The Future of Post-Human Geology

The Future of Post-Human Geology

A Preface to a New Theory of Statics and Dynamics

Peter Baofu

Viva Books

New Delhi | Mumbai | Chennai | Kolkata | Bengaluru | Hyderabad | Kochi | Guwahati

First Indian Edition 2012

VIVA BOOKS PRIVATE LIMITED

- 4737/23 Ansari Road, Daryaganj, New Delhi 110 002
  E-mail: vivadelhi@vivagroupindia.net, Tel. 42242200
- 76, Service Industries, Shirvane, Sector 1, Nerul, Navi Mumbai 400 706
  E-mail: vivamumbai@vivagroupindia.net, Tel. 27721273, 27721274
- Jamals Fazal Chambers, 26 Greams Road, Chennai 600 006
  E-mail: vivachennai@vivagroupindia.net, Tel. 28290304, 28294241
- B-103, Jindal Towers, 21/1A/3 Darga Road, Kolkata 700 017
  E-mail: vivakolkata@vivagroupindia.net, Tel. 22836381, 22816713
- 7, Sovereign Park Apartments, 56-58, K. R. Road, Basavanagudi, Bangalore 560 004
  E-mail: vivabangalore@vivagroupindia.net, Tel. 26607409, 26607410
- 101-102 Mughal Marc Apartments, 3-4-637 to 641, Narayanguda, Hyderabad 500 029
  E-mail: vivahyderabad@vivagroupindia.net, Tel. 27564481, 27564482
- Beevi Towers, First Floor, SRM Road, Kaloor, Kochi 682 018, Kerala
  E-mail: vivakochi@vivagroupindia.net, Tel: 0484-2403055, 2403056
- 232, GNB Road, Beside UCO Bank, Silpukhuri, Guwahati 781 003
  E-mail: vivaguwahati@vivagroupindia.net, Tel: 0361-2666386

www.vivagroupindia.com

Published by arrangement with
Cambridge International Science Publishing
7, Meadow Walk, Great Abington
Cambridge CB1 6AZ, UK

ISBN : 978-81-309-1996-6

Published by Vinod Vasishtha for Viva Books Private Limited, 4737/23, Ansari Road, Daryaganj, New Delhi-110 002. Printed and bound by Syndicate Binders, Noida.

To Those in the Future World beyond Earth Geology

BOOKS ALSO BY PETER BAOFU

Category I: Social Sciences and Related Fields

- 33. *The Future of Post-Human War and Peace* (2010)
- 32. *The Future of Post-Human Law* (2010)
- 31. *The Future of Post-Human Organization* (2009)
- 30. *The Future of Post-Human Mass Media* (2009)
- 29. *The Rise of Authoritarian Liberal Democracy* (2007)
- 28. *Beyond the World of Titans, and the Remaking of World Order* (2007)
- 27. *Beyond Capitalism to Post-Capitalism* (2005)
- 26. Volume 1: *Beyond Democracy to Post-Democracy* (2004)
- 25. Volume 2: *Beyond Democracy to Post-Democracy* (2004)
- 24. *The Future of Capitalism and Democracy* (2002)

Category II: Natural Sciences and Related Fields

- 23. *Beyond Cosmology to Post-Cosmology* (2010)
- 22. *The Future of Post-Human Personality* (2010)
- 21. *The Future of Post-Human Sexuality* (2010)
- 20. *The Future of Post-Human Urban Planning* (2009)
- 19. *The Future of Post-Human Engineering* (2009)
- 18. *The Future of Post-Human Unconsciousness* (2008)
- 17. *The Future of Complexity* (2007)
- 16. *Beyond Nature and Nurture* (2006)
- 15. *The Future of Post-Human Space-Time* (2006)
- 14. *The Future of Post-Human Consciousness* (2004)

Category III: Formal Sciences and Related Fields

- 13. *The Future of Post-Human Formal Science* (2010)
- 12. *The Future of Post-Human Language* (2009)
- 11. *The Future of Post-Human Geometry* (2009)
- 10. *The Future of Post-Human Mathematical Logic* (2008)
- 9. *The Future of Information Architecture* (2008)

Category IV: The Humanities and Related Fields

- 8. *Beyond Ethics to Post-Ethics* (2010)
- 7. *The Future of Post-Human Martial Arts* (2009)
- 6. *The Future of Post-Human Creative Thinking* (2009)
- 5. *The Future of Post-Human Knowledge* (2008)
- 4. *The Future of Aesthetic Experience* (2007)
- 3. *Beyond Civilization to Post-Civilization* (2006)
- 2. Volume 1: *The Future of Human Civilization* (2000)
- 1. Volume 2: *The Future of Human Civilization* (2000)

CONTENTS

Part Three: Dynamics

Part Four: Conclusion

TABLES

Category III: Visions on the Mind

Category IV: Visions on History

Category VII. Visions on Society (Socio-Political)

Category VIII. Visions on Society (Socio-Economic)

Category IX: Visions on Culture

ACKNOWLEDGMENTS

This book, like all other previous ones of mine, is written with the critical spirit to challenge conventional wisdom with a novel way of thinking, or more specifically in the current context, a new theory to understand the future fate of Earth geology.

In light of this political incorrectness, this book receives no external funding nor help from any formal organization or institution—as this is something that I often stressed in all of my previous books.

My only reward is that joyful wonder of something new not conceived before in history.

In any event, and as always, I bear the sole responsibility for the ideas presented in this book.

ABBREVIATIONS

ALD = Peter Baofu. 2007. *The Rise of Authoritarian Liberal Democracy: A Preface to a New Theory of Comparative Political Systems.* Cambridge, England: Cambridge Scholars Publishing, Ltd.

BCIV = Peter Baofu. 2006. *Beyond Civilization to Post-Civilization: Conceiving a Better Model of Life Settlement to Supersede Civilization.* NY: Peter Lang Publishing, Inc.

BCPC = Peter Baofu. 2005. *Beyond Capitalism to Post-Capitalism: Conceiving a Better Model of Wealth Acquisition to Supersede Capitalism.* NY: The Edwin Mellen Press.

BCOS = Peter Baofu. 2010. *Beyond Cosmology to Post-Cosmology: A Preface to a New Theory of Different Worlds.* Cambridge, England: Cambridge International Science Publishing, Ltd.

BDPD1 = Peter Baofu. 2004. Volume 1. *Beyond Democracy to Post-Democracy: Conceiving a Better Model of Governance to Supersede Democracy.* NY: The Edwin Mellen Press.

BDPD2 = Peter Baofu. 2004. Volume 2. *Beyond Democracy to Post-Democracy: Conceiving a Better Model of Governance to Supersede Democracy.* NY: The Edwin Mellen Press.

BEPE = Peter Baofu. 2010. *Beyond Ethics to Post-Ethics: A Preface to a New Theory of Morality and Immorality.* Charlotte, NC: Infomration Age Publishing.

BNN = Peter Baofu. 2006. *Beyond Nature and Nurture: Conceivng a Better Way to Understand Genes and Memes.* Cambridge, England: Cambridge Scholars Publishing, Ltd.

BWT = Peter Baofu. 2007. *Beyond the World of Titans, and the Renaking of World Order: A Preface to a New Logic of Empire-Building.* Cambridge, England: Cambridge Scholars Publishing, Ltd.

FAE = Peter Baofu. 2007. *The Future of Aesthetic Experience: Conceiving a Better Way to Understand Beauty, Ugliness and the Rest.* Cambridge, England: Cambridge Scholars Publishing, Ltd.

FC = Peter Baofu. 2007. *The Future of Complexity: Conceiving a Better Way to Understand Order and Chaos.* London, United Kingdom: World Scientific Publishing Co.

FCD = Peter Baofu. 2002. *The Future of Capitalism and Democracy.* MD: The University Press of America.

FHC1 = Peter Baofu. 2000. Volume 1. *The Future of Human Civilization.* NY: The Edwin Mellen Press.

FHC2 = Peter Baofu. 2000. Volume 2. *The Future of Human Civilization.* NY: The Edwin Mellen Press.

FIA = Peter Baofu. 2008. *The Future of Information Architecture: Conceiving a Better Way to Understand Taxonomy, Network, and Intelligence.* Oxford, England: Chandos Publishing (Oxford) Limited.

FPHC = Peter Baofu. 2004. *The Future of Post-Human Consciousness.* NY: The Edwin Mellen Press.

FPHCT = Peter Baofu. 2009. *The Future of Post-Human Creative Thinking: A Preface to a New Theory of Invention and Innovation.* Cambridge, England: Cambridge Scholars Publishing, Ltd.

FPHE = Peter Baofu. 2009. *The Future of Post-Human Engineering: A Preface to a New Theory of Technology.* Cambridge, England: Cambridge Scholars Publishing, Ltd.

FPHFS = Peter Baofu. 2010. *The Future of Post-Human Formal Science: A Preface to a New Theory of Abstraction and Application.* Cambridge, England: Cambridge Scholars Publishing, Ltd.

FPHG = Peter Baofu. 2009. *The Future of Post-Human Geometry: A Preface to a New Theory of Infinity, Symmetry, and Dimensionality.* Cambridge, England: Cambridge Scholars Publishing, Ltd.

FPHGEOL = Peter Baofu. 2010. *The Future of Post-Human Geology: A Preface to a New Theory of Statics and Dynamics.* Cambridge, England: Cambridge International Science Publishing, Ltd.

FPHK = Peter Baofu. 2008. *The Future of Post-Human Knowledge: A Preface to a New Theory of Methodology and Ontology.* Oxford, England: Chandos Publishing (Oxford) Limited.

FPHL = Peter Baofu. 2009. *The Future of Post-Human Language: A Preface to a New Theory of Structure, Context, and Learning.* Cambridge, England: Cambridge Scholars Publishing, Ltd.

FPHLAW = Peter Baofu. 2010. *The Future of Post-Human Law: A Preface to a New Theory of Necessity, Contingency, and Justice.* Cambridge, England: Cambridge Scholars Publishing, Ltd.

FPHMA = Peter Baofu. 2009. *The Future of Post-Human Martial Arts: A Preface to a New Theory of the Body and Spirit of Warriors.* Cambridge, England: Cambridge Scholars Publishing, Ltd.

FPHML = Peter Baofu. 2008. *The Future of Post-Human Mathematical Logic: A Preface to a New Theory of Rationality.* Cambridge, England: Cambridge Scholars Publishing, Ltd.

FPHMM = Peter Baofu. 2009. *The Future of Post-Human Mass Media: A Preface to a New Theory of Technology.* Cambridge, England: Cambridge Scholars Publishing, Ltd.

FPHO = Peter Baofu. 2009. *The Future of Post-Human Organization: A Preface to a New Theory of Communication, Decision-Making, and Leadership.*.Cambridge, England: Cambridge Scholars Publishing, Ltd.

FPHP = Peter Baofu. 2010. *The Future of Post-Human Personality: A Preface to a New Theory of Normality and Abnormality.* Cambridge, England: Cambridge International Science Publishing, Ltd.

FPHS = Peter Baofu. 2010. *The Future of Post-Human Sexuality: A Preface to a New Theory of the Body and Spirit of Love-Makers.* Cambridge, England: Cambridge Scholars Publishing, Ltd.

FPHST = Peter Baofu. 2006. *The Future of Post-Human Space-Time: Conceivng a Better Way to Understand Space and Time.* New York: Peter Lang Publishing, Inc.

FPHU = Peter Baofu. 2008. *The Future of Post-Human Unconsciousness: A Preface to a New Theory of Anomalous Experience.* Cambridge, England: Cambridge Scholars Publishing, Ltd.

FPHUP = Peter Baofu. 2009. *The Future of Post-Human Urban Planning: A Preface to a New Theory of Density, Void, and Sustainability.* Cambridge, England: Cambridge Scholars Publishing, Ltd.

FPHWP = Peter Baofu. 2010. *The Future of Post-Human War and Peace: A Preface to a New Theory of Aggression and Pacificity.* Cambridge, England: Cambridge Scholars Publishing, Ltd.

• PART ONE •

Introduction

CHAPTER 1
INTRODUCTION—THE USE OF GEOLOGY

> [W]ith *Time's Arrow, Time's Cycle*, [Stephen Jay] Gould has turned to the history of geology… In this…work he offers a revisionist historical account of the discovery of geological time…. [W]hat underlies the discovery of deep time is by no means fieldwork, as the myths of geology textbooks would have us believe. Rather, Gould pinpoints a powerful pair of metaphors—time's arrow and time's cycle—by which humankind has always tried to grasp the concept of time.
> —Frank J. Sulloway (1987)

Two Views in Contention

Is the history of geology really so metaphorical in its foundation, according to some critics like Stephen Jay Gould, that "what underlies the discovery of deep time is by no means fieldwork, as the myths of geology textbooks would have us believe"? (F. Sulloway 1987)

This revisionist account by the critics like Gould can be contrasted by an opposing view in the older days, when Charles Lyell, for example, "portrayed the previous history of his discipline as a gradual overcoming of primitive superstitions, wild speculations, and biblical allegiances. In doing so he created his own legend…, as an archempiricist free of all bias and preconception." (F. Sulloway 1987)

Contrary to these two opposing views (and others as will be discussed in the rest of the book), the history of geology is neither so metaphorical nor so empirical to the extent that the respective defenders would like us to believe.

But this rejection of the opposing views in contention does not logically follow that geology is therefore not worth studying, or that it is useless. Neither to say, neither of these two extreme views is reasonable either.

Instead, this book offers an alternative (better) way to understand the future of geology, especially in the context of statics and dynamics—while learning from different approaches in the literature but without favoring any one of them (nor integrating them, since they are not necessarily compatible with each other). Thus, this book offers a new theory to go beyond the existing approaches in the literature on geology in an original way.

If successful, this seminal project is to fundamentally change the way that we think about geology, from the combined perspectives of the mind, nature, society, and culture, with enormous implications for the human future and what I originally called its "post-human" fate.

Geology and Geosciences

A good starting point for the inquiry is to clarify a confusion between the term "geology" and a related term known as "geosciences" (or "the geosciences," or even the "Earth sciences").

The term "geology," which is from "Greek γῆ, gê, meaning 'earth' and λόγος, logos, meaning 'study of.'" (WK 2010; S. Winchester 2001) More to the point, "geology" refers to the study of "the rocky parts of the Earth's crust (or lithosphere) and its historic development." (WK 2010q)

By contrast, the term "geosciences" (or "the geosciences," or even the "Earth sciences") is much broader in scope and refers to different sciences about the Earth, with "geology" as only one of them.

For instance, the term "geosciences" also includes "geology" as part of the sciences about the Earth but also has other concerns, as shown below (and summaried in *Table 1.1*). (WK 2010q; G. Smith 2006)

- "Geophysics and Geodesy" (for the "Lithosphere/Geosphere")—These geosciences "investigate the shape of the Earth, its reaction to forces and its magnetic and gravity fields. Geophysicists explore the Earth's core and mantle as well as the tectonic and seismic activity of the lithosphere."
- "Soil Science" (for the "Pedosphere")—This geoscience "covers the outermost layer of the Earth's crust that is subject to soil formation

processes (or pedosphere). Major subdisciplines include edaphology and pedology."

- "Oceanography and Hydrology" (for the "Hydropshere")—These goesciences "describe the marine and freshwater domains of the watery parts of the Earth (or hydrosphere). Major subdisciplines include hydrogeology and physical, chemical, and biological oceanography."
- "Glaciology" (for the "Cryosphere")—This geoscience "covers the icy parts of the Earth (or cryosphere)."
- "Atmospheric Sciences" (for the "Atmosphere")—This geosciences "cover the gaseous parts of the Earth (or atmosphere) between the surface and the exosphere (about 1000 km). Major subdisciplines are meteorology, climatology, atmospheric chemistry and atmospheric physics."
- "Biology" (for "the Biosphere")—This geoscience studies the biosphere, which "consists of all forms of life, from single-celled organisms to pine trees to people. The interactions of Earth's other spheres—lithosphere/geosphere, hydrosphere, atmosphere and/or cryosphere and pedosphere—create the conditions that can support life."

In other words, geology is concerned mostly (though not exclusively) with the lithosphere, although it also incorporates the knowledge from other geosciences in regard to other spheres—whereas the geosciences (or the Earth sciences) are concerned with multiple spheres (including the lithosphere), such as the atmosphere, the hydrosphere, the cryosphere, the pedosphere, and the biosphere.

In general, the resources in the biosphere tend to be "biotic" (like forests, animals, birds, fish, fuels from decayed organic maerials, etc.), whereas the resources in other spheres tend to be "abiotic" (like air, water, ice, soil, rocks, metals, etc.). (WK 2010o)

Statics and Dynamics in Geology

With this clarification of the relationship between geology and the geosciences in mind—the term geology "was first used by Jean-André Deluc in 1778 and introduced as a fixed term by Horace-Bénédict de Saussure in 1779." (WK 2010; S. Winchester 2001)

More specifically, geology is "the study of the composition..., properties, and history of the planet's physical material, and the processes by which it is formed, moved, and changed." (WK 2010) Or alternatively, it is "a science that deals with the history of the earth and its life especially as recorded in rocks." (MWD 2010) And its major sub-fields include "mineralogy and petrology, geochemistry, geomorphology, paleontology, stratigraphy, structural geology, engineering geology and sedimentology," for instance. (WK 2010q)

More generally, geology is concerned with both the *structure* (or *statics*) and *process* (or *dynamics*) of the planet's physical matter.

Geology is important enough, since it has been used commercially "for mineral and hydrocarbon extraction, as well as for predicting and understanding earthquakes and volcanic eruptions. It is also an important foundation for many other science and engineering disciplines, including evolutionary biology, paleoclimatology and geotechnical engineering." (WK 2010)

With this use of geology in mind—it is thus important to elaborate further what the "structure" (or "statics") and "process" (or "dynamics") of the planet's physical matter (as indicated above and summarized in *Table 1.2*) are, in what follows, in that order.

Statics

The word "statics" is formally defined as a branch of physical science (or mechanics) which deals with "the relations of forces that produce equilibrium among material bodies." (MWD 2010a)

"Statics" is used here, in the context of geology, to refer to the *structure* of the planet's physical matter (like its composition, properties, etc.) in such a way that it is in a state of equilibrium, in the absence of any change. (MWD 2010a)

For illustration, there are two major categories of the statics (or structure) of the planet's physical matter, namely, (a) "geologic materials" and (b) "whole-Earth structure"—to be illustrated hereafter, respectively. (WK 2010)

Geologic Materials

Geologic materials can be of two types, namely, (a) "rock" and (b) "unconsolidated material" (e.g., soils, etc.) (WK 2010)

(a) In relation to "rock"—there are three major sub-types classified in the literature, namely, "igneous, sedimentary, and metamorphic." (WK 2010)

These three major sub-types of rock can be understood by way of the "rock cycle," which "illustrates the relationships between these three sub-types of rock, and magma." (WK 2010)

Firstly, "[w]hen a rock crystallizes from melt (magma and/or lava), it is an igneous rock." (WK 2010)

Secondly, "[igneous] rock can be weathered and eroded, and then redeposited and lithified into a sedimentary rock." (WK 2010) The term "sediment" in "sedimentary rock" here refers to "a naturally-occurring material that is broken down by processes of weathering and erosion, and is subsequently transported by the action of fluids such as wind, water, or ice, and/or by the force of gravity acting on the particle itself." (WK 2010a)

Thirdly, a rock can also "be turned into a metamorphic rock due to heat and pressure that change the mineral content of the rock and give it a characteristic fabric." (WK 2010)

In fact, the relationships among the three sub-types of rock can be more complicated, because "the sedimentary rock can then be subsequently turned into a metamorphic rock due to heat and pressure, and the metamorphic rock can be weathered, eroded, deposited, and lithified, becoming a sedimentary rock. Sedimentary rock may also be re-eroded and redeposited, and metamorphic rock may also undergo additional metamorphism. All three types of rocks may be re-melted; when this happens, a new magma is formed, from which an igneous rock may once again crystallize." (WK 2010)

(b) And in relation to "unconsolidated material"—there is "unlithified material, which typically comes from more recent deposits. Because of this, the study of such material is often known as Quaternary geology, after the recent Quaternary Period. This includes the study of sediment and soils, and is important to some (or many) studies in geomorphology, sedimentology, and paleoclimatology." (WK 2010)

Whole-Earth Structure

Another category of the statics (or structure) of the planet's physical matter concerns "whole-Earth structure," which can be illustrated by way of two types, namely, (a) "plate tectonics" and (b) "Earth structure"—as shown below, in that order.

(a) In relation to "plate tectonics"—there is "the Earth's lithosphere, which includes the crust and rigid uppermost portion of the upper mantle" and "is separated into a number of tectonic plates that move across the plastically deforming, solid, upper mantle, which is called the asthenosphere." (WK 2010; H. Hess 1962; J. Kious 1996)

What is interesting here is that "[t]here is an intimate coupling between the movement of the plates on the surface and the convection of the mantle: oceanic plate motions and mantle convection currents always move in the same direction, because the oceanic lithosphere is the rigid upper thermal boundary layer of the convecting mantle. This coupling between rigid plates moving on the surface of the Earth and the convecting mantle is called plate tectonics." (WK 2010)

The understanding of plate tectonics serves as a physical explanation of "many observations of the solid Earth"; for instance "[m]id-ocean ridges, high regions on the seafloor where hydrothermal vents and volcanoes exist, were explained as divergent boundaries, where two plates move apart. Arcs of volcanoes and earthquakes were explained as convergent boundaries, where one plate subducts under another. Transform boundaries, such as the San Andreas fault system, resulted in widespread powerful earthquakes. Plate tectonics also provided a mechanism for Alfred Wegener's theory of continental drift, in which the continents move across the surface of the Earth over geologic time." (WK 2010; J. Kious 1996)

In the end, "[t]he power of the theory of plate tectonics lies in its ability to combine all of these observations into a single theory of how the lithosphere moves over the convecting mantle." (WK 2010)

(b) And in relation to "Earth structure"—there is "the internal composition and structure of the Earth," thanks to the "advances in seismology, computer modeling, and mineralogy and crystallography at high temperatures and pressures." (WK 2010)

For instance, "seismologists can use the arrival times of seismic waves in reverse to image the interior of the Earth. Early advances in this field showed the existence of a liquid outer core (where shear waves were not able to propagate) and a dense solid inner core. These advances led to the development of a layered model of the Earth, with a crust and lithosphere on top, the mantle below (separated within itself by seismic discontinuities at 410 and 660 kilometers), and the outer core and inner core below that." (WK 2010)

Dynamics

The above examples about geologic materials and whole-Earth structure also reveal the fascinating dynamics of the planet's physical matter, such that the structure (or statics) and process (or dynamics) of the planet's physical matter are intricately related.

For instance, the above example of the "rock cycle" describes "the dynamic transitions through geologic time among the three main rock types: sedimentary, metamorphic, and igneous," such that "each type of rock is altered or destroyed when it is forced out of its equilibrium conditions." (WK 2010b)

With this intricate relationship between statics and dynamics in mind—the word "dynamics" is formally defined as a branch of physical science (or mechanics) which deals with "the process of change, growth, or activity" in material bodies. (MWD 2010b)

"Dynamics" is therefore used here, in the context of geology, to refer to the *process* of change in relation to the planet's physical material (like its change, growth, etc.) in such a way that it is not in a state of equilibrium.

An excellent illustration of the dynamics of the planet's physical material concerns "the geological evolution of an area," as described hereafter. (WK 2010) WK 2010)

The Geological Evolution of an Area

The geological evolution of an area refers to the process of change when "rock units are deposited and inserted and deformational processes change their shapes and locations." (WK 2010)

In general, "[r]ock units are first emplaced either by deposition onto the surface or intrusion into the overlying rock. Deposition can occur when sediments settle onto the surface of the Earth and later lithify into sedimentary rock, or when as volcanic material such as volcanic ash or lava flows blanket the surface. Igneous intrusions such as batholiths, laccoliths, dikes, and sills, push upwards into the overlying rock, and crystallize as they intrude." (WK 2010)

Later, "[a]fter the initial sequence of rocks has been deposited, the rock units can be deformed and/or metamorphosed. Deformation typically occurs as a result of horizontal shortening, horizontal extension, or side-to-side (strike-slip) motion. These structural regimes broadly relate to convergent boundaries, divergent boundaries, and transform boundaries, respectively, between tectonic plates." (WK 2010)

For instance, "[w]hen rock units are placed under horizontal compression, they shorten and become thicker. Because rock units, other than muds, do not significantly change in volume, this is accomplished in two primary ways: through faulting and folding. In the shallow crust, where brittle deformation can occur, thrust faults form, which cause deeper rock to move on top of shallower rock. Because deeper rock is often older, as noted by the principle of superposition, this can result in older rocks moving on top of younger ones." (WK 2010)

In addition, "[m]ovement along faults can result in folding, either because the faults are not planar, or because the rock layers are dragged along, forming drag folds, as slip occurs along the fault. Deeper in the Earth, rocks behave plastically, and fold instead of faulting. These folds can either be those where the material in the center of the fold buckles upwards, creating 'antiforms,' or where it buckles downwards, creating 'synforms.' If the tops of the rock units within the folds remain pointing upwards, they are called anticlines and synclines, respectively. If some of the units in the fold are facing downward, the structure is called an overturned anticline or syncline, and if all of the rock units are overturned or the correct up-direction is unknown, they are simply called by the most general terms, antiforms and synforms." (WK 2010)

Surely, "all of these processes do not necessarily occur in a single environment, and do not necessarily occur in a single order," and there can be other processes not mentioned above.

But the examples here suffice for the illustrative purpose at hand.

The Wide Variety of Geological Methods

But this understanding of the statics and dynamics of the planet's physical matter raises another closely related issue concenring the methods which have been used to discover them in the first place.

In geology, there are two major types of methods to be summarized below (and in *Table 1.2*), namely, (a) "field methods" and (b) "laboratory methods." (WK 2010)

Field Methods

The first type of geological methods is known as "field methods," which require fieldwork and "varies depending on the task at hand," as shown in some examples below. (WK 2010; R. Compton 1985; USGS 2009; R. Burger 2006; W. Krumbein 1978; I. McDougall 1999; B. Hubbard 2005)

- "Geological Mapping"
  - "Structural mapping: the locations of the major rock units and the faults and folds that led to their placement there"
  - "Stratigraphic mapping: the locations of sedimentary facies (lithofacies and biofacies) or the mapping of isopachs of equal thickness of sedimentary rock"
  - "Surficial mapping: the locations of soils and surficial deposits"
- "Surveying of Topographic Features"
  - "Creation of topographic maps"

- –"Work to understand change across landscapes, including"
    - "Patterns of erosion and deposition"
    - "River channel change through migration and avulsion"
    - "Hillslope processes"
- "Subsurface Mapping through Geophysical Methods"
  - –"These methods include"
    - "Shallow seismic surveys"
    - "Ground-penetrating radar"
    - "Electrical resistivity tomography"
  - –"They are used for"
    - "Hydrocarbon exploration"
    - "Finding groundwater"
    - "Locating buried archaeological artifacts"
- "High-Resolution Stratigraphy"
  - –"Measuring and describing stratigraphic sections on the surface"
  - –"Well drilling and logging"
- "Biogeochemistry and Geomicrobiology"
  - –"Collecting samples to"
    - "Determine biochemical pathways"
    - "Identify new species of organisms"
    - "Identify new chemical compounds"
  - –"And to use these discoveries to"
    - "Understand early life on Earth and how it functioned and metabolized"
    - "Find important compounds for use in pharmaceuticals"
- "Paleontology: Excavation of Fossil Material"
  - –"For research into past life and evolution"
  - –"For museums and education"
- "Collection of Samples for Geochronology and Thermochronology"
- "Glaciology: Measurement of Characteristics of Glaciers and their Motion"

Surely, there can be other field methods, so the examples above are illustrative for the purpose at hand.

Laboratory Methods

By contrast, geological methods which are conducted in the laboratory (not in the field) are known as "laboratory methods"; the three most often used ones are, namely, (a) "petrology," (b) "structural geology," and (c) "stratigraphy," to be summarized hereafter, in that order.

Petrology

The laboratory method known as "petrology" studies "rock samples in the laboratory." (WK 2010)

It should be stressed that, although both "petrology" and "lithology" study "rocks," there is a distinction between the two terms, in that "lithology is a subdivision of petrology focusing on macroscopic hand-sample or outcrop-scale description of rocks, while petrography is the speciality that deals with microscopic details." (WK 2010d)

With this clarification in mind—"two of the primary methods for identifying rocks in the laboratory are through optical microscopy and by using an electron microprobe." (WK 2010)

In relation to "optical microscopy"—the focus here is on the analysis of "thin sections of rock samples…through a petrographic microscope, where the minerals can be identified through their different properties in plane-polarized and cross-polarized light, including their birefringence, pleochroism, twinning, and interference properties with a conoscopic lens." (WK 2010; W. Nesse 1991)

And in relation to the "electron microprobe"—the focus here is on the analysis of "individual locations…for their exact chemical compositions and variation in composition within individual crystals. Stable and radioactive isotope studies provide insight into the geochemical evolution of rock units." (WK 2010; A. Morton 1985; Y. Zheng 2003; M. Condomines 1995)

Structural Geology

In the laboratory method known as "structural geology," geologists use "thin sections of geologic samples to observe the fabric within the rocks which gives information about strain within the crystal structure of the rocks. They also plot and combine measurements of geological structures in order to better understand the orientations of faults and folds in order to reconstruct the history of rock deformation in the area. In addition, they perform analog and numerical experiments of rock deformation in large and small settings." (WK 2010)

Some of "the most well-known experiments in structural geology" concerns "those involving orogenic wedges, which are zones in which mountains are built along convergent tectonic plate boundaries. In the analog versions of these experiments, horizontal layers of sand are pulled along a lower surface into a back stop, which results in realistic-looking patterns of faulting and the growth of a critically tapered (all angles remain the same) orogenic wedge. Numerical models work in the same way as these analog models, though they are often more sophisticated and can include patterns of erosion and uplift in the mountain belt. This helps to show the relationship between erosion and the shape of the mountain range. These studies can also give useful information about pathways for metamorphism through pressure, temperature, space, and time." (WK 2010; F. Dahlen 1990; M. Gutscher 1998; P. Koons 1995)

Stratigraphy

And in the laboratory method known as "stratigraphy" (for the study of "rock layers and layering"—or simply "stratification"), geologists "analyze samples of stratigraphic sections that can be returned from the field, such as those from drill cores. Stratigraphers also analyze data from geophysical surveys that show the locations of stratigraphic units in the subsurface. Geophysical data and well logs can be combined to produce a better view of the subsurface, and stratigraphers often use computer programs to do this in three dimensions. Stratigraphers can then use these data to reconstruct ancient processes occurring on the surface of the Earth, interpret past environments, and locate areas for water, coal, and hydrocarbon extraction." (WK 2010 & 2010c; D. Hodell 1994; A. Bally 1987; O. Fernández 2004; C. Poulsen 1998)

Other stratigraphers focus on different issues. For instance, "biostratigraphers analyze rock samples from outcrop and drill cores for the fossils found in them. These fossils help scientists to date the core and to understand the depositional environment in which the rock units formed. Geochronologists precisely date rocks within the stratigraphic section in order to provide better absolute bounds on the timing and rates of deposition. Magnetic stratigraphers look for signs of magnetic reversals in igneous rock units within the drill cores. Other scientists perform stable isotope studies on the rocks to gain information about past climate." (WK 2010; D. Hodell 1994; M. Toscano 1999)

The Theoretical Debate

However, this introduction to the wide variety of geological methods (and the intricate relationship between statics and dynamics in regard to the planet's physical matter, as shown in the earlier section) thus leads us to a more important theoretical question concerning the extent to which geology is possible and desirable in the first place.

For illustration, there are four main theoretical approaches in the debate to be summarized hereafter, and they can be called, in the absence of better words, namely, (a) the *catastrophe* argument, (b) the *uniformity* argument, (c) the *revision* argument, and (d) the *resettlement* argument (which is my original contribution to the debate)—to be elabroated hereafter, respectively.

Lest misunderstanding occurs, two clarifications should be made here, as this is something that I regularly emphasize in my previous books, whenever a theoretical debate is summarized for a topic, *almost verbatim*.

Firstly, there can be other variants within each argument (besides the ones to be summarized hereafter). The advantage to select them here has to do with their illustrative representation of the diverse theories in the literature (which are deemed sufficient for the purpose at hand in this book).

And secondly, the first three arguments are not necessarily mutually exclusive. For instance, those who advocate the "revision" argument can also consider the "uniformity" argument, although they do not focus on the latter.

And the reverse is also true, in that those who make the "uniformity" argument can also consider the "revision" argument, although they do not focus on the latter.

For instance, a version of the "revision" argument as proposed by Stephen Jay Gould does not totally ignore the importance of empiricism in doing geological work, although it does not focus on the latter (as will be clear later).

In other words, their disagreement is often one in degree, not in kind.

The Catastrophe Argument

With these two clarifications in mind—the first major theoretical approach is called, in the absence of better words, the *catastrophe* argument, which focuses, relatively speaking of course, more on some "sudden, short-lived, violent events, possibly worldwide in scope," that affected the history of the physical development of the Earth. (WK 2010e)

Therefore, the word "catastrophe" in the title of the argument is suggestive, because it refers to "abrupt" changes, rather than "incremental" changes, in the history of the physical development of the Earth.

Thus, the catastrophe argument holds that "the distant past 'consisted of epochs of paroxysmal and catastrophic action interposed between periods of comparative tranquility.'" (WK 2010f; W. Whewell 1832)

For illustration, consider two main positions in the catastrophe argument, namely, (a) creationist catastrophism and (b) scientific catastrophism—as explained below, respectively.

Creatonist Catastrophsim

For much of the pre-modern history of geology, for instance, "[u]ntil the 19th century, the dominant scientific beliefs in Europe were founded on the biblical narratives of Creation and the universal deluge. Other ancient deluge myths have been discovered since then, explaining why the flood story was 'stated in scientific methods with surprising frequency among the Greeks,' an example being Plutarch's account of the Ogygian flood." (WK 2010e; C. King 1877)

In this way, "Earth's history was viewed as the result of an accumulation of catastrophic events over a relatively short time period, before the depth of geological time was appreciated. In this way they were able to explain the observations of early geologists within the framework of a short Earth history." (WK 2010e)

But for the critics, there are some problems with the "creationist" variant of the catastrophe argument. Consider two examples hereafter, for illustration.

Firstly, one criticism of the "creationist" variant of the catastrophe argument is that it ignores opposing arguments, like the "uniformity" one, which can shed further light on the history of the Earth.

And secondly, another criticism of the "creatonist" variant of the catastrophe argument is that it does not give sufficient attention to other variants of the catastrophe argument itself (like scientific catastrophism, as explained below).

Scientific Catastrophsim

Unlike the "creationist" variant—the "scientific" variant of the catastrophe argument tries to do without the religious indoctrination while maintaining the catastrophe thesis.

For instance, Georges Cuvier, who was "the leading scientific proponent of catastrophism in the early nineteenth century," tried "to explain the patterns of extinction and faunal succession that he and others were observing in the fossil record." (WK 2010e)

What is distinctive about Cuvier's approach is that, "[w]hile he did speculate that the catastrophe responsible for the most recent extinctions in Eurasia might have been the result of the inundation of low lying areas by the sea, he never made any reference to Noah's flood. Nor did he ever make any reference to divine creation as the mechanism by which repopulation occurred following the extinction event." (WK 2010e; C. McGowan 2001)

Instead, Cuvier lel himself "influenced by the ideas of the Enlightenment and the intellectual climate of the French revolution" and "avoided religious or metaphysical speculation in his scientific writings. Cuvier also believed that the stratigraphic record indicated that there had been several of these revolutions, which he viewed as recurring natural events, amid long intervals of stability during the history of life on earth. This led him to believe the Earth was several million years old." (WK 2010e; M. Rudwick 1972)

But for the critics, there are also some problems with the "scientific" variant of the catastrophe argument.

For instance, one often cited criticism of the "scientific" variant of the catastrophe argument is that it does not avoid its own bias in ignoring opposing arguments, like the "uniformity" one, which proposes a contrastive view about the history of the Earth.

The Uniformity Argument

The second theoretical approach is called, in the absence of better words, the *uniformity* argument, which focuses, relatively speaking of course, more on some "gradual" changes that affected the history of the physical development of the Earth over a long period of time. (WK 2010e)

Therefore, the word "uniformity" in the title of the argument is suggestive, because it assumes that "the same natural laws and processes that operate in the universe now, have always operated in the universe in the past and apply everywhere in the universe. It is frequently summarized as 'the present is the key to the past,' because it holds that all things continue as they were from the beginning of the world," over a long period of time. (WK 2010f)

For illustration, consider two main positions in the uniformity argument, namely, (a) the work of James Hutton and (b) the work of Charles Lyell—as explained below, respectively.

The Work of James Hutton

In 1785, James Hutton "proposed a self-maintaining infinite cycle," in which "[t]he solid parts of the present land appear in general, to have been composed of the productions of the sea, and of other materials similar to those now found upon the shores." (WK 2010f; P. Bowler 2003)

Hutton then made three interesting conclusions, as shown below: (WK 2010f)

- "That the land on which we rest is not simple and original, but that it is a composition, and had been formed by the operation of second causes."
- "That before the present land was made, there had subsisted a world composed of sea and land, in which were tides and currents, with such operations at the bottom of the sea as now take place."
- "That while the present land was forming at the bottom of the ocean, the former land maintained plants and animals; at least the sea was than inhabited by animals, in a similar manner as it is at present."

With these observations, "Hutton then sought evidence to support his idea that there must have been repeated cycles, each involving deposition on the seabed, uplift with tilting and erosion then undersea again for further layers to be deposited....In the Spring of 1788, he took a boat trip along the Berwickshire coast with John Playfair and the geologist Sir James Hall, and found a dramatic unconformity showing the same sequence at Siccar Point....Hutton concluded a 1788 paper he presented at the Royal Society of Edinburgh, later rewritten as a book, with the phrase 'we find no vestige of a beginning, no prospect of an end.'" (WK 2010f; K. Thomson 2001)

But for the critics, there are some problems with Hutton's idea. For instance, one criticism of the uniformity argument by Hutton is that it does not give sufficient attention to opposing arguments like the "catastrophe" one.

The critic Georges Cuvier did his "paleontological work in the 1790s, which established the reality of extinction" and "explained this by local catastrophes, after which other fixed species repopulated the affected areas. In Britain, geologists adapted this idea into 'diluvial theory' which proposed repeated worldwide annihilation and creation of new fixed species adapted to a changed environment, initially identifying the most recent catastrophe as the biblical flood." (WK 2001f; P. Bowler 2003)

The Work of Charles Lyell

Trying to improve the work of Hutton—Charles Lyell in the 1830's published his multi-volume *Principles of Geology*, with the subtitle "An attempt to explain the former changes of the Earth's surface by reference to causes now in operation." (WK 2010f)

Lyell "drew his explanations from field studies…and developed Hutton's idea that the earth was shaped entirely by slow-moving forces still in operation today, acting over a very long period of time. The terms uniformitarianism for this idea, and catastrophism for the opposing viewpoint, were coined by William Whewell in a review of Lyell's book." (WK 2010f; L. Wilson 1973)

For Reijer Hooykaas (1963), "Lyell's uniformitarianism is a family of four related propositions, not a single idea," as shown below: (WK 2010f)

- "Uniformity of Law"—This holds that "the laws of nature are constant."
- "Uniformity of Methodology"—This holds that the appropriate hypotheses for explaining the geological past are those with analogy today."
- "Uniformity of Kind" (or "Rate")—This holds that "past and present causes are all of the same kind, have the same energy, and produce the same effects."
- "Uniformity of Degree" (or "State")—This holds that "geological circumstances have not changed over time."

But for the critics, there are some problems with the uniformity argument by Lyell.

One of the most vocal critics is Stephen Jay Gould, as explained in the following sub-section on the "revision" argument.

The Revision Argument

The third theoretical approach is called, in the absence of better words, the *revision* argument, which focuses, relatively speaking of course, more on a critical reinterpretation of science as "mental constructs," not "empirical discoveries." (F. Sulloway 1987)

Many of these critics are influenced by Thomas Kuhn's *Structure of Scientific Revolutions* (1962), in that "science is a social activity and that theories are intellectual constructions imposed on data, not demanded by them….'Facts' are so embedded in theory that they simply do not have the kind of independent probative power they were once supposed to possess." (F. Sulloway 1987)

In this sense, the word "revision" in the title of the argument is suggestive, because it refers to the revision of conventionally well accepted scientific views.

For illustration, therefore, consider the revison argument by Stephen Jay Gould in his critique of Lyell's uniformitarianism (as discussed above).

The Work of Stephen Gould

The work of Gould "should be seen as part of the generation of historians who have been affected by T. S. Kuhn's *Structure of Scientific Revolutions* (1962). Kuhn argued, in part, that science is a social activity and that theories are intellectual constructions imposed on data, not demanded by them. The views of Kuhn and other philosophers and sociologists of science have helped historians of science to recognize, as Gould emphatically does, that mental constructs (metaphors, analogies, personal philosophies, imaginative leaps)—not empirical discoveries—are what bring about scientific advance." (F. Sulloway 1987)

In the context of Lyell's four related propositions in uniformitarianism (as indicated above), Gould offered his critique in *Time's Arrow, Time's Cycle* (1987), in that "Lyell conflated two different types of propositions," namely, (a) "a pair of methodological assumptions" (i.e., in relation to the first two propositions) and (b) "a pair of substantive hypotheses" (i.e., in relation to the last two propositions). (WK 2010f)

(a) In regard to "methodological assumptions"—Gould claimed that the first two propositions by Lyell (i.e., "uniformity of law" and "uniformity of methodology") are not really empirical hypotheses but only working methodological assumptions.

For Gould, "the methodological assumptions are universally acclaimed by scientists, and embraced by all geologists. Gould further states that these philosophical propositions must be assumed before you can proceed as a scientist doing science. 'You cannot go to a rocky outcrop and observe either the constancy of nature's laws or the working of unknown processes. It works the other way around.' You first assume these propositions and 'then you go to the out crop of rock.'" (WK 2010f)

(b) In regard to "substantive hypotheses"—Gould rejected Lyell's last two propositions (i.e., "uniformity of kind" or "rate," and "uniformity of degree" or "state") "as an unjustified limitation on scientific inquiry, as it constrains past geologic rates and conditions to those of the present." (WK 2010f)

However, the view by Gould and his followers is not without some problems. For illustration, consider a few criticisms below.

Firstly, one major criticism of the view held by Gould and his followers is that the same historicist critique against Lyell can also be directed against Gould himself.

In other words, when Gould told us, according to Sulloway (1987), that "scientists are actually influenced by history all the time,"—this same critique can also be directed against Gould himself, because in this postmodernist age of ours, there is the obsession with revisionist (deconstructivist) critcisms, to the extent that there is also a need to go beyond the "postmodernist" discourse.

This is something that I already originally proposed as "after-postmodernity" in *The Future of Human Civilization* (2000), in that the postmodernist project is to be transcended too.

And secondly, another major criticism of the view held by Gould and his followers is that the truth may lie somewhere between the opposing views.

For instance, for many geologists nowadays, "Earth's history is a slow, gradual process punctuated by occasional natural catastrophic events that have affected Earth and its inhabitants. In practice it is reduced from Lyell's conflation to simply the two philosophical assumptions. This is also known as the principle of actualism (geology), which states that all past geological action was like all present geological action. The principle of actualism is the cornerstone of paleoecology." (WK 2010f; CE 2007)

The Resettlement Argument

And fourthly, unlike the previous three arguments (viz., the catastrophe, uniformity, and revision arguments)—the fourth one to be introduced here can be called, in the absence of better words again, the *resettlement* argument, which learns from all other theoretical approaches (which include the three previous arguments and also other examples as will be introduced in the rest of this book), without, however, favoring any of them, nor trying to integrate them (as they are not necessarily compatible with each other, as the introduction of the theoretical debate in the previous sub-sections has shown).

The Resettlement Theory of Geology

My *resettlement* argument can be more precisely called *the resettlement theory of geology*, since it is to propose a new theory of chess.

My argument is related to "resettlement" because one of its theses has to do with my vision of the post-human resettlement outside Earth in future eras, with profound implications for the need to go beyond "Earth geology."

With this clarification in mind—the resettlement theory of geology has four distinctive features to remember.

Firstly, my theory makes good use of all theoretical approaches in the literature on geology, be they about the catastrophe argument, the uniformity argument, and the revision argument—especially from the most comprehensive combined perspectives of the mind, nature, society, and culture (as will be clear in Chapter Two and Chapter Three).

Secondly, just like many other theories of mine in my previous books, my theory here does not heavily favor any specific theory over others in the literature, nor trying to integrate them (as they are not necessarily compatible with each other).

Thirdly, my theory treats the issue of geology in the distinctive dialectic context of statics and dynamics in geology—especially in the larger dialectic context of no statics without dynamics, and vice versa, together with the subsequent transcendence of both.

And fourthly, it contains sixteen major theses, namely, (a) the first thesis: the absoluteness-relativeness principle, (b) the second thesis: the predictability-unpredictability principle, (c) the third thesis: the explicability-inexplicability principle, (d) the fourth thesis: the preciseness-vagueness principle, (e) the fifth thesis: the simpleness-complicatedness principle, (f) the sixth thesis: the openness-hiddenness principle, (g) the seventh thesis: the denseness-emptiness principle, (h) the eighth thesis: the slowness-quickness principle, (i) the nineth thesis: the expansion-contraction principle, (j) the tenth thesis: the theory-praxis principle, (k) the eleventh thesis: the convention-novelty principle, (l) the twelveth thesis: the evolution-transformation principle, (m) the thirteenth thesis: the symmetry-asymmetry principle, (n) the fourteenth thesis: the regression-progression principle, (o) the fifteenth thesis: the sameness-difference principle, and (p) the sixteenth thesis: the post-human resettlement, in relation to the three great future transformations of geology—to be elaborated in the rest of the book, with a summary in the concluding chapter.

Of course, as this is something that I often emphasized in my previous books, other principles (besides the 15 as cited above) are also relevant, but these 15 are the most relevant in the current case study (in terms of the number of citations of each principle in the book).

Even then, in some cases, the difference between any two given principles, for instance, in terms of the number of citations in a book, is rather small, so the reason in those cases is more aesthetic (than otherwise), because it looks nicer to list only 15 theses for 15 principles (than 21 theses for all of the 21 principles) in the Table of Contents.

This is true, even if different studies of the same kind can yield different views about the degree of relevance for each principle, depending on the specific nature of a rcsearch in question, needless to say. So, if a different author analyzes the same subject matter in a different way, the relevance of the principles will be different.

With this clarification in mind—the seminal project is to fundamentally change the way that we think about geology, from the combined perspectives of the mind, nature, society, and culture, with enormous implications for the human future and what I originally called its "post-human" fate.

Theory and Meta-Theory

The *resettlement* argument, or the resettlement theory of geology, presupposes some kind of meta-theory, in special relation to methodology and ontology—just as all theories, either explicitly or implicitly, require their own versions of meta-theory.

In my case, as this is something that I often explained in my previous books, I already proposed my distinctive approach to methodology (known as "sophisticated methodological holism") and ontology (known as "existential dialectics").

The next two sections are to introduce my meta-theory, or existential dialectics and sophisticated methodological holism, respectively—to be repeated, *often verbatim* (with only a few updated revisions), from my previous works, as this is something that I often do in each new book of mine as an introduction to some background information for the convenience of the readers.

A Unified Theory of Everything

In the end, my distinctive approach to meta-theory serves as a foundation to unify all domains of knowledge for an overarching understanding of all forms of reality (by way of some ontological principles at the meta-theoretical level and the comprehensive perspectives of the mind, nature, society, and culture at the theoretical level—as will be explained in the next two sections).

This is so, without commiting the sins of reductionism and reverse-reductionism (as often seen in many holistic approaches, with the fad of systems approach as a most recent notorious example, as already critically explained in *The Future of Post-Human Formal Science*, or *FPHFS*), since my distinctive approach makes good use of different schools of thought without favoring any of them nor trying to integrate them (as they are not necessarily compatible with each other), so as to include (or adjust for) heterogeneity, conflict, subjectivity, and complexity, for example.

In this sense, the word "unified" does not have to mean an integrative approach, in a narrow sense, nor a systems approach, in a broad one, both of which I have critically rebuked (in my previous books), while learning from them.

This overarching project (consisting of numerous theories of mine, as will be summarized in the next section), thus, constitutes my grand unifed theory of everything in the world and beyond.

The Logic of Existential Dialectics

To start, the ontology of existential dialectics can be summarized in this section—*often verbatim* (with only a few updated revisions), from my previous works, as this is something that I regularly do in each new book of mine as an introduction to some background information for the convenience of the readers.

After all, this book, like all others of mine, is in conversation with all previous ones of mine, for the final aim to converge all of them into a larger grand project on the future of intelligent life, both here on earth and there in deep space unto multiverses.

The summary can be organized in relation to (a) the conception of existential dialectics (or its ontics), (b) the syntax of existential dialectics (or its ontomethodology and ontologic), (c) the semantics of existential dialectics (or its ontosemantics), and (d) the pragmatics of existential dialectics (or its ontopragmatics), respectively hereafter—again *often verbatim*.

This means that, for those readers who had read some of my previous books, the summary below is really not much new (*almost verbatim*, with only some updated revisions).

But for those who have never read my previous books, it is a good review of them.

The Conception of Existential Dialectics (or Its Ontics)

This ontology was originally designated as "existential dialectics" in *Beyond Capitalism to Post-Capitalism* (herein abbreviated as *BCPC*), although it was already analyzed in other books of mine like the 2 volumes of *The Future of Human Civilization* (*FHC*), *The Future of Capitalism and*

Democracy (*FCD*), *The Future of Post-Human Consciousness* (*FPHC*), and the 2 volumes of *Beyond Democracy to Post-Democracy* (*BDPD*).

Lest any confusion occurs, it is important to stress at the outset that the word 'existential' in "existential dialectics" has nothing to do with Existentialism, which I rebuked in *FHC*, *FCD*, and also *FPHC*. Instead, the word here merely refers to the existence of intelligent life (both primitive and advanced) in a broad sense.

Some Basic Concepts

The conception of existential dialectics (or its ontics) makes use of different concepts (like "sets," "elements," "relations," "operations," "functions," "truth values," "axioms," "postulates," and "principles"—as shown in *Table 4.3*), which are important for the understanding of any logic of ontology.

That clarified—existential dialectics, as a language of ontology, can be analyzed in different ways that a language, as an analogy, is often analyzed, namely, in relation to phonology ("the study of patterns of a language's sounds"), phonetics ("the study of the physical aspects of sounds of human language"), morphonology ("the study of the internal structure of words"), syntax ("the study of how words combine to form grammatical sentences"), semantics ("the study of the meaning of words [lexical semantics] and fixed word combinations [phraseology]"), and pragmatics ("the study of how utterances are used…literally, figuratively, or otherwise…in communicative acts"). (WK 2007)

But since my theory of existential dialectics makes use of the English language for communication here, it does not propose a new way to make the patterns of a language's sounds (as in phonology), to study the physical aspects of sounds (as in phonetics), or even to strive for a new organization of the internal structure of words for the English language (as in morphonology).

For this reason, the language of existential dialectics to be analyzed hereafter is not concerned with the phonological, phonetic, and morphonological aspects of the English language. Instead, the analysis will explore the syntax, semantics, and pragmatics of existential dialectics as a language of ontology to improve our understanding of reality.

The Ontic Dispute

But first, there is an ontic dispute to be addressed, which is related to what I called "the dilemma of ontology" as introduced earlier in *Sec. 1.4* of *The Future of Post-Human Knowledge* (*FPHK*).

A good point of departure concerns what I want to call *the ontic dispute*, on how to understand the very nature of ontology.

Traditionally, scholars in the field used to study the conception of ontology in terms of its entities (e.g., material objects, abstract numbers), properties (e.g., duration, plurality), and relationships (e.g., causation).

This is a good starting point, but it suffers from different problems. Consider three examples of bias below, which are quite common in the literature, and the recent version (of treating concepts and principles in ontology on the basis of "plurality," "dynamism," "duration," "interaction," "life," "consciousness," and "volition") by Reginald Firehammer (2005) only constitutes a latest addition to this old (impoverishing) tradition.

Firstly, the classification of entities, properties, and relationships is too material-centric, in often focusing more on the existence of material entities than on non-material ones.

Secondly, the classification is too anthropo-centric, in often favoring the factor of human agency in depicting reality.

And lastly, the classification is too confused among the three categories relating to each other, in often lumping them together without adequate consideration of their intricated interrelationships (both in relation to each and in relation to all).

A good alternative is precisely what I want to offer in this debate, that is, a new classification of ontological entities, properties, and their relationships in terms of *method*, *structure*, *process*, *agency*, and *outcome* instead. Let me explicate what these words mean below.

In relation to "method"—it concerns how ontology is to be studied. In this sense, the word 'method' here should not be confused with the different usage of the same term 'method' as a technique of investigation in the context of methodology (as already introduced in *Sec. 1.2* of *FPHK*).

In relation to "structure"—it concerns what the general patterns of entities, properties, and relationships are.

In relation to "process"—it concerns how these general patterns change over time.

In relation to "agency"—it concerns how agency affects any change of these general patterns over time (without putting, however, too much emphasis on the role of human agency).

And in relation to "outcome"—it concerns what the outputs of this change over time are.

The Syntax of Existential Dialectics (or Its Ontomethodology and Ontologic)

The syntax of existential dialectics, analogously speaking, refers here to the structure of an ontology which can be studied by way of combining ideas into complex relationships like ontological principles to depict reality in the world—and can be analyzed in seven sub-sections, namely, (a) the selection criteria, (b) further clarifications on classification, (c) ontomethodology, (d) ontologic, (e) the principles as short cuts, (f) the principles as family resemblances, and (g) the dialectic constraints imposed by the principles, respectively hereafter.

The Selection Criteria

To start, not any pair of relationship can be chosen as an ontological principle in existential dialectics; otherwise, there could be billions of them under the sky.

Four illustrations below suffice to clarify this point.

(a) Firstly, an ontological principle cannot *overlap* with other principles, in that it is too closely related with other ones.

Of course, one cannot totally rule out any relationship between two entities, especially in a complex world of everything being linked to everything else—but the comparison here is relative, not absolute.

In this sense, the flexibility-inflexibility pair cannot be an ontological principle, since it has something closely in common with other ones (like the simpleness-complicatedness principle, for instance).

As an illustration, something which is flexible in interpretation is more likely to allow multiple viewpoints and thus makes the overall picture more complicated than otherwise (simple).

(b) Secondly, an ontological principle cannot be *redundant* in relation to other principles, in that it somehow duplicates other principles.

In this sense, for instance, the directness-indirectness pair cannot be an ontological principle, because it somehow duplicates other ontological principles (like the preciseness-vagueness principle, for instance).

As an illustration, something which is direct means to be right to the point and does not need to go to an unnecessarily long loop—so it is less vague, in being clearer about the thing in question.

(c) Thirdly, an ontological principle cannot be *derived* from other ones—as if it is a child-parent relationship.

In this sense, the convenience-inconvenience pair cannot be an ontological principle, because it can be derived from other ones (like the slowness-quickness principle, for instance).

For instance, something which is convenient already implies that it can be obtained quickly (rather than slowly)—all things considered.

(d) And fourthly, an ontological principle cannot be *trivial* (or *parochial*)—as if it is only one component competing with other ones for inclusion in a set.

In this sense, the consistency-inconsistency pair cannot be an ontological principle, because its validity is limited (or parochial), as it can be easily challenged and replaced by opposing major (not minor) theories (e.g., the correspondence theory of truth, the pragmatic theory of truth, the semantic theory of truth, and whatnot), not just in accordance to the consistency theory of truth (for consistency).

In this sense, the consistency argument is only one among different opposing arguments in relation to the logic of what constitutes "truth."

Although these examples are not exhaustive, they illustrative the selectiveness of any ontological principle to be existentially dialectic.

Further Clarifications on Classification

Lest any misunderstanding carelessly occurs, some further clarifications on classification should be made here (and summarized in *Table 4.8*).

(a) Firstly, the total number of ontological principles is unknown, to be discovered later, as our knowledge of the world becomes more advanced.

Consequently, the principles as introduced in my books are not exhaustive, with new ones being added, whenever more of them are discovered in later research.

At least, future generations can pick up where I leave off and continue the discovery.

In this sense, my work should be treated as a pioneering effort for the development of a systematic, comprehensive analysis of a new general ontology for the future of knowledge.

(b) Secondly, the principles are not rigidly classified, as they can be reclassified in a different way.

For instance, the preciseness-vagueness principle is classified under the category of "structure" but can be reclassified under the category of "method," although in so doing, it has a different meaning in the context of method.

The same logic applies to the same-difference principle under the category of "outcome," which can be reclassified under the category of "structure," although in so doing, once more, it has a different meaning in the context of structure.

(c) Thirdly, the principles are to be understood together, not that each prinicple is to be treated independently of others. After all, the principles are mutually constraining, in that they work together as a whole.

(d) Fourthly, the principles are relevant to all subject matters, but some principles are, relatively speaking of course, more useful to some subject matters than others under certain conditions—as implied in the symmetry-asymmetry principle.

(e) And fifthly, the principles constitute only two levels of analysis, this time, at the ontological and methodological levels—while other levels of analysis (from the perspectives of the mind, nature, society, and culture) are also needed, in order to understand reality in its totality (as explained in the section on "sophisticated methodological holism," so as to avoid the dual dangers of reductionism and reverse-reductionism).

Ontomethodology

With these selection criteria and clarifications in mind—the syntax of existential dialectics can be divided into two main smaller sections for analysis, since they are related to each other in the context of some fundamental principles, that is, (a) ontomethodology and (b) ontologic, to be addressed hereafter, respectively.

In the context of ontomethodology, some ontological principles are proposed here in relation to the nature of methodology, which however is tied up with the general issue of ontology in the context of knowledge.

In this sense, ontomethodology (in the context of these ontological principles in the category of method) are related with ontologic in general (in the context of ontological principles in general).

Some good instances of ontological principles involving ontomethodology include the formalness-informalness principle, the predictability-unpredictability principle, the explicability-inexplicability principle, the partiality-totality principle, and the absoluteness-relativeness principle, which will be introduced in conjunction with ontological principles in the ontologic of existential dialectics in general hereafter.

Ontologic

With this clarification of the relationship between ontomethodology and ontologic in mind (which is comparable to the relationship between methodology and ontology as described in *Sec. 1.3* of *FPHK*)—the first three principles in existential dialectics were introduced in *BCPC*. And more principles were later added to the logic in my subsequent books (like *FC*, *FAE*, and others).

More specifically, in *BCPC*, I proposed three principles (based on previous works of mine, not just something out of the blue) for the ontological logic, namely, (a) the regression-progression principle on the direction of history, (b) the symmetry-asymmetry principle on the relationships among existents, and (c) the change-constancy principle on the alteration of things—or in a more elegant term, the dynamics of space-time (in *FC*).

In *The Future of Complexity* (*FC*), I added three more principles, on top of the three aforementioned, namely, (d) the partiality-totality principle on the relationships between the parts and the whole, (e) the order-chaos principle on the pattern of things, and (f) the predictability-unpredictability principle on the occurrence of things—as part of the ontological logic of existential dialectics.

In *The Future of Aesthetic Experience* (or simply *FAE*), another principle is added, namely, (g) the evolution-transformation principle (on the multiple kinds of agency).

Also in *FAE*, I then reclassified the 7 principles into four categories, namely, (i) in relation to method, for the partiality-totality principle and the predictability-unpredictability principle, (ii) in relation to process, for the change-constancy principle and the order-chaos principle, (iii) in relation to agency, for the symmetry-asymmetry principle and the evolution-transformation principle, and (iv) in relation to outcome, for the regression-progression principle.

In *The Rise of Authoritarian Liberal Democracy* (*ALD*), I added two more principles, namely, (h) the softness-hardness principle on the force of agency (which is to be classified under the category about agency) and (i) the same-difference principle on the metamorphosis of change (which is to be classified under the category about outcome).

In *The Future of Information Architecture* (*FIA*), I introduce three additional principles, that is, (j) the simpleness-complicadness principle on the interconnection among things, (k) the preciseness-vagueness principle on the refinement of things, and (l) the slowness-quickness principle on the speed of change—with the first two in relation to structure and the third in relation to process.

In *The Future of Post-Human Unconsciousness* (*FPHU*), I added three more principles, that is, (m) the openness-hidden principle on the detection of things, in relation to structure, (n) the explicability-inexplicability principle on the underlying mechanisms of things, in relation to method, and (o) the expansion-contraction principle on the growth of things, in relation to process.

In *The Future of Post-Human Knowledge* (*FPHK*), I added one more principle, that is, (p) the absoluteness-relativeness principle on the multiplicity of entities, in relation to method.

In *The Future of Post-Human Mathematical Logic* (*FPHML*), I added one more principle, that is, (q) the formalness-informalness principle on the formal requirements of systems, in relation to method.

In *The Future of Post-Human Engineering* (*FPHE*), I added one more principle, that is, (r) the theory-praxis principle on the duality of knowledge, in relation to agency.

In *The Future of Post-Human Creative Thinking* (*FPHCT*), I added one more principle, that is, (s) the convention-novelty principle on the nature of creative thinking, in relation to agency.

In *The Future of Post-Human Geometry* (*FPHG*) here, I added one more principle, that is, (t) the finiteness-transfiniteness principle on the nature of numbers, in relation to structure.

And in *The Future of Post-Human Urban Planning* (*FPHUP*), I added one more principle, that is, (u) the denseness-emptiness principle on the distribution of entities in space, in relation to structure.

With this update, there are twenty-one principles so far in existential dialectics, that is, five principles for the category in relation to method (viz., the formalness-informalness principle, the partiality-totality principle, the predictability-unpredictability principle, the explicability-inexplicability principle, and the absoluteness-relativeness principle), five principles for the category in relation to structure (viz., the finiteness-transfiniteness principle, the simpleness-complicatedness principle, the preciseness-vagueness principle, the openness-hiddenness principle, and the denseness-emptiness principle), four principles for the category in relation to process (viz., the change-constancy principle, the slowness-quickness principle, the order-chaos principle, and the expansion-contraction principle), five principles for the category in relation to agency (viz., the theory-praxis principle, the convention-novelty principle, the symmetry-asymmetry principle, the evolution-transformation principle, and the softness-hardness principle), and two principles for the category in relation to outcome (viz., the regression-progression principle and the same-difference principle).

There are thus twenty-one principles in existential dialectics—so far (as summarized in *Table 4.4*).

The Principles as Short Cuts

Yet, these principles should be treated with caution, lest misunderstanding occurs, since they do not constitute rigid dualities (or dichotomies).

The reason is that each pair in an ontological principle consists of two opposites, which are, however, merely short cuts both for multiple variations and degrees, as well as for different interactions with multiple other entities unlike them.

In this light, each pair can end up having hundred (or even thousand, if not more) different versions, which interact with hundred (or even thousand, if not more) other entities—as summarized in *Table 4.5*.

There are two clarifications here.

Firstly, this conception of shortcuts is not mutually exclusive nor absolute, in that the opposites can come in all shapes and sizes, with different degrees. For instance, by analogy, just as there are different degrees of the two colors "white" and "black"—there are likewise different degrees of the opposites in each ontological pair, to the extent that there can be multiple entities (not only two) interacting with each other in each pair.

And secondly, this conception of shortcuts do not ignore other possible entities in interacting with the two opposites (with their different versions) in each pair. For instance, by analogy, there are not only the two colors "white" and "black" as opposites, since there are other colors too besides them like "yellow," "green," "purple," or else. The same logic can be applied to each ontological pair, in that they also interact with other entities, not with only two of them (with their different versions).

Therefore, with these two clarifications in mind—each pair in an ontological principle serves only as an abbreviation for something more complicated and, therefore, although it contains two opposites, it should not be confused as a duality (dualism).

Existential dialectics rejects any dualism (or dichotomy) as too rigid and instead allows the multiplicity of entities, to the extent that between the two opposites in each pair exist many other alternatives to choose from. They are named in that short form for aesthetic elegance only, instead of listing all possible entities between the two opposites in the title (which would be too cumbersome and sound extremely awkward).

To be dialectic is to go beyond any rigid dichotomy and transcend into something different altogether in the long haul. One may be tempted to call the dialectic logic here with a different name like existential "multilectics" (instead of "dialectics"), but this naming is incorrect (or even misleading), for the two reasons aforestated.

After all, all the possible entities between any two opposites in an ontological pair are still about the two opposites, although each ontological pair come in all shapes and sizes (with different degrees) and do not exist by themselves but also interact with other entities unlike them. The virtue of revealing an ontological pair is to show how they relate within themselves (in multiple versions) and also interact with others unlike them (also in multiple versions).

This will be clear in a later sub-section titled "the dialectic constraints imposed by the principles."

The Principles as Family Resemblances

In addition, each ontological principle is generic, with some other comparable ontological pairs to be put in the same family (like a *family resemblance*)—as summarized in *Table 4.6.*

Ludwig Wittgenstein (1953) once suggested the idea of "family resemblance" in explaining different games classified under the same family called *games.*

Why should, for instance, playing football and chess as playing "games," when it is well understood that football is not the same as chess? The answer is that, although each game is different and has different rules, many of them (though not all) share, more or less, some commonalites (e.g., scoring as necessary for winning).

And this is so, even though some games share more than some others in any given selection of critieria, and no two games are exactly identical. So, Wittgenstein's point here is that there is no essential core which is common to all games, and the best that one can look for is some characteristics which are common to many (but not all) games. (A. Biletzki 2006)

(a) Appropriate Family Resemblances

In the context of existential dialectics, I want to distinguish two types of family resemblances, and, in the absence of better words, let me call them (a) *appropriate* family resemblances and (b) *inappropriate* family resemblances.

In regard to appropriate family resemblances—good examples include some of the pairs as already explained in an earlier section on selection criteria, like the flexibility-inflexibility pair, which can be put in the family resemblance of the simpleness-complicatedness principle, although the two pairs are not exactly identical. Likewise, the directness-indirectness pair can be put in the family resemblance of the preciseness-vagueness principle, although, again, the two pairs are not exactly identical.

In this sense, which specific pair in a family should be used to designate the name of the family can be at times a bit arbitrary, but with good reason.

For illustration, in the context of method, the partiality-totality principle can take the different form like individualisticness-holisticness, just as the explicability-inexplicability principle can take the different form like underlyingness-regularness—although each two pairs are not exactly identical and have slightly different meanings and usages.

A more comprehensive listing of this family resemblance for all other principles is shown in *Table 4.6* on the syntax of existential dialectics.

In fact, this section on family resemblances overlaps with the semantics of existential dialectics, which will be introduced later in the next section (and summarized in *Table 4.11*).

(b) Inappropriate Family Resemblances

However, there are some family resemblances which are not appropriate, and let me distinugish two of them by calling them, in the absence of better words, (b1) *inadequate* family resemblances and (b2) *compound* family resemblances (as summarized in *Table 4.10*).

(b1) In regard to "inadequate" family resemblances—there are some pairs which are not adequate, because they fail the test of selection criteria (as already pointed out in an earlier section), especially in relation to the criterion of not being *trivial* (or *parochial*)—as if it is only one component competing with other ones for inclusion in a set.

For instance, the *consistency-inconsistency* pair is inadequate, because it is only one component competing with other ones for inclusion in the larger set of the formalness-informalenss principle (like the *completeness-incompleteness* pair and the *soundness-unsoundness* pair, as discussed in my book titled *The Future of Post-Human Mathematical Logic*, or *FPHML* in short).

The same is true for the *cognition-noncognition* pair, which overlaps with both the completeness-incompleteness pair and the soundness-unsoundness pair in the larger set of the formalness-informalenss principle.

(b2) In regard to "compound" (or non-atomic) family resemblances—there are some pairs which are not appropriate for a different reason, because they fail the test of selection criteria (as already pointed out in an earlier section), especially in relation to the criterion of not *overlapping* with other principles, in that it is too closely related with other ones.

For instance, the *normality-abnormality* pair is compound (and thus inappropriate), because it overlaps with several other principles, like the symmetry-asymmetry principle, the density-emptiness principle, the convention-novelty principle, and the same-difference principle, for instance (as already discussed in *FPHP*).

The normality-abnormality pair overlaps with the symmetry-asymmetry principle, because being "abnormal" (say, for those with mental illnesses) is often treated not equally by others who are "normal." So, the relationship reveals the symmetry-asymmetry principle.

The normality-abnormality pair also overlaps with the density-emptiness principle, because being "abnormal" (say, for those with mental illnesses) is often not as dense in the overall (not local) distribution of a population as those who are "normal." So, the relationship reveals the density-emptiness principle.

The normality-abnormality pair also overlaps with the convention-novelty principle, because being "abnormal" (say, for those with mental illnesses) often exhibits non-conventional (non-conformist) ideas and behaviors when contrasted with those who are "normal." So, the relationship reveals the convention-novelty principle.

The normality-abnormality pair also overlaps with the same-difference principle, because being "abnormal" (say, for those with mental illnesses) shares some similarities in outcome when compared with those who are "normal" (in that both contribute to the control and regulation of human thoughts and behaviors in society, albeit in different ways). Yet, being "abnormal" also has something different from being "normal," in that abnormal thoughts and behaviors follow a different pattern when contrasted with those which are normal. So, the relationship reveals the same-difference principle.

These illustrations of indequate and compound family resemblances show how difficult it is to come up with an ontological principle in existential dialectics out of the blue, because many of them violate the selection criteria in one way or the other. It is all too easy to join two opposing words and simply make them into a pair, but it is all too difficult to pass the test of the selection criteria to be qualified as an ontological principle in existential dialectics.

Other examples of compound (and inadequate) family resemblances are shown in *Table 4.10* on "The Syntax of Existential Dialectics VII: Types of Inappropriate Family Resemblances."

The Dialectic Constraints Imposed by the Principles

The principles, as they constitute the syntax of existential dialectics, are dialectic in character, such that, when they are applied, they impose dialectic constraints on how reality is to be understood.

Consider, say, the symmetry-asymmetry principle as an illustration here, in order to summarize two main characters of the dialectic constraints in question (as summarized in *Table 4.7*).

Firstly, to be dialectic here is to go beyond the narrow dichotomies (and, for that matter, any rigid multi-dimensional classificatory scheme), be they about "self" vs. "world," "freedom" vs. "unfreedom," "barbarity" vs. "civilization," "individuality" vs. "communality," and so on.

One way to do so (to go beyond) is to consider them all in terms of co-existence (without favoring one over the rest). For instance, my theory of "post-civilization" (to be summarized later in the section on the pragmatics of existential dialectics) is to go beyond barbarity and civilization in terms of understanding barbarity and civilization as being co-existent.

And the same logic can be said in relation to my theories of "post-democracy," "post-capitalsm," and others (also to be introduced later in the section on the pragmatics of existential dialectics), in regard to freedom vs. unfreedom, equality vs. inequality, communality vs. individuality, spirituality vs. materiality, and so on.

But to consider them all (in the dichotomies—and, for that matter, in any rigid multi-dimensional classifactory scheme) as co-existent is not the same as to imply that the opposing categories in any classificatory scheme are all equal, since, in accordance to the symmetry-asymmetry principle (as an illustration here), if they are equal in terms of being considered as co-existent, they are asymmetric in terms of being unequal in dialectic interaction (e.g., X can be more dominant than Y in case A, or Y is more dominant than X in case B), although in some special cases, they can be relatively equal (e.g., X and Y are relatively equal in case C).

For this reason, there are different versions of "post-democracy" and "post-capitalism" in my theories. As an illustration, in version I of the theory of post-democracy, freedom is more dominant than equality, whereas in version II of the theory of post-democracy, equality is more so than freedom.

But this "X more than Y" has to be understood in the context of dialectic logic (not in symbolic logic, as conventionally understood), in that both "X" and "Y" are important in post-democracy (in the context of dialectic logic), but in an asymmetry way.

By contrast, in symbolic logic, it often favors one over the other—be it in regard to privileging freedom over equality (in Fascism), favoring freedom relatively more than equality (in Liberal Democracy), or favoring equality relatively more than freedom (in Socialist Democracy). In regard to the latter two cases (about Liberal Democracy and Socialist

Democracy), the difference between dialectic logic and symbolic logic can be one in degree, not in kind—in this sense, albeit not in other senses.

The same reasoning can be said about the relatonships between individuality and communality, between spirituality and materiality, and between formal legalism and informal legalism in the different versions of my theory of "post-capitalism."

Secondly, to be dialectic is to go beyond the narrow dichotomies (and, for that matter, any rigid multi-dimensional classificatory scheme) in another way, this time, in a transcendent way, that is, in exploring other possibilities or even other issues not considered within the narrow confines of narrow dichotomies (and, for that matter, any rigid multi-dimensional classificatory scheme).

As an analogy, in this second meaning, to go beyond the narrow dichotomy of "black" vs. "white" is not just to choose both "black" and "white" (as in the first meaning) but also to explore other options (e.g., "green," "purple," "blue," etc.—and, alternatively, "shade," "line," "curve," etc.).

By the same reasoning, to go beyond "democracy" is to transcend democracy (as in version III of my theory of "post-democracy") and to explore other possibilties of lifeforms (e.g., "floating consciousness," "hyper-spatial consciousness," etc., to live beyond the narrow obsession with freedom and equality).

This dialectic character of the principles in existential dialectics has important implications for the pragmatics of existential dialectics (as will be clear shortly, in the section on the pragmatics of existential dialectics).

The Semantics of Existential Dialectics (or Its Ontosemantics)

The syntax of existential dialectics so understood in terms of ontological principles only gives us the structure of ontology in the world, in an abstract (general) sense. These principles by themselves do not tell us the specific meanings in a given context.

In order to grasp the specific meanings of the principles in a given context, it is necessary to study the semantics of existential dialectics (as summarized in *Table 4.11*).

And the analysis of the ontological principles as family resemblances in the previous section on syntax is only a starting point (and thus overlaps a bit with the semantics of existential dialectics here).

For this reason, I have often gone to great lengths in my previous books on different subjects to explain the specific meanings of the principles when applied in different contexts.

Consider two examples below for illustration, which are not exhaustive.

In *FPHST*, for instance, I used the first three principles (i.e., the change-constancy principle, the regression-progression principle, and the symmetry-asymmetry principle) to propose "the perspectival theory of space-time," for a better way to understand space and time—especially, though not exclusively, in relation to future post-human history (as summarized in *Table 1.3*, *Table 1.4*, *Table 1.5*, and *Table 1.6*).

In so doing, I had to introduce concepts and theories specific to the field of physics and other related fields (e.g., "absolute space" and "absolute time" in "classical mechanics" and "relative space-time" in "the theory of relativity").

And in *BNN*, I also exploited the three principles to propose the "transcendent" approach to the study of genes and memes as a new way to understand the interaction between nature and nurture.

In so doing, I had to explore concepts and theories in the world of evolutionary biology (e.g., "mutation," "variation," "adaptation," "selection," and "inheritance" in classical evolutionary theory) and neural biology (e.g., "chromosome," "gene," "DNA," "RNA," "protein," "neuron," "neural network," and "behavior").

Other examples, besdies these two here, are given in other books of mine on different topics.

Therefore, this semantic dimension of existential dialectics is important, since, in each of my books when the ontological principles are used, each principle becomes relevant in a different meaning when contextualized within a given case study in question.

In this book, for instance, the principles are relevant in relation to the subject matter of cosmology (as summarized in the concluding chapter), and they can be shown in all shapes and sizes (since they are also to be treated as "short cuts," as indicated before), without, however, being reduced to "specific ontology" (as opposed to "general ontology"), as explained in *Table 4.9*.

The Pragmatics of Existential Dialectics (or Its Ontopragmatics)

Even the semantics of existential dialectics is not enough to understand reality, since semantics by itself does not tell us the nature of intentionality and interests in relation to the matrix of power formations and conflicting interests when existential dialectics as an ontology is applied.

Surely, one cannot completely separate the pragmatics of existential dialectics from its semantics (and for that matter, its syntax), but the point here is to give each a more elaborated analysis.

For this reason, the pragmatics of existential dialectics needs to be studied too, in special relation to the inclusion of power formation and conflicting interests in the application of the principles of existential dialectics (as summarized in *Table 4.12*).

It is thus no wonder that all my previous books have shown in different ways how and why these ontological principles and their theoretical applications can reveal the future world to come in a way that humans have never known, in a totally different battleground for competing human (and later post-human) interests by myriad groups fighting for their own versions of hegemony.

Perhaps a summary of my previous works in relation to existential dialectics and their contributions to my new theories as proposed over the years is deemed revealing here, for the further understanding of the pragmatics of existential dialectics (in close relation to the semantics of existential dialectics too).

That stressed—the following summary in four sub-sections is something that I often do in my previous books too, as a way to introduce the intellectual background of my ideas.

So, for those readers who had read my previous works, the following sub-sections serve as a reminder, as they are repeated hereafter *often verbatim* (with only some updated changes, with the rest kept intact).

These sub-sections are, namely, (a) the two-way street connecting theory with meta-theory, (b) the theoretical application of existential dialectics, (c) direct and indirect applications of existential dialectics, and (d) multiple applications of existential dialectics—to be summaried herefafter, respectively.

The Two-Way Street Connecting Theory and Meta-Theory

However, lest any misunderstanding occurs, it is important to remember that the pragmatics of existential dialectics is not a one-way street (that is, using the ontological principles for theoretical insights in praxis) but a two-way one, namely, (a) from meta-theory to theory, and (b) from theory to meta-theory.

(a) On one side of the street, the ontological principles can inspire some theoretical insights in praxis, that is, in relation to some specific fields.

(b) On the other (opposing) side of the street, however, the study of a subject matter in the specific fields in turn reveals some more hitherto unknown ontological principles to be discovered and identified.

For this reason, three new principles were added in *FC*, one in *FAE*, two in *ALD*, and three in *FIA*, on top of the original three in *BCPC*, for example (as there are more)—after some research on the specific subject matters.

Both theory and meta-theory enrich each other in all my works over the years.

The Theoretical Application of Existential Dialectics

This two-way street connecting theory and meta-theory has contributed to my original construction of 44 specific theories in 34 books (including this one here, up to now, in early 2010) on numerous fields in all domains of knowledge, which can be classified in terms of specific "meta-theories" (about epistemology, methodology, logic, ontology, etc.) and specific "theories" (about the mind, nature, culture, society, and the rest—as will be further explained later in the sub-section on "the holistic organization of an inquiry" in the section about "sophisticated methodological holism").

The final aim of my numerous new specific theories (on both theory and meta-theory) is to converge all of them (as proposed in my numerous books) into a grand project about the future of intelligent life, both here on earth and in deep space.

In fact, all these theories of mine serve as a foundation to unify all domains of knowledge for an overarching understanding of all forms of reality (by way of some ontological principles and the comprehensive perspectives of the mind, nature, society, and culture), while allowing for heterogeneity, conflict, subjectivity, and complexity, for example.

With this grand project in mind—the summary of my specific theories on meta-theory are given in this section (on "existential dialectics") and the next (on "sophisticated methodological holism"), whereas the specific theories on theory will be summarized in the following categories of (a) society, (b) culture, (c) the mind, (d) nature, and (e) the rest, as shown below, in that order.

(a) Society

The category of "society" here can be divided into four sub-categories, i.e., (a1) social organizations, (a2) social institutions, (a3) social structure, and (a4) social systems.

(a1) Social Organizations

In relation to social organizations, I proposed a new thesis on the ambivalent nature of virtual organizations in the future, known as "the theory of e-civic alienation" in *FCD* (and summarized in *Table 4.19*).

And in *FPHO*, another theory called "the combinational theory of organization" is worked out, although it was first proposed in Ch.6 of *ALD*.

(a2) Social Institutions

There are different social institutions, be they about (a1.1) politics, (a2.2) economics, (a2.3) military affairs, and (a2.4) mass media, for instance.

(a2.1) In relation to politics, I offered different visions on (a2.1.1) political theory, (a2.1.2) comparative politics, and (a2.1.3) international relations, to cite three main ones.

(a2.1.1) In the context of political theory, my vision of a path-breaking political system in future times concerns the different forms of "post-democracy" to supersede democracy unto the post-human age.

My vision of post-democracy was called "the theory of post-democracy" (as summarized in *Table 4.22*, *Table 4.23*, and *Table 4.24*), whereas *Table 4.25* explains the distinctions among democracy, non-democracy, and post-democracy.

And *Table 4.26* stresses the multiple causes of the emergence of post-democracy, while *Table 4.27* clarifies some possible misunderstandings in regard to post-capitalism and post-democracy.

(a2.1.2) In the context of comparative politics, I also showed, in *ALD*, how and why authoritarianism can be an advanced form of liberal democracy (just as it has its own counterpart in non-liberal democracy), or what I originally called "authoritarian liberal democracy" (as summarized in *Table 4.21*), before it is to be superceded by "post-democracy."

(a2.1.3) In the context of international relations (both at the institutional and structural levels), I provided, in *BWT*, "the theory of cyclical progression of empire-building," for a better way to understand the logic of empire-building on earth and beyond—with the rise of what I originally envisioned as "hyper-empires" (like "the Chinese Union" and "the Indian Union"), "meso-empires" (like "the European Union" and "the North American Union") and "micro-empires" (like "the Middle Eastern Union" and "the Latin American Union") in the coming "post-post-Cold War era" as a visionary illustration, as summarized in *Table 4.20*.

In addition, in *FPHWP*, I proposed "the multifaceted theory of war and peace" to understand war and peace, in a small sense—or aggression and pacificity, in a broad sense.

(a2.2) In relation to economics, I also offered another vision, this time to understand the future forms of economic systems, that is, the different forms of "post-capitalism" to supersede capitalism unto the post-human age.

My vision of post-capitalism was referred to (in *FCD*, *BDPD* and *BCPC*) as "the theory of post-capitalism" (as summarized in *Table 4.28*, *Table 4.29*, *Table 4.30*, *Table 4.31*, and *Table 4.32*). And *Table 4.33* clarifies the differences among capitalism, non-capitalism, and post-capitalism, whereas *Table 4.34* shows multiple causes of the emergence of post-capitalism.

This vision on "post-capitalism" constitutes the third theoretical application of existential dialectics, this time, in that there is no wealth without poverty (or the wealth/poverty dialectics).

(a2.3) In relation to military affairs, I also proposed "the multifaceted theory of war and peace" (in *FCD*, although it was first so labeled in *FPHK*) for a new way to understand the causes of war—and, for that matter, the rise and fall of different forms of warfare over time.

This theory was further expanded in *FPHWP*, where I suggested a new way to understand war and peace, in a small sense—or aggression and pacificity, in a broad sense.

(a2.4) In relation to mass media, I also proposed "the totalistic theory of communication" in *The Future of Post-Human Mass Media* (or in short, *FPHMM*).

(a3) Social Structure

In relation to social structure, I proposed two additional theories, that is, "the theory of the cyclical progression of hegemony" in *FCD* (and also in *FPHC* and *BDPD*), and "the theory of the cyclical progression of empire-building" (in *BWT*, as already indicated above)—with important implications for the symmetry-asymmetry princple in existential dialectics, just to cite one example.

Also, in *BNN*, I also proposed "the theory of contrastive advantages" to understand the persistence of social structure by gender, race, ethnicity, or any other groupings.

(a4) Social Systems

In relation to social systems, there are several sub-categories for illustration, which can be about (a4.1) urbanization, (a4.2) technology, (a4.3) demography, (a4.4) the environment, and (a4.5) social trends as a whole.

(a4.1) In relation to urbanization, I proposed "the contingent theory of urban planning" in *FPHUP* for a new way to understand density and void.

(a4.2) In relation to technology, I worked out "the ambivalent theory of technology" in *FPHE* for a new way to understand the nature of technology.

Some other books (like *FPHMM*, *FAE*, *FCD*, and *FHC*) also deal with technology, and a good example concerns "the synthetic theory of information architecture" in *FIA* on taxonomy, network, and intelligence, in conjunction with the rise of technology.

(a4.3) In relation to demography, I proposed, first in *FHC* (and later in other books like *FCD*, *FPHC*, *FPHU*, etc.), "the theory of post-humanity," in relation to the "post-human" age at some distant point of "after-postmodernity," long after human extinction, to be eventually superseded by "post-humans" of various forms.

Some good candidates of the post-human lifeforms include, for instance, "thinking robots," "thinking machines," "cyborgs," "genetically altered superior beings," "floating consciousness," "hyper-spatial consciousness," "unfolding unconsciousness," "the hyper-sexual body," and "the hyper-martial body." This post-human vision of mine was first originally worked out in *FHC* and further elaborated in both *FCD*, *FPHC*, *FPHU*, *FPHS*, and *FPHMA*, for instance.

Lest any confusion occurs, I need to emphasize two clarifications here about the term "post-human" as a neologism in my works.

Firstly, the word "post-human" here should *not* be confused with another term which looks similar but has a totally different meaning in the literature of postmodernism, namely, "post-humanism"—which constitutes a critique of "humanism" as tradionally understood (especially, though not exclusively, in relation to the idea of progress in science and reason in the Enlightenment project). (WK 2008)

And secondly, the word "post-human" here should also *not* be confused with a similar term which is used to champion the ideology of technology for the future co-existence between humans and cyborgs in "trans-humanism." (WK 2008a)

Instead, my term "post-human" in relation to "posthuman-ism" also rejects "transhumanism" (especially, though not exclusively, in relation to the promises of technology) and refers to something else altogether, that is, the future extinction of humans and its post-human successors in deep space and beyond unto multiverses.

Later, in *BEPE*, I went further and provided a more comprehensive critique of "transhumanism," together with other issues.

With these two clarifications in mind (as summarized in *Table 1.31*)—in the end, a most fundamental question about intelligent life now has an answer, in that, if asked, "What is the future of human civilization?"—my answer in *FCD* (89) is thus: "As addressed in Ch.7 of *FHC*, a later epoch of the age of after-postmodernity (that is, at some point further away from after-postmodernity) will begin, as what I called the 'post-human' history (with the term 'post-human' originally used in my doctoral dissertation at M.I.T., which was finished in November 1995, under the title *After Postmodernity*, still available at M.I.T. library, and was later revised and published as *FHC*). The post-human history will be such that humans are nothing in the end, other than what culture, society, and nature (with some luck) have shaped them into, to be eventually superseded by post-humans (e.g., cyborgs, thinking machines, genetically altered superior beings, and others), if humans are not destroyed long before then."

(a4.4) In relation to the environment, I also addressed the issue of sustainability, especially in *FHC*, *FCD*, and *FPHUP*. For instance, in *FPHUP*, I specifically proposed a new theory, "the contingent theory of urban planning," on the need to go beyond the contemporoary obsession with sustainability, in the context of not only the environment but also other issues.

(a4.5) In relation to social trends as a whole, I proposed in Ch.9 of *FCD* (367-8), that "civilizational history will continue into the following cyclical progression of expansion, before it is to be superseded (solely as a high probability, since humans might be destroyed sooner either by themselves or in a gigantic natural calamity) by posthumans at some distant point in after-postmodernity (which I already discussed in *FHC*)" unto multiverses (different constellations of universes):

Local → Regional → Global →Solar →
Galactic → Clustery...→ Multiversal

In *BDPD*, this thesis of mine was referred to as "the theory of the cyclical progression of system integration and fragmentation," at the systemic level—in close relation to another thesis analyzed in Ch.9 of *FCD*, which was called, in the absence of better words, "the multifaceted theory of war and peace" in *FPHK* (as already indicated above).

In *FC*, I also proposed "the dialectic theory of complexity" to account for the factor of uncertainty, which can have a major role to play in the world.

(b) Culture

The category of "culture" here can also be divided into different sub-categories like (b1) aesthetics, (b2) religion, (b3) popular (mass) culture, (b4), logic, (b5) civilization as a whole, (b6) ethics, (b7) epistemology (methodology), (b8) metaphysics (ontology), and (b9) ideology.

However, (b7) epistemology (methodology), (b8) metaphysics (ontology), and (b9) ideology are already discussed elsewhere, be they about "meta-theory" (e.g., methodology and ontology under the categories of "existential dialectics" here and "sophisticated methodological holism" in the next section) or "theory" (e.g., ideology under the categories of "social institutions," as described earlier, and also of "civilization as a whole," as will be addressed shortly).

(b1) Aesthetics

In relation to aesthetics, I proposed "the transformative theory of aesthetic experience" in *FAE* about some great transformations of aesthetic experience in future history.

(b2) Religion

In relation to religion, I proposed "the comparative theory of religion" in *FHC* (and also in *FCD*, although it was first so labeled in *FPHK*).

(b3) Popular (Mass) Culture

In relation to popular culture, I proposed, in *FHC*, a new way to understand mass culture in terms of some fundamental dilemmas, which is called, in the absence of better words, "the dualistic theory of mass culture" in *FPHO*.

(b4) Logic

In relation to logic, I also worked out, in *FPHML*, "the contrastive theory of rationality" for a new way to understand rationality and knowledge (in the context of mathematical logic).

(b5) Civilization as a Whole

In relation to civilization as a whole, I also argued, in *FHC*, no freedom without unfreedom (as shown in *Table 4.13*) and no equality without inequality (as shown in *Table 4.14*), especially in relation to the seven dimensions of life existence first worked out in *FHC* (i.e., the technological, the everyday, the true, the holy, the sublime/beautiful, the good, and the just).

Even more interestingly, in *BDPD* and later in *BCIV*, I further revealed a theoretical refinement of existential dialectics, this time, in arguing that there is no civilization without barbarity, with human civilization to be eventually superseded by what I originally analyzed as "post-human post-civilization" (which should not be confused with "post-human civilization"), in the context of the freedom/unfreedom and equality/inequality dialectics.

In the final analysis, civilization cannot live without barbarity and has to learn to co-exist with it in ever new ways. Preposteorus as this may seem to many contemporaries—it is no more imperative to preserve civilization than necessary to destroy barbarity, and the ideal of civilization is essentially bankrupt, to be eventually superseded by "post-civilization" (as summarized in *Tables 4.35-4.43*).

This freedom/unfreedom and equality/inequality dialectics holds in pre-moderny, modernity, postmodernity, and what I orignally called "after-postmodernity" in *FHC* and *FCD*—especially, though not exclusively, in the context of the emergence of "post-humans."

There are two specific theories about "after-postmodernity" here. The first theory about "after-postmodernity" is about the trinity of pre-modernity, modernity, postmodernity, and after-postmodernity at the cultural level—and was called, in *BDPD*, "the theory of the trinity of modernity to its after-postmodern counterpart," although it was first analyzed in *FHC* and later in *FCD* and also *BCIV* (as summarized in *Table 1.34*, *Table 1.35*, *Table 1.36*, and *Table 1.37* and will be elaborated further in the later sub-section on "the rest" about future history).

The second theory about "after-postmodernity" is about pre-modernity, modernity, postmodernity, and after-postmodernity at the historical level. For instance, a conclusion in *FHC*, which is shocking indeed for many contemporaries, is none other than that "[t]he post-human history will therefore mark the end of human history as we know it and, for that matter, the end of human dominance and, practically speaking, the end of humans as well. The entire history of human civilization, from its beginning to the end, can be summarized by four words, linked by three arrows (as already discussed in *FHC*)":

Pre-Modernity → Modernity →
Post-Modernity → After-Postmodernity

In *BDPD*, this thesis of mine was specifically called "the theory of the evolution from pre-modernity to after-postmodernity," at the historical level—as shown in *Table 1.34*, *Table 1.35*, *Table 1.36*, and *Table 1.37* (and will be elaborated further in the later sub-section on "the rest" about future history).

Therefore, "[t]he end of humanity in the coming human extinction is the beginning of post-humanity. To say an untimely farewell to humanity is to foretell the future welcome of post-humanity." (P. Baofu 2002: 89) This thesis of mine was known in *BDPD* as "the theory of post-humanity," at the systemic level—as already indicated in (a4.3) above.

(b6) Ethics

And in relation to ethics, I also proposed, in *BEPE*, "the theory of post-ethics" for a new way to understand morality and immorality.

(c) The Mind

This important vision about "post-humans" (as summarized above) brings us to the category of "the mind" here, which, as is often understood in everyday's usage, refers to the "brain" and the "body" (together with their interactions) and has different levels of consciousness which can be both "spiritual" and non-spiritual.

Therefore, the "mind" already presupposes the "brain," "body," and "spirit" (or "soul"), so that one does not fall into the intellectual trap of the "mind-body dualism" as debunked in modern (and especially, postmodern) philosophy (which I already went to great lengths to explain in *The Future of Human Civilization*).

That said—the category of "the mind" can be divided into two sub-categories, namely, in relation to (c1) biology and (c2) psychology.

(c1) Biology

In relation to biology, there are two concerns here, namely, (c1.1) genes and intelligence and (c1.2) the human body.

(c1.1) In relation to genes and intelligence, I worked out "the theory of contrastive advantages" (which was originally proposed in *FCD* and further worked out in *BNN*, as summarized in *Table 1.24*), to show the nature-nurture interactions of multiple levels in action (e.g., the biological, the psychological, the structural, the systemic, the cultural, and others) for humans and post-humans on earth and beyond, to the extent that different groups, be they on the basis of race, gender, ethnicity, class, age, or else, are not equal, on average, and have never been, nor will they be, even if there can be exceptions.

(c1.2) In relation to the human body, for instance, in *FPHMA*, I suggested "the expansive-contractive theory of martial arts," in which two great future transformations of the martial body were envisioned, that is, both "with the hyper-martial body" and "without the body" (as summarized in *Table 1.32*)—although this issue can also be classified under the category of "culture."

And in *FPHS*, I proposed "the theory of virtual sexuality," in which three great future transformations of the sexual body were envisioned, that is, "virtual pleasure," "the hyper-sexual body," and "the post-sexual floating mind" (as summarized in *Table 1.33*).

(c2) Psychology

In relation to psypchology (which also overlaps with biology too, however), I also proposed some greater transformations of the mind to come in the future, especially though not exclusively in the post-human age.

These greater transformations of the mind in psychology can be further sub-divided into different categories, namely, (c2.1) consciousness, (c2.2) unconsciousness (together with subconsciousness), (c2.3) creativity, (c2.4) learning, (c2.5) aggression (and pacificity), and (c2.6) personality.

(c2.1) In relation to consciousness, I analyzed, in *FPHC*, the nature of consciousness in relation to the conceptual dimensions of consciousness (as shown in *Table 1.12*), the theoretical levels of consciousness (as shown in *Table 1.13*), the thematic issues of consciousness (as shown in *Table 1.14*), the different dimensions of human existence (viz., having-ness, belonging-ness, and being-ness) in relation to consciousness (as shown in *Table 1.15*, *Table 1.16*, *Table 1.17*, and *Table 1.18*).

In *BCPC*, the nature of consciousness was further explored in relation to cognition, emotion, and behavior, which can be in consciousness and other mental states (e.g., subconsciousness and unconsciousness), in the context of existential dialectics, as shown in *Table 1.19*, *Table 1.20*, *Table 1.21*, *Table 1.22,* and *Table 1.23*.

In fact, in *FCD*, I already proposed three different theories in relation to the limits of cognition, emotion, and behavior, namely, "the theory of cognitive partiality," "the theory of emotional non-neutrality," and "the theory of behavioral alteration."

Then, I argued that the existence of human consciousness will be superseded one day with "floating consciousness" (as first proposed in *FCD*) and "hyper-spatial consciousness" (as first proposed in *FPHC*) as a climax of evolution in consciousness, after the future extinction of human consciousness (as shown below):

Primordial consciousness → Human consciousness →
Post-human consciousness (with floating consciousness and hyper-spatial consciousness as a climax in the evolution of consciousness)

In *BDPD*, these latest theses of mine were called "the theory of floating consciousness" and "the theory of hyper-spatial consciousness," both at the cosmological and psychological levels (as summarized in *Table 1.26* and *Table 1.25*).

(c2.2) In relation to unconsciousness, for instance, in *FPHU*, the unconsious mind will undergo tremendous change, to the rise of what I originally called "unfolding unconsciousess" in relation to anomalous experience, as a climax of evolution in unconsciousness, in conjunction with the conscious mind (both human now and post-human later), especially at the psychological level, as shown below:

Primordial unconsciousness → Human unconsciousness →
Post-human unconsciousness (with unfolding unconsciousness
as a climax in the evolution of unconsciousness)

In *FPHU*, this thesis of mine is called "the theory of unfolding unconsciousness" in a small sense, or "the unfolding theory of anomalous experience," in a larger one, as shown in *Table 1.27* and *Table 1.28*.

(c2.3) In relation to creativity, for instance, in *FPHCT*, I also proposed "the comprehensive theory of creative thinking" to explore the role of creative thinking, in relation to technology and other factors, together with my original proposal of important creative techniques and traits (as shown in *Table 1.29*) and its double nature of desirability and undesirability (as shown in *Table 1.30*).

(c2.4) In relation to learning, for instance, in *FPHL*, I worked out "the multilogical theory of learning" for a new way to understand the nature of structure and context in language, in relation to learning and its future development in the context of the mind.

(c2.5) In relation to aggression (and pacificity), I proposed "the multifaceted theory of war and peace" in *FPHWP* (as mentioned above).

(c2.6) And in relation to personality, I worked out "the contrarian theory of personality" in *FPHP* for a new way to understand normality and abnormality.

(d) Nature

The category of "nature" here refers to the state of nature, which can be divided into some sub-categories like (d1) physics, (d2) cosmology, (d3) biology, (d4) geography, and (d5) geology.

But since biology in (d3) overlaps with biology in (c1)—only phyiscs, cosmology, geography, and geology are addressed hereafter.

(d1) Physics

In relation to physics, I also proposed "the perspectival theory of space-time" in *FPHST* (based on *FPHC*) for a new way to understand space and time, as shown in *Table 1.3*, *Table 1.4*, *Table 1.5*, and *Table 1.7*.

In *FPHG* (based on *FPHC* and *FPHST*), I proposed "the selective theory of geometry" for a new way to understand geometry, with implications for the understanding of space-time in the context of infinity, symmetry, and dimensionality for future lifeforms that our world has never known (as shown in *Table 1.6*, *Table 1.7*, *Table 1.8; Table 1.9*, *Table 1.10,* and *Table 1.11*).

And in *FC*, I proposed "the dialectic theory of complexity" for a new way to understand order and chaos in the state of nature and beyond.

(d2) Cosmology

In relation to cosmology, my theories on physics (as indicated above) already have important implications for understanding the nature and the future of the universe (which was also addressed in *FHC*, among other works aforcited).

I also proposed "the theory of hyper-spatial consciousness" in *FPHC* (and also in *FPHG*) for a new way to understand the future evolution of the mind to enter into different dimensions of space-time that our current world has never known.

In addition, in *BCOS*, I proposed "the theory of post-cosmology" for a new way to understand the beginnings and ends of the cosmos and beyond.

(d4) Geography

In relation to geography, I also proposed "the theory of the geopower of nature" in *ALD* for a different way to understand the relationships between geography and other entities (like the mind, culture, and society).

(d5) Geology

And in relation to geology, I also propose "the resettlement theory of geology" in *FPHGEOL* for a different way to understand the relationships between statics and dynamics.

(e) The Rest

The category of "the rest" here refers to what cannot be strictly classified in any of the previous categories (i.e., society, culture, the mind, and nature). A good illustration of "the rest" refers to that which is, namely, (e1) historical.

(e1) Historical

Precisely here, I proposed "the theory of the evolution from pre-modernity to after-postmodernity" (as first worked out in *FHC* and later in *FCD* and *FPHC*), which overlaps—as a reminder from (b5)—with the issue concerning "civilization as a whole" in (b5).

For instance, in both *FCD* and *FPHC*, I worked out the structure of "post-human civilization" in terms of the trinity of after-postmodernity (i.e., "free-spirited after-postmodernity," "post-capitalist after-postmodernity," and "hegemonic after-postmodernity").

Both conceptually and theoretically, the trinity of after-postmodernity is a sequential extension to the trinity of modernity (i.e., "free-spirited modernity," "capitalist modernity,"and "hegemonic modernity") and the trinity of postmodernity (i.e., "free-spirited postmodernity," "capitalist postmodernity," and "hegemonic postmodernity") as first proposed in *FHC*.

And the trinity of pre-modernity (i.e., "pre-free-spirited pre-modernity," "pre-capitalist pre-modernity" and "hegemonic pre-modernity") was later conceived in *BCIV* to complete the historical set from pre-modernity to after-postmodernity.

In *BDPD*, this thesis about the trinity of pre-modernity, modernity, postmodernity, and after-postmodernity was collectively known as "the theory of the trinity of modernity to its after-postmodern counterpart," at the cultural level—as already so mentioned in (b5) earlier, and summarized in *Table 1.34*, *Table 1.35*, *Table 1.36*, and *Table 1.37*.

At the structural level, all these trinities are subject to the existential constraints (e.g., the freedom/unfreedom and equality/inequality dialectics in the context of "the cyclical progression of hegemony"), be the historical epoch in pre-modernity, modernity, postmodernity, or after-postmodernity in future times. Each of the historical epochs has its ever new ways of coming to terms with the ever new (different) mixtures of freedom/unfreedom and equality/inequality.

This is importantly so, not because, as is falsely assumed in conventional wisdom, one certain way is superior (or better) than another in terms of achieving more freedom and less unfreedom, or more equality with less inequality.

On the contrary, indeed, in each of the historical epochs, each increase of unfreedom greets each freedom achieved, just as each increase of inequality welcomes each equality achieved, albeit in ever new (different) ways. In *BDPD*, this thesis of mine was labeled as "the theory of the cyclical progression of hegemony," at the structural level, though it was first analyzed in *FCD*.

In *BDPD*, more theoretical applications of existential dialectics were further examined, in relation to five main features, in the context of the duality of oppression, namely, (a) that each freedom/equality achieved is also each unfreedom/inequality created, (b) that the subsequent oppressiveness is dualistic, both by the Same against the Others and itself and by the Others against the Same and themselves, (c) that both oppression and self-oppression can be achieved by way of downgrading differences (between the Same and the Others) and of accentuating them, (d) that the relationships are relatively asymmetric among them but relatively symmetric within them, even when the Same can be relatively asymmetric towards itself in self-oppression, and the Others can be likewise towards themselves, and (e) that symmetry and asymmetry change over time, with ever new players, new causes, and new forms, be the locality here on Earth or in deep space unto multiverses—as summarized in *Table.4.13*, *Table 4.14*, *Table 4.15*, and *Table 4.16*.

The same logic also holds both in relation to wealth and poverty (as addressed in *BCPC* and summarized in *Table 4.17* on the wealth/poverty dialectics) and in relation to civilization and barbarity (as addressed in *BCIV* and summarized in *Table 4.18* on the civilization/barbarity dialectics).

In *BDPD*, this thesis on existential dialectics was labeled as "the theory of existential dialectics."

Direct and Indirect Applications of Existential Dialectics

A different way to appreciate the usefulness of existential dialectics is by way of the analysis of its (a) direct and (b) indirect applications.

(a) In direct applications, on the one hand, the logic of existential dialectics can shed some theoretical insights on diverse phenomena in the world, and good instances are the usage of the principles of existential dialectics for the theoretical insights on the freedom/unfreedom dialectics, the equality/inequality dialectics, and the wealth/poverty dialectics—as introduced earlier in (b5) and (e) in the previous sub-section.

For instance, my books like *FPHST*, *BNN*, *FAE*, *FC*, *FIA*, *FPHU*, *FPHK*, *FPHML*, *FPHE*, *FPHMM*, *FPHCT*, *FPHG*, *FPHUP*, *FPHL*, *FPHO*, *FPHMA*, *FPHS*, *FPHLAW*, *FPHFS*, *FPHWP*, *FPHP*, *BEPE*, *BCOS*, and *FPHGEOL* use the principles to reveal some theoretical insights on the perspectives of space and time (as in *FPHST*), of nature and nurture (as in *BNN*), of beauty and ugliness (as in *FAE*), of order and chaos (as in *FC*), of taxonomy and network (as in *FIA*), of the unconscious mind (as in *FPHU*), of the nature of knowledge (as in *FPHK*), of mathematical logic (as in *FPHML*), of engineering (as in *FPHE*), of mass media (as in *FPHMM*), of creative thinking (as in *FPHCT*), of geometry (as in *FPHG*), of urban planning (as in *FPHUP*), of language (as in *FPHL*), of organization (as in *FPHO*), of martial arts (as in *FPHMA*), of sexuality (as in *FPHS*), of law (as in *FPHLAW*), of formal science (as in *FPHFS*), of aggression and pacificity (as in *FPHWP*), of normality and abnormality (as in *FPHP*). of morality and immorality (as in *BEPE*), of the beginnings and ends of the universe (as in *BOCS*), and of statics and dynamics (as in *FPHGEOL*).

(b) In indirect applications, on the other hand, the theoretical insights can further be used to reveal other phenomena directly from them (viz., the theoretical insights) and therefore indirectly from the principles themselves.

A good illustration is of course the use of the theoretical insights on the freedom/unfreedom and equality/inequality dialectics for the understanding of the civilization/barbarity dialectics.

This distinction between direct and indirect applications may sound a bit academic, since even in indirect applications, the phenomena under study can still be directly related back to the principles themselves.

In the previous example, as an illustration, the civilization/barbarity dialectics can be directly related to the principles of existential dialectics without the intermediate role of the freedom/unfreedom and equality/inequality dialectics.

Multiple Levels of Application

There is another issue to be clarified, however. In other words, the theoretical insights can be applied to multiple levels of analysis—even though, in a given example, it may refer to one level or a few only.

For instance, in the example concerning the freedom/unfreedom dialectics, it can be used at the structural level (e.g., in relation to the theory of the cyclical progression of hegemony), but it can also be exploited for other levels (e.g., the theory of post-capitalism at the institutional level).

All these levels of application should not be misleadingly construed, as I stress this before, as a one-way street (that is, to use the ontological principles for theoretical application at multiple levels of specific analysis) but a two-way one, in which theoretical insights in praxis, when studied in more specific contexts, can in turn refine the nature of existential dialectics (for example, with the addition of new principles).

Sophisticated Methodological Holism

The summary of my original meta-theory on methodology is provided below, again, to be repeated *often verbatim* from my previous works—as this is something I regularly do in each new book to introduce my past works, which are in conversation with this current one.

This means that, for those readers who had read some of my previous books, the summary below is really not much new (*almost verbatim*, with only some updated revisions).

That clarified—this original contribution of mine to the study of methodology takes the form of an approach known as "the theory of methodological holism" or "methodological holism" in short, as already worked out in *FPHC* (2004).

I have made good use of this methodological approach of mine for all of my previous works.

The Confusion of Holism with Systems Approach

Yet, it is imperative to stress at the outset, as I often emphasized this in all my previous books, that my approach of "methodological holism" does not oppose or exclude "methodological individualism" (as some readers may be tempted to assume, as is conventionally understood) but actually includes it.

The confusion here has to do with equating holism with systems approach in general (be it about systems theory, chaos theory, complexity theory, or else), which I debunked, while learning something from it, as already elaborated in *The Future of Complexity* (2007)—and *The Future of Post-Human Formal Science* (2010) is another addition to this critique of the intellectual fad of systems approach.

In other words, my meta-theory of holism, or sophisticated methodological holism in short, makes a distinction between doing holistic work and doing systems work, for two reasons.

The first reason is that holism, in my theory, does not imply systems approach, since it rejects systems approach because of the myriad problems that the literature on systems approach has been unable to resolve, while learning something from it.

And the second reason is that my theory is related to the other theory of mine, that is, existential dialectics, which does not exclude methodological individualism (but includes it, because it simply goes beyond both reductionism and reverse-reductionism in a dialectic way).

For these reasons (and others too, as summarized in *Table 4.1*), my version of methodological holism is *sophisticated*—not *vulgar* as sometimes used by inapt scholars employing the same term.

The Ontological Constraints

Sophisticated methodological holism is subject to some ontological constraints, and five good examples include "the partiality-totality principle," "the predictability-unpredictability principle," "the explicability-inexplicability principle," "the absoluteness-relativeness principle," and "the formalness-informalness principle" in existential dialectics.

Of course, there can be more than five ontological principles in relation to method, but the five examples here are sufficient to illustrate the point in question.

With this caveat in mind—the first two were already analyzed in previous books of mine like *FC* and *FAE*, whereas the third one was addressed in *FPHU*, and the last two were accounted for in *FPHK* and *FPHML*, respectively.

For now, it suffices to show that sophisticted methodological holism, because of these ontological constraints on methodology, targets two major sins of methodology, namely, what I call, in the absence of better words, (a) *reductionism* and (b) *reverse-reductionism*—both of which come in all shapes and sizes, to be summarized below (and also shown in *Table 4.2*).

Against the Varieties of Reductionism

There are four versions of reductionism to be summarized here, which sophisticated methodological holism rejects, namely, (a1) conceptual, (a2) theoretical, (a3) methodological, and (a4) ontological forms of reductionism.

(a1) In conceptual reductionism, a good case in point concerns myriad dualities like mind vs. body, self vs. world, democracy vs non-democracy, and the like (as already addressed in *FHC*, *FPHC*, and *BDPD*, for instance).

(a2) In theoretical reductionism, an illuminating case study is best exemplified by what I originally called "the foundation fallacy" in *FPHST*, in any attempt to naively understand space-time from the physical perspective as the foundation and, consequently, to dangerously dismiss (or belittle) other perspectives.

In *FAE*, I elaborated further these versions of reductionism in the literature on aesthetics (e.g., form vs. content, representation vs. expression, critics vs. artists, and externalism vs. internalism).

In *FIA*, I revealed other forms of reductionism in the literature on information architecture (e.g., the constructivist argument and the representational argument).

In *FPHU*, I showed the persistent legacy of reductionism, this time, in the literature on anomalous experience (e.g., the obsession with physics, chemistry, and biology for explaining anomalous experience).

In *FPHE*, I examined another case of reductionism in action, in the context of engineering (e.g., technical constraints vs. normative constraints).

In *FPHMM*, I elaborated one more version of reductionism, in relation to the three domains of communication (e.g., the competing views on sending, connecting, and receiving).

In *FPHCT*, I explored another version of reductionism, in relation to invention and innovation (e.g., the bio-psychological argument vs. the socio-cultural arguments).

In *FPHG*, I identified another version of reductionism, in relation to infinity, symmetry, and dimensionality (e.g., the Euclidean argument vs. the non-Euclidean arguments).

In *FPHUP*, I analyzed another version of reductionism, in relation to density and void (e.g., the engineering argument and the ecology argument).

In *PFHL*, I examined another version of reductionism, in relation to structure and context (e.g., the structuralist argument and the contextualist argument).

In *PFHO*, I explored another version of reductionism, in relation to communcation, decision-making, and leadership (e.g., the rational-system argument vs. the natural-system argument vs. the open-system argument).

In *PFHMA*, I revealed another version of reductionism, in relation to the martial body and spirit (e.g., the spiritual argument vs. the materialist argument vs. the defensive argument).

In *PFHS*, I analyzed another version of reductionism, in relation to the sexual body and spirit (e.g., the naturalist argument vs. the constructivist argument).

In *PFHLAW*, I examined another version of reductionism, in relation to law (e.g., the necessity argument vs. the contengency argument).

In *FPHWP*, I show another version of reductionism, in relation to war and peace (e.g., the aggressivist argument vs. the pacifist argument).

In *BEPE*, I analyzed another version of reductionism, in relation to morality and immorality (e.g., the objectivist argument vs. the non-objectivist argument vs. the skeptical argument).

In *BCOS*, I examined another version of reductionism, in relation to the contested beginnings and speculative ends of the universe (e.g., the scientific argument vs. the religious argument vs. the esoteric argument vs. the metaphysical argument).

In *FPHP*, I showed another version of reductionism, in relation to normality and abnormality (e.g., the natural argument vs. the social argument vs. the cultural argument vs. the mental argument).

And in *FPHGEOL*, I scrutinize another version of reductionism, in relation to statics and dynamics (e.g., the catastrophe argument vs. the uniformity argument vs. the revision argument).

(a3) In methodological reductionism, a good illustration can be the debate between different versions of qualitative and quantitative methods (as already analyzed in *FC* and also *FHC*).

In *FPHML*, I examined similar reductionism, this time, in the literature on mathematical logic (e.g., the obsession with consistency, soundness, and completeness).

And in *FPHFS*, I explored the problems of reductionism in the context of formal science (e.g., the analytical argument, as opposed to the synthetic argument).

(a4) In ontological reductionism, an excellent instance involves another debate, this time between emergentism and reductionism in complexity theory (as addressed in *FC*) and also in psychology (as elaborated in *FPHC*, in the context of Being and Belonging).

But to target against the varieties of reductionism constitutes only a side of the same coin, and the other side concerns the varieties of reverse-reductionism.

Against the Varieties of Reverse-Reductionism

The other side of the same coin is the reverse version of reductionism, which is what I want to call, in the absence of better words, *reverse-reductionism*.

My sophisticated version of methodological holism targets against the varieties of reverse-reductionism (just as it also rejects the ones of reductionism).

Perhaps nothing expresses better the popularity of reverse-reductionim than the "anything-goes" mentality in postmodernism of our time, as shown in the following four versions.

(b1) In conceptual reverse-reductionism, any concept of "art" (e.g., fine art, applied art, outsider art, junk art) is welcome in postmodernism (as already analyzed in Ch.4 of *FHC*—and also in *FAE*).

(b2) In theoretical reverse-reductionism, a variety of art and literary theories co-exist. Take the case of literary studies, as there are now Literary Structuralism, Marxist Literary Criticism, New Criticism, Phenomenology, Hermeneutics, Language-Game Literary Criticism, Feminist Literary Criticism, Reception Theory, Reader Response Criticism, Poststructuralism, Semiotics, Pyschoanalytic Literary Criticism, just to cite some well-known ones, with no one being said to be better than any others (as detailedly analyzed in Ch.4 of *FHC*). (S. Raman 1997)

In *BNN*, I even introduced "the compromise fallacy" as another good illustration of theoretical reverse-reductionism, in misleadingly treating both genetic and environmental approaches as equally valid.

(b3) In methodological reverse-reductionism, multiple methodologies are deemed as acceptable in postmodernism (e.g., doing art without praxis, doing art with praxis, and doing art by sublation), as analyzed in Ch.4 of *FHC*.

And in *FPHFS*, I exposed the problems of reverse-reductionism in the context of formal science in relation to systems theory (e.g., the synthetic argument).

(b4) In ontological reverse-reductionism, no privileged ontology is allowed, and the door is open practically for anything in postmodernism (e.g., the equal status of the ontology of Being vs. that of Becoming, as already addressed in Ch.4 of *FHC*—and also in *FPHC*).

In *FAE*, I also introduced another version of reverse-reductionism, that is, "the pluralist fallacy," in the context of understanding aesthetic experience, for instance—although this fallacy has been committed not exclusively in relation to the ontological level (but also at the conceptual, theoretical, and methodological ones).

These dual dangers against reverse-reductionism (in this sub-section) and reductionism (in the previous sub-section) are something that sophisticated methodological holism rejects. And in *FPHFS*, I addressed these two problems further.

This point about the dual dangers is important enough, since many scholars often favor one against the other, but it is vital to target both.

The Holistic Organization of an Inquiry

With these dual dangers against reductionism and reverse-reductionism in mind—sophisticated methodological holism suggests that an inquiry of any given phenomenon is more complete, if treated in the context of a comprehensive analysis at all relevant levels, which challengingly encompass all the fields of human knowledge, ranging from the natural sciences and formal sciences through the social sciences to the humanities.

This section is something that I had already stressed in all my previous books and repeat (*often verbatim*) hereafter. But, for those readers who had read my previous books, this serves as a reminder.

With this reminder in mind—there are multiple ways to engage in a holistic inquiry with all relevent levels of analysis. Over the years, I have proposed different ways to fulfill this holistic methodological requirement, as repeated in my previous books.

Hereafter is a summary of four major ways, namely, (a) by discipline, (b) by domain, (c) by subject, and (d) by meta-analysis—all of which fulfill the holistic requirement of sophisticated methodological holism on a given issue.

(a) In a holistic organization by discipline, a good illustration is the multiple levels of analysis, namely, (a1) the micro-physical, (a2) the chemical, (a3) the biological, (a4) the psychological, (a5) the organizational, (a6) the institutional, (a7) the structural, (a8) the systemic, (a9) the cultural, (a10) the macro-physical (cosmological), and (a11) other relevant levels, which are either a combination of all these levels or the practical applications with a combination of them.

This holistic organization by discipline is used in *FCD* and *BCPC*, for instance.

(b) In a holistic organization by domain, some good candidates include the classification of different perspectives of analysis in relation to (b1) nature, (b2) the mind, (b3) society, and (b4) culture, as already worked out in *BCIV*, *FPHST*, *FC*, *FAE*, *FPHU*, *FPHK*, *FPHE*, and *FPHFS*.

Culture in (b4), in this re-classification, is the same as culture in (a9) and can be further divided into different sub-categories, with good examples like (a9i) epistemology (methodology), (a9ii) aesthetics, (a9iii) ethics, (a9iv) metaphysics (ontology), and (a9v) religion. For clarification, epistemology (methodology) in (a9i) and metaphysics (ontology) in (a9iv) overlap with the category of "meta-theory."

Society in (b3) includes the organizational in (a5), the institutional in (a6), the structural in (a7), and the systemic in (a8).

The mind (b2) has more to do with the chemical in (a2), the biological in (a3), and the psychological in (a4), although it can overlap with (a1), (b3) and (b4), for instance.

It should be stressed that the "mind" here also includes the categories like the "brain," "body," and the "spirit" (or the "soul")—whenever needed in a given context, so that one does not fall into the intellectual trap of the "mind-body dualism" in the history of modern (and especially, postmodern) philosophy, which I already went to great lengths to explain in *The Future of Human Civilization*. After all, in everyday's usage, the "mind" already implies the "brain" and the "body" (together with their interactions) and has different levels of consciousness which can be both "spiritual" and non-spiritual.

And nature in (b1) refers to the micro-physical in (a1) and the macro-physical (cosmological) in (a10), although it can also overlap with (a2), (a3) and (a4).

(c) In a holistic organization by subject, some excellent examples concern the analysis of consciousness in *FPHC* (i.e., on Having, Belonging, and Being); of civilization in *FHC* (i.e., the True, the Holy, the Everyday, the Technological, the Beautiful/Sublime, the Good, and the

Just); of international politics in *BWT* (i.e., Hyper-Empires, Meso-Empires, and Micro-Empires); of nature and nurture in *BNN* (i.e., genes and memes); of information architecture in *FIA* (i.e., taxonomy and network); of mathematical logic in *FPHML* (i.e., consistency, soundness, and completeness); of communication in *FPHMM* (i.e., sending, connecting, and receiving); of creative thinking in *FPHCT* (i.e., invention and innovation); of geometry in *FPHG* (i.e., infinity, symmetry, and dimensionality); of urban planning in *FPHUP* (i.e., density and void), of language in *FPHL* (i.e., structure and context), and of organization in *FPHO* (i.e., communication, decision-making, and leadership), in *FPHMA* (i.e., the martial body and the martial spirit), in *FPHS* (i.e., the sexual body and the sexual spirit), in *FPHLAW* (i.e., necessity and contingency), in *FPHWP* (i.e., aggression and pacificity), in *FPHP* (i.e., normality and abnormality), in *BEPE* (i.e., morality and immorality), and in *BCOS* (i.e., the beginnings and ends of the universe)—by incorporating all the relevent levels of analysis as cited above in each.

(d) And in a holistic organization by meta-analysis, two good cases in point involve the works on post-capitalism in *BCPC* and on authoritarian liberal democracy in *ALD*, with the classification of analysis in theory and meta-theory.

The essential point to remember here, as I thus stress in all my previous books, is that the multiple levels of analysis can be reorganized in many different ways, insofar as none of the levels (if relevant to an inquiry in question) is ignored or dismissed, to avoid the danger of reductionism (and for that matter, the one of reverse-reductionism, in the opposite direction).

Three Clarifications

Three clarifications are needed here, however, to avoid any hasty misunderstanding.

Firstly, the entities in each classification are not mutually exclusive. For instance, in the holisitc organization by domain, nature in (b1) can also be linked to the chemical in (a2), the biological in (a3), and the systemic in (a8). And the mind (b3) can alternatively be related to the micro-physical in (a1) and the chemical in (a2), for example.

Secondly, there is also the important factor of luck (or uncertainty in conventional usage), but it is already implied or allowed in each of the organizations of an inquiry—especially in relation to the predictability-unpredictability principle and the order-chaos principle in existential dialectics. This topic about uncertainty was already addressed in *FHC*—and more extensively, in *FC*, in the context of order and chaos.

And lastly, the comparison in a classification is not absolute, but relative, as there are often some shades of gray, not exactly black or white (figuratively speaking).

Some further clarifications and qualifications of sophisticated methodological holism are shown in *Table 4.2*.

Some Distinctive Usefulness

In the end, sophisticated methodological holism—when applied, especially though not exclusively, as illustrated in my numerous works—can enrich the understanding of reality in some distinctive ways. Here are four examples (as revealed in each of my books).

Firstly, it provides a comprehensive analysis of a subject matter, from which much can be learned about reality, both in relation to the perspectives of the mind, nature, society, and culture.

Secondly, it offers a new classification of the subject matter.

Thirdly, it suggests some visions of the future in relation to the subject matter in question.

And fourthly, it proposes some insights on meta-theory (e.g., methodology and ontology) in general—with the clear understanding, however, of the dilemma of specific vs. general ontology (as shown in *Table 4.9* on the syntax of existential dialectics in the context of the dilemma of ontology).

For this very reason, all of these ways are important, without reducing one into the analysis of another. (*FPHCT*)

Chapter Outline

With this summary of my two original meta-theories (on methodology and ontology, *almost verbatim* from my previous works) in mind—this book is organized in four main parts, namely, (a) Part I on the introduction to this book, (b) Part II on statics, (c) Part III on dynamics, and (d) Part IV on the conclusion of the book.

The book is thus organized in four chapters.

The introductory chapter here, that is, Chapter One titled "Introduction—The Use of Geology," begins by first introducing the views in contention, geology and the geosciences, the relationship between statics and dynamics in geology, and the wide variety of geological methods—together with the theoretical debate in the literature, my resettlement theory of geology, the relationship between theory and meta-theory, existential dialectics, sophisticated methodological holism, and some further clarifications.

Chapter Two titled "Statics and Its Contingency" analyzes the nature of statics in geology from the four perspectives of the mind, nature, society, and culture, with a constructive critique of each.

Chapter Three titled "Dynamics and Its Restriction" explores the nature of dynamics in geology from the four perspectives of the mind, nature, society, and culture, with a deconstructive analysis of each.

The last chapter titled "Conclusion—The Future of Geology" summarizes the analysis in the book in the dialectic context of statics and dynamics, with sixteen major theses, namely, (a) the first thesis: the absoluteness-relativeness principle, (b) the second thesis: the predictability-unpredictability principle, (c) the third thesis: the explicability-inexplicability principle, (d) the fourth thesis: the preciseness-vagueness principle, (e) the fifth thesis: the simpleness-complicatedness principle, (f) the sixth thesis: the openness-hiddenness principle, (g) the seventh thesis: the denseness-emptiness principle, (h) the eighth thesis: the slowness-quickness principle, (i) the nineth thesis: the expansion-contraction principle, (j) the tenth thesis: the theory-praxis principle, (k) the eleventh thesis: the convention-novelty principle, (l) the twelveth thesis: the evolution-transformation principle, (m) the thirteenth thesis: the symmetry-asymmetry principle, (n) the fourteenth thesis: the regression-progression principle, (o) the fifteenth thesis: the sameness-difference principle, and (p) the sixteenth thesis: the post-human resettlement, in relation to the three great future transformations of geology—to be elaborated in the rest of the book, with a summary in the concluding chapter.

This seminal project is to fundamentally alter the way that we think about geology, from the combined perspectives of the mind, nature, society, and culture, with enormous implications for the human future and what I originally called its "post-human" fate.

Some Clarifications

But some additional clarifications are needed, before a formal analysis of the nature of geology is to begin in Chapter Two.

These clarifications, lest any potential misunderstanding by the readers occurs, refer to (a) the conversation with my previous books, (b) the illustrative selection of case studies and examples, (c) detailed analysis versus overall analysis, (d) two distinctive features of using quotations, and (e) the use of neologisms.

These clarifications are something that I also regularly repeat in my previous books too, so these points are repeated hereafter, *often verbatim*—albeit in a different context this time.

In Conversation with My Previous Books

Firstly, my theory of post-ethics is constructed on the theoretical foundation of my previous books (i.e., *FHC*, *FCD*, *FPHC*, *BDPD*, *BCPC*, *BCIV*, *FPHST*, *BNN*, *BWT*, *FC*, *FAE*, *ALD*, *FIA*, *FPHU*, *FPHK*, *FPHML*, *FPHE*, *FPHMM*, *FPHCT*, *FPHG*, *FPHUP*, *FPHL*, *FPHO*, *FPHMA*, *FPHS*, *FPHLAW*, *FPHFS*, *FPHWP*, *FPHP*, *BCOS*, and *BEPE*—as explicated in "The List of Abbreviations").

Just consider the following two illustrations, namely, (a) in relation to other related theoretical debates and (b) in relation to different visions as worked out in my previous books.

(a) The first illustration is that the theoretical debate on the nature of morality and immorality also involves a few other related theoretical issues.

These other related theoretical issues have to do with some deeply contested theoretical debates, which, however, I had already gone to great lengths to analyze in my numerous (previous) books, as summarized below. It should also be stressed, lest misunderstanding occurs, that each of the theoretical debates (as listed below) does not exhaust the issues covered in the books, as each book covers more issues than what is listed below (for illustration only).

- On Space and Time
  - –Ex: *The Future of Post-Human Space-Time*
  - –Ex: *The Future of Post-Human Geometry*
- On Science
  - –Ex: 2 volumes, *The Future of Human Civilization*
  - –Ex: *The Future of Complexity*
  - –Ex: *The Future of Post-Human Knowledge*
  - –Ex: *The Future of Post-Human Space-Time*
  - –Ex: *Beyond Cosmology to Post-Cosmology*
  - –Ex: *The Future of Post-Human Geometry*
- On Infinity, Symmetry, and Dimensionality
  - –Ex: *The Future of Post-Human Geometry*
- On the Future of the Universe
  - –Ex: *The Future of Post-Human Space-Time*
  - –Ex: *The Future of Post-Human Geometry*

–Ex: 2 volumes, *The Future of Human Civilization*
–Ex: *The Future of Capitalism and Democracy*
- On Uncertainty
–Ex: *The Future of Complexity*
–Ex: 2 volumes, *The Future of Human Civilization*
- On Systems Thinking
–Ex: *The Future of Post-Human Formal Science*
- On Epistemology and Ontology
–Ex: *The Future of Post-Human Knowledge*
–Ex: *The Future of Post-Human Consciousness*
–Ex: 2 volumes, *The Future of Human Civilization*
- On Consciousness and Unconsciousness
–Ex: *The Future of Post-Human Consciousness*
–Ex: *The Future of Post-Human Unonsciousness*
–Ex: *The Future of Capitalism and Democracy*
- On Creativity
–Ex: *The Future of Post-Human Creative Thinking*
–Ex: *The Future of Aesthetic Experience*
- On Technology and Communication
–Ex: 2 volumes, *The Future of Human Civilization*
–Ex: *The Future of Post-Human Engineering*
–Ex: *The Future of Post-Human Mass Media*
- On Cognition
–Ex: *The Future of Post-Human Unconsciousness*
–Ex: *The Future of Post-Human Consciousness*
–Ex: *The Future of Capitalism and Democracy*
–Ex: *The Future of Post-Human Creative Thinking*
- On Intuition
–Ex: *The Future of Post-Human Unconsciousness*
–Ex: *The Future of Post-Human Consciousness*
–Ex: *Beyond Capitalism to Post-Capitalism*
–Ex: *The Future of Capitalism and Democracy*
- On Being and Belonging
–Ex: *The Future of Post-Human Consciousness*
–Ex: 2 volumes, *The Future of Human Civilization*
- On Different Dimensions of Human Existence
–Ex: 2 volumes, *The Future of Human Civilization*
- On Modernity, Post-Modernity, and After-Postmodernity
–Ex: 2 volumes, *The Future of Human Civilization*
–Ex: *Beyond Civilization to Post-Civilization*

- On Pre-Modernity
  - –Ex: *Beyond Civilization to Post-Civilization*
- On No Freedom Without Unfreedom
  - –Ex: 2 volumes, *The Future of Human Civilization*
  - –Ex: *The Future of Capitalism and Democracy*
  - –Ex: *Beyond Capitalism to Post-Capitalism*
  - –Ex: *Beyond Democracy to Post-Democracy*
  - –Ex: *Beyond Civilization to Post-Civilization*
- On Spirituality
  - –Ex: 2 volumes, *The Future of Human Civilization*
  - –Ex: *The Future of Capitalism and Democracy*
  - –Ex: *Beyond Capitalism to Post-Capitalism*
  - –Ex: *The Future of Post-Human Sexuality*
  - –Ex: *The Future of Post-Human Martial Arts*
  - –Ex: *The Future of Post-Human Unconsciousness*
  - –Ex: *The Future of Post-Human Consciousness*
  - –Ex: *Beyond Ethics to Post-Ethics*

Therefore, those who are interested in these other theoretical issues (and debates) which are related to the current context can consult these other books of mine for more details.

(b) And the second illustration is that one of the major theses of this book concerns the future, be it human and later post-human, which is closely linked with the different visions that I had worked out in my previous books, in relation to the mind, nature, society, and culture.

Of course, for the convenience of the readers, a summary of some of these visions will be provided whenever necessary, as shown in the two sections on meta-theory above (*often verbatim*) and in different tables at the end of this chapter and the concluding chapter, for instance.

Yet, I still expect the readers to read my previous books directly for more details—as this book is written in conversation with my previous ones.

Case Studies and Examples

Secondly, the case studies and examples in the chapters are not exhaustive but solely illustrative. Nor are they necessarily mutually exclusive, as they can be reclassified in a different way. And exceptions are allowed.

One is tempted to assume, however, that, since the case studies and their examples are not exhaustive, more case studies and examples are needed. But the problem here is that just adding more of them does not necessarily change the conclusions to be drawn and may even lead to redundancy.

After all, the case studies and examples used in this book are chosen with care and deemed sufficient for the purpose at hand, even though they are solely illustrative.

Detailed Analysis versus Overall Synthesis

Thirdly, there is a distinction between detailed analysis and overall synthesis, in that the former is, especially though not exclusively, more for the specialized viewpoint of a specific discipline, whereas the latter is, especially though not exclusively, more for the broad horizon of an interdisciplinary approach—relatively speaking, of course.

Or, by analogy, there is a distinction between depicting a "tree" (with tiny details) in a "forest" and viewing the entire "forest" itself (with the broader horizon).

This book, like all previous ones of mine, tries to strike a balance between the two forms of understanding, although this does not imply that the two forms should be exactly equally offered, in an exact 50-50 balance.

Instead, the book, more often than not, does not allow the tiny details to obscure the larger picture in a given case study, because the entire book is inter-disciplinary in nature and tries to look at the entire "forest," not being bogged down with the detailed analysis of an individual "tree" in the forest, by analogy.

In this sense, although detailed analysis can be provided whenever needed for a given case study, the important point to remember is the overall horizon with its broadness of scope.

Two Distinctive Features of Using Quotations

Fourthly, the way that quotes are used in this book (and others of mine) has two distinctive features, as explained below.

The first distinctive feature is that, at the end of each quotation, sometimes there are more than one reference—in which case the first reference is for the original citation, and the rest of the references are for the convenience of the reader for more information about the issue under quotation.

And the second feature is that quotations are used as often as possible, so as to let others speak for themselves, without the nuisance of putting my own words into their mouths. Even more importantly, this allows me to use the original text as an evidence for critical analysis at the end of each section.

The Use of Neologisms

And finally, I use different neologisms in my books, mostly to introduce my original concepts and theories, and good instances include those here or elsewhere in my previous books (e.g., "the resettlement theory of geology," "post-democracy," "hyper-spatial consciousness," "post-capitalism," and whatnot). Needless to say, they are used here solely for our current intellectual convenience, as they will be renamed differently in different ways in future history.

As I thus wrote in *FCD* (508-9), "all these terms 'post-capitalism,' 'post-democracy'...and other ones as introduced in...[the] project (e.g.,...'posthuman elitists,' and 'posthuman counter-elitists,' just to cite a few of them) are more for our current intellectual convenience than to the liking of future humans and post-humans, who will surely invent more tasteful neologisms to call their own eras, entities, and everything else, for that matter. But the didactic point here is to use the terms to foretell what the future might be like, not that its eras and entities must be called so exactly and permanently. After all, William Shakespeare (1995: Act II, Scene II, Line 47) well said long ago: 'What is in a name? That which we call a rose by any other name would smell as sweet.'"

As I had also stressed time and again before, each of the neologisms can be re-written as a different "X," only to be re-named differently by the powers that be in different eras of future history.

With these clarifications for thought in mind—I now proceed to Chapter Two on statics and its contingency.

Table 1.1. Geology and Geosciences

• **Geology**

—The term "geology," which is from "Greek γῆ, gê, meaning 'earth' and λόγος, logos, meaning 'study of.'" (WK 2010; S. Winchester 2001) More to the point, "geology" refers to the study of "the rocky parts of the Earth's crust (or lithosphere) and its historic development." (WK 2010q)

• **Geosciences**

—By contrast, the term "geosciences" (or the "geosciences," or even the "Earth sciences") is much broader in scope and refers to different sciences about the Earth, with "geology" as only one of them. For instance, the term "geosciences" also includes "geology" as part of the sciences about the Earth but also has other concerns, as shown below. (WK 2010q; G. Smith 2006)

- Ex: "Geophysics and Geodesy" (for the "Lithosphere/Geosphere")
- Ex: "Soil Science" (for the "Pedosphere")
- Ex: "Oceanography and Hydrology" (for the "Hydropshere")
- Ex: "Glaciology" (for the "Cryosphere")
- Ex: "Atmospheric Sciences" (for the "Atmosphere")
- Ex: "Biology" (for "the Biosphere")

Notes: The examples in the categories are solely illustrative (not exhaustive), and the comparison is relative (not absolute), nor are they necessarily mutually exclusive. And some can be easily re-classified elsewhere. As generalities, they allow exceptions.
Source: From *Sec. 1.2* of *FPHGEOL*. See book for citations.

Table 1.2. The Nature of Geology

- **Statics and Dynamics**
  - —Statics
    - Ex: "geologic materials" (e.g., rocks, unsolidated materials)
    - Ex: "whole-Earth structure" (e.g., plate tectonics, Earth structure)
  - —Dynamics
    - Ex: "the geological evolution of an area"

- **The Varieties of Methods**
  - —Field Methods
    - Ex: "Geological Mapping"
    - Ex: "Surveying of Topographic Features"
    - Ex: "Subsurface Mapping through Geophysical Methods"
    - Ex: "High-Resolution Stratigraphy"
    - Ex: "Biogeochemistry and Geomicrobiology"
    - Ex: "Paleontology: Excavation of Fossil Material"
    - Ex: Sample Collection "for Geochronology &Thermochronology"
    - Ex: "Measurement of Characteristics of Glaciers and their Motion"
  - —Lab Methods
    - Ex: "petrology"
    - Ex: "structural geology"
    - Ex: "stratigraphy"

Notes: The examples in the categories are solely illustrative (not exhaustive), and the comparison is relative (not absolute), nor are they necessarily mutually exclusive. And some can be easily re-classified elsewhere. As generalities, they allow exceptions.
Source: From *Sec. 1.3* and *Sec. 1.4* of *FPHGEOL*. See book for citations.

Table 1.3. The Theoretical Debate on Space-Time (Part I)

• **Isaac Newton's Absolutist (Substantivist) Theory of Space-Time**

—space and time are independent from each other. The structure of space-time is $E3 \times E1$ (with the structure of space, *P*, as the set of spatial locations in a three-dimensional Euclidean space, *E3*, and the structure of time as the set of temporal moments, *T*, in the one-dimensional real time, *E1*).

—space and time are also independent from the effects of matter and events. The existence of space and time is possible even in a world absent of matter (and, for that matter, even in a world absent of events), as if they were material objects but with their total unchangingness thorough time.

• **Albert Einstein's Relativist Theory of Space-Time**

—space and time are interchangeable (not absolute), just as matter and energy are equivalent (not independent) with the famous equation, $E = mc^2$ (as in the special theory of relativity in 1905).

—space-time and matter-energy are also relative in a grand union (as in the general theory of relativity in 1915). Thus, each pair affects the other pair, as "matter 'tells' spacetime how to curve [in a non-Euclidean geometry] and curved spacetime 'tells' matter how to behave….Space contracts near mass and dilates away from it. Time dilates near mass and contracts away from it. Clocks positioned farther away from the mass of the earth run faster than clocks closer to the earth." (L. Shlain 1991: 328-330)

(continued on next page)

Table 1.3. The Theoretical Debate on Space-Time (Part II)

- **Peter Baofu's Perspectival Theory of Space-Time**
  —space and time can be understood from multiple perspectives, be they in relation to culture, society, nature, and the mind, with each perspective revealing something about the nature of space-time and simultaneously delimiting its view. This is subject to "the regression-progression principle" in existential dialectics.
  —each perspective of space and time exists in society and culture with good reasons, with some being more successful and hegemonic (dominant) than others. This is subject to "the symmetry-asymmetry principle" in existential dialectics.
  —space and time will not last, to be eventually superseded (altered) by post-humans in different forms (e.g.,stretching/shrinking space-time, engineering more dimensions of space-time, and manipulating multiverses), be they here in this universe or in multiverses. Thus, even the physical existence of space-time cannot last forever, with ever more transformations in the process. This is subject to "the change-constancy principle" in existential dialectics.
  —the conventional wisdom (especially by physicists) of treating the physical perspective of space and time as the foundation of all other perspectives (of space and time) and of regarding them as much less important is a form of reductionism,committing what I call *the foundation fallacy*, in misleadingly dismissing the multiple perspectives of space and time in relation to culture, society, nature, and the mind.

Notes: The examples in each category are solely illustrative (not exhaustive), and the comparison is relative (not absolute), nor are they necessarily mutually exclusive. Some can be easily re-classified elsewhere. As generalities, they allow exceptions.
Source: A summary of *Sec.1.1*, *Sec.1.2*, and *Sec.1.3* of *FPHST*—and, for that matter, the rest of *FPHST*

Table 1.4. Main Reasons for Altering Space-Time

- **The Need to Make New Energy-Matter**
  —Ex: manipulating molecular bonds for new materials
  —Ex: creating nanotechnologies on the atomic scale
  —Ex: engineering the atomic nucleus
  —Ex: restructuring most elementary particles
  —Ex: inventing new forms of matter and energy

- **The Need to Create New Space-Time**
  —Ex: creating "warp drive" (as in science fiction) for space travel
  —Ex: creating "pocket universes"

- **The Need to Conquer the Cosmos unto Multiverses**
  —Ex: spreading floating consciousness and hyper-spatial consciousness, besides other forms that humans have never known, in the cosmos and beyond unto multiverses for ultimate conquest

Notes: The examples in each category are solely illustrative (not exhaustive) nor necessarily mutually exclusive, and the comparison is relative (not absolute). As generalities, they allow exceptions. Also, it should be stressed that the three reasons are all related, in that they all contribute to the evolution of intelligent life in the cosmos unto multiverses in the most distant future beyond our current knowledge.
Sources: A summary of *Sec.6.2* of *FPHST*. See also *FHC*, *FCD*, *FPHC*, *FC*, *FPHU*, and *FPHG*, for example.

Table 1.5. The Technological Frontiers of the Micro-World

- **Type I-Minus**
  —Ex: building structures and mining

- **Type II-Minus**
  —Ex: playing with the genetic makeups of living things

- **Type III-Minus**
  —Ex: manipulating molecular bonds for new materials

- **Type IV-Minus**
  —Ex: creating nanotechnologies on the atomic scale

- **Type V-Minus**
  —Ex: engineering the atomic nucleus

- **Type VI-Minus**
  —Ex: restructuring most elementary particles

- **Type Ω-Minus**
  —Ex: altering the structure of space-time

Notes: As already indicated in *Sec.4.4.2.2* of *FPHC*, the problem with this micro-classification (from Barrow's work) is that the civilization types (with the exception of Type Ω-Minus, for example) are not quite distinct, since many of them can be achieved more or less in a civilization, to the extent that Type II-minus and Type III-minus, just to cite two plausible types, can be historically contemporaneous, relatively speaking, unlike the vast historical distance between, say, Type O and Type I (or Type I and Type II) civilizations. In other words, the micro-classification here is not very useful to understand civilization types but is revealing to see the technological frontiers of the micro-world.
Sources: A reconstruction from J.Barrow (1998:133), as originally shown in *Table 4.7* of *FPHC*. See *FPHC* for more info.

Table 1.6. Finity, Transfinity, and Infinity (Part I)

• **Infinity**

—The term “infinity,” which is “symbolically represented with ∞,” derives from the Latin *infinitas* (or “unboundedness”), to refer to “several distinct concepts—usually linked to the idea of 'without end'—which arise in philosophy, mathematics, and theology.” (WK 2008c)

—In the context of mathematics, “infinity” is often “treated as if it were a number (i.e., it counts or measures things: 'an infinite number of terms') but it is a different type of 'number' from the real numbers. Infinity is related to limits,...large cardinals,...projective geometry, extended real numbers and the absolute Infinite,” for instance. (WK 2008c)

• **Transfinity**

—To avoid confusion, my usage of the word “transfinite” (as used here in the ontological principle) is not the same as the one which was originally coined by Georg Cantor to refer to “cardinal numbers or ordinal numbers that are larger than all finite numbers, yet not necessarily absolutely infinite” (in the mind of God), as opposed to “relative” infinity (in the mind of Man). (WK 2008d) In other words, for Cantor, there are two kinds of infinity, “absolute” and “relative”—with “relative” infinity or “transfinity” (in the mind of Man) being dependent on “absolute” infinity (in the mind of God) for its existence.

—By contrast, my usage of the word “transfinite” here differs radically from the one used by Cantor (and other mathematicians) and is more limited, in light of the problems confronting any attempt to understand the idea of infinity, be it by intuition, imagination, and conception (as already shown in the prevous three sub-sections).

—With this clarification in mind, I allow numbers which can be many times larger—or smaller, for that matter—than the finite things that we encounter in daily life, but they do not have to be related to the idea of infinity at all (which may not exist).

(continued on next page)

Table 1.6. Finity, Transfinity, and Infinity (Part II)

- **Transfinity** ***(cont'd)***
  —Of course, there may be some *borderline* cases, in which it is not clear whether the number in question is transfinite (in my usage) or simply a mathematical convenience. A good example of a borderline case is the Planck unit of length for "the smallest space possibly measured in nature," which is "less than billionths of trillionths of trillionths of an inch" (or something like 1.6×10^{-35} meters). (P. Baofu 2006a; N. McAleer 1987: 219; WKV 2008; D. Corbett 2008)
  —The ideas concerning symmetry and dimensions in the examples above were addressed in Chapter Three and Chapter Four of *FPHG*. For now, it suffices to further add that it is mind-boggling to even conceive or imagine extreme numbers like the Planck unit of length (or others). Does it really exist at all? Or is it merely a mathematical convenience to describe a physical situation that the human mind does not understand properly?
  —In any event, "transfinity" (in my usage) can be used as a further extension (of number) to the finite numbers in daily life—albeit with some borderline cases in mind.

- **Finity**
  —It refers to numbers which are bounded (that is, with an end), especially (though not exclusively) in relation to things in everyday life.

Source: From *Sec. 1.4* and *Sec. 2.2.3* of *FPHG*

Table 1.7. Theoretical Speculations of Multiverses

- **"Baby Universes" (Ex: Andre Linde and others)**
  —Ex: In a flat universe theory, "even if our part of it eventually collapses,...some spots in the cosmos would suddenly start inflating on their own, creating brand-new 'baby universes.'" (P. Baofu 2000: 623)

- **"Parallel Universes" (Ex: Stephen Hawking and others)**
  —Ex: In quantum cosmology, there allows the existence of infinite numbers of parallel universes, with tunneling among them. (M. Kaku 1994: 256) Hawking later revised his views on this.

- **"Pocket Universes" (Ex: Alan Guth)**
  —Ex: "As the pocket universes live out their lives and recollapse or dwindle away, new universes are generated to take their place....While life in our pocket universe will presumably die out, life in the universe as a whole will thrive for eternity." (A. Guth 1997: 248; P. Baofu 2002: 482)

- **"Brane Worlds" (Ex: Warren Siegel, Lisa Randall, and others)**
  —Ex: Our universe is stuck on a membrane of space-time embedded in a larger cosmos, with different brane worlds connecting and/or colliding with each other.

Notes: These examples are solely illustrative (not exhaustive), and some of the items can be reclassified somewhere else. Nor are they always mutually exclusive. Since they are generalities, exceptions are expected.
Source: From *Table 4.8* of *FPHC*

Table 1.8. The Confusion between "Many Worlds" and "Multiverse" (Part I)

- **First Problem**

—The first problem in this second confusion concerns a debate between the "many-worlds" interpretation of quantum mechanics by Hugh Everett (1957) and others, and the Copenhagen interpretation by Niels Bohr and Werner Heisenberg on the effects of quantum states after interacting with an external environment (e.g., by a measuring observer).

—On the one hand, the Copenhagen interpretation of quantum mechanics is more indeterministic in its concern with the probabilities of the outcomes of a wave function after interacting with an external environment (e.g., by a measuring observer using light to detect, for example, "an electron in a particular region around the nucleus at a particular time." (WK 2008e) Each wave function for the state of a particle refers to "a mathematical representation used to calculate the probability for it to be found in a location, or state of motion." (WK 2008f) And the "wavefunction collapse" means that "the act of measurement causes the calculated set of probabilities to 'collapse' to the value defined by the measurement." (WK 2008f)

—On the other hand, the Everett's interrpetation is more deterministic, in rejecting "the objective reality of wavefunction collapse" and "instead explaining the subjective appearance of wavefunction collapse with the mechanism of quantum decoherence." (WK 2008g) The words "quantum decoherence" here refers to "the mechanism by which quantum systems interact with their environments to exhibit probabilistically additive behavior," in such a way that "the quantum nature of the system is simply 'leaked' into the environment." (WK 2008g) For example, "[a]s a result of an interaction, the wave functions of the system and the measuring device become entangled with each other. Decoherence happens when different portions of the system's wavefunction become entangled *in different ways* [italic added] with the measuring device...." (WK 2008g) Thus, for Everett, the appearance of "wavefunction collapse" can be explained by "quantum decoherence" for quantum states to branch out in different independent worlds after interacting with an external environment.

(continued on next page)

Table 1.8. The Confusion between "Many Worlds" and "Multiverse" (Part II)

- **First Problem *(cont'd)***
  —Yet, a major criticism against Everett's interpretation is that these different independent worlds in Everett's sense "will never be accessible to us," because of the interference by the measuring observer in an external environment. (WK 2008g)
  —For illustration, "once a measurement is done, the measured system becomes entangled with both the physicist who measured it and a huge number of other particles, some of which are photons flying away towards the other end of the universe; in order to prove that the wave function did not collapse one would have to bring all these particles back and measure them again, together with the system that was measured originally. This is completely impractical, but even if one can theoretically do this, it would destroy any evidence that the original measurement took place (including the physicist's memory)." (WK 2008g)
  —But then, the critics ask, If these independent worlds in Everett's sense are inaccessible to us, how do we know much of anything really about them?

- **Second Problem**
  —What is relevant at the quantum mechanical level may not necessarily apply to the world at the physicsl level.
  —In fact, the persistent inability among contemporoary physicists to unifiy the theory of relativity for large objects at the physical level and quantum mechanics for tiny objects at the quantum mechanical level should warn us about the danger of reductionism, by reducing the understanding of different universes ("parallel universes") at the physical level from a generalization of "many worlds" at the quantum mechanical level.

- **Third Problem**
  —There is no empirical evidence to suggest that there are infinite (or extremely many) possible worlds in the cosmos as the many-worlds model by Everett would have us to believe.

Source: From *Sec.4.4.1* of *FPHG*

Table 1.9. Hyperspace and Its Challenge (Part I)

• **First Problem**

—Hyperspace for intra- and inter-universal travel is assumed to be possible, on the basis of four main assumptions; although there may be others, these are the often cited ones.

—The first assumption concerns the existence of "entry" and "exit" points in hyperspace, which is easier said than done to enter, go through, and then exit hyperspace.

—For instance, using hyperspace "requires complex calculations…of single or multiple hyperspace jumps and the control of the jump….Peturbations such as those experienced by ship in space from the gravitional field around an object such as a planet or even a star are exacerbated in hyperspatial travel, since mass in real space distorts hyperspace in an equal measure. 'Jumping' near to a gravitational mass is likely to make resulting exit from hyperspace to be highly uncertain, with the level of improbability *i* increasing with the square of the distance to the nearest gravitional 'well.'" (WK 2008h)

—Besides, "[o]nly spaceships equipped with a special force field can enter hyperspace, because exposure to hyperspace even for short period of time is hazardous to unprotected humans." (WK 2008h)

• **Second Problem**

—The second assumption concerns the faster-than-light speed. It is well known that Einstein argued for the speed of light as the ultimate limit, but nowadays phyiscists do allow the violation of this limit under certain conditions (like the hypothetical existence of tachyons, as will be explained further in a later sub-section on time travel).

—Even then, there is a challenge to explain in science fiction "why ships can travel faster than light in hyperspace," and two good arguments are that "hyperspace may be smaller than real space and therefore a star ship's propulsion seems to be greatly multiplied, or else the speed of light in hyperspace is not a barrier as it is in real space. Whatever the reasoning, the general effect is that ships traveling in hyperspace seem to have broken the speed of light, appearing at their destinations much more quickly and without the shift in time that the Theory of Relativity would suggest." (WK 2008h)

(continued on next page)

Table 1.9. Hyperspace and Its Challenge (Part II)

• **Third Problem**

—The third assumption concerns the speed of time (to be measured, say, by a clock), but this is elaborated in a different section on the related issue of "time travel" (as shown in a different table on time travel).

• **Fourth Problem**

—The fourth assumption concerns spatial shortcut, as it is possible, because of the curvature of space, in that "the manifold of ordinary three-dimensional space is curved in four or more 'higher' spacial dimensions (a 'hyperspace' in the geometric sense...). This curvature causes certain widely separated points in three-dimensional space to nonetheless be 'adjacent' to each other four-dimensionally. Creating an aperture in 4D space (a wormhole) between these locations can allow instantaneous transit between the two locations; a common comparison is that of a folded piece of paper, where a hole punched through two folded sections is more direct than a line drawn between them on the sheet. This idea probably arose out of certain popular descriptions of General Relativity and/or Riemannian manifolds, and may be the original form from which later concepts of hyperspace arose." (WK 2008h)

—But this does not mean that hyperpsace is always safe. In fact, "in some science fiction, the danger of hyperspace travel is due to the chance that the route through hyperspace may take a ship too close to a celestial body with a large gravitational field, such as a star. In such scenarios, if a starship passes too close to a large gravitational field while in hyperspace, the ship is forcibly pulled out of hyperspace and reverts to normal space. Therefore, certain hyperspace 'routes' may be mapped out that are safe, not passing too close to stars or other dangers." (WJ 2008h)

Source: From *Sec.4.4.1* of *FPHG*

Table 1.10. The Problems of Time Travel into the Future (Part I)

• **First Problem**

—The first problem concerns "time dilation under the Theory of Special Relativity" (which separates space-time from mass-energy as special cases). Suppose you are on a moving train and an observer outside the train is watching you along the tracks. As your train moves along at the speed of light (or something close to it), "time, as measured by your watch, ticked along at a slower pace than time measured by the observer. Not only that, distance changed, too. For the observer, a one-foot ruler whizzing by on the train would have measured less than a foot." (NO 2000)

—In fact, you may not even notice that this slower clicking of the clock; in fact, everything looks normal to you inside the train as it was before: "The weird thing is that, for you on the train, time wouldn't seem to be moving slower and your ruler wouldn't be shorter—all would appear normal." (NO 2008) By contrast, you instead think that "time on the rest of the Earth would appear to be ticking along slower and its rulers would be shorter." (NO 2000) So, both of you insist that the other clock (not his) is clicking more slowly—assuming, of course, an inertial frame of reference.

—Now, change the story a bit, in that, this time, instead of an "inertial reference frame," acceleration is allowed, and you are on your way to board a spaceship. As an illustration, suppose "[y]ou board a spaceship and take off for deep space. The ship approaches the speed of light. Time for you seems to pass as it always has. It takes you about five seconds to tie your shoe. But to an observer on Earth (assuming he or she could watch you), you are moving at a snail's pace. It takes hours to tie your shoe." (NO 2000) In any event, "you continue on your journey. You slow down, stop, and accelerate back to Earth. You arrive home. You have aged two years during your flight. Two hundred years have passed on Earth. You have successfully travelled forward through time." (NO 2000) In this case, both the observer (if he is still alive) and you agree that you have aged less.

(continued on next page)

Table 1.10. The Problems of Time Travel into the Future (Part II)

- **First Problem** ***(cont'd)***

—In other words, when you are back to Earth, a lot of things have changed since, and you may find rather strange to be in a new environment which you hardly recognize, when compared with what you used to think as your city, or the Earth (two years ago, well, in accordance to your time of measurement). This means that you are in the future (two hundred years later), even though you are still relatively young (only two years older). But, can you go back? Well, no. The reason is that "[a]ccording to relativity, you can only move through time in one direction." (NO 2000)

- **Second Problem**

—The second problem concerns "time dilation under the Theory of General Relativity" (which unifies space-time with mass-energy in a general theory), the same result of time dilation occurs.

—For instance, when "one clock is deeper in a gravity well than the other," a result is that "the clock deeper in the well" will "tick…more slowly; this effect must be taken into account when calibrating the clocks on the satellites of the Global Positioning System, and it could lead to significant differences in rates of aging for observers at different distances from a black hole." (WK 2008i)

—More precisely, by way of another example, "it has been calculated that, under general relativity, a person could travel forward in time at a rate four times that of distant observers by residing inside a spherical shell with a diameter of 5 meters and the mass of Jupiter. For such a person, every one second of their 'personal' time would correspond to four seconds for distant observers. Of course, squeezing the mass of a large planet into such a structure is not expected to be within our technological capabilities in the near future." (WK 2008i; R. Gott 2002)

—The point here, however, is that time travel into the future, in this sense, is allowed, theoretically speaking, but in a very limited way, from the theory of relativity. But one can also ask, What if the theory of relativity were wrong? Well, in that case, there would be a need to find a better theory to explain time travel into the future.

Source: From *Sec.4.4.2* of *FPHG*

Table 1.11. The Problems of Time Travel into the Past (Part I)

- **First Problem**
  - —The first problem concerns the faster-than-light-speed, since, in accordance to the theory of special relativity, when an object goes faster than the speed of light, something weird can happen, in that, instead of having the clock clicking more slowly, it is moving backward, in relation to some inertial frame of reference—which then raisies the issue of whether time travel into the past is possible at all.
  - —But the problem here is about the violation of the law of casuality. In everyday language, an excellent illustration of causality violation is that, suppose you travel back into the past and then kill your parents, this then means that you could not have been born. But this is not true, since you are alive.
  - —Yet, "in the case of a hypothetical signal moving faster than light, there would always be some frames in which the signal was received before it was sent, so that the signal could be said to have moved backwards in time." The violation of casuality exists in this case.
  - —The first part of the problem is that, in accordance to the theory of special relativity, "it would take an infinite amount of energy to accelerate a slower-than-light object to the speed of light"—let alone the energy needed to propel an object to go faster than the speed of light. (WK 2008i) Although some suggest the possibility of "negative energy," it remains to be seen to what extent this is true.
  - —And the second part of the problem is that, "although relativity does not forbid the theoretical possibility of tachyons which move faster than light at all times, when analyzed using quantum field theory it seems that it would not actually be possible to use them to transmit information faster than light and there is no evidence for their existence." (WK 2008i; S. Chase 1993)

(continued on next page)

Table 1.11. The Problems of Time Travel into the Past (Part II)

• **Second Problem**

—The second problem concerns cosmic string and black holes, since it is the theory of general relativity which "extends the special theory to cover gravity, illustrating it in terms of curvature in spacetime caused by mass-energy and the flow of momentum." (WK 2008i)

—On the other hand, "[g]eneral relativity describes the universe under a system of field equations...and there exist solutions to these equations that permit what are called 'closed time-like curves,' and hence time travel into the past....The first of these was proposed by Kurt Gödel, a solution known as the Gödel metric, but his (and many others') example requires the universe to have physical characteristics that it does not appear to have." (WK 2008i; K. Thorne 1994)

—But then, the problem now becomes: "Whether general relativity forbids closed time-like curves for all realistic conditions is unknown." (WK 2008i)

• **Third Problem**

—The third problem concerns wormholes and related ones, and "[w]ormholes are a hypothetical warped spacetime which are also permitted by the Einstein field equations of general relativity, although it would be impossible to travel through a wormhole unless it was what is known as a traversable wormhole." (WK 2008i; M. Visser 1995) In other words, "a wormhole is a hypothetical topological feature of spacetime that is fundamentally a 'shortcut' through space and time. Spacetime can be viewed as a 2D surface, and when 'folded' over, a wormhole bridge can be formed. A wormhole has at least two mouths which are connected to a single throat or tube. If the wormhole is traversable, matter can 'travel' from one mouth to the other by passing through the throat." (WK 2008j)

—But the problem here is three-fold, as shown in what follows.

(continued on next page)

Table 1.11. The Problems of Time Travel into the Past (Part III)

- **Third Problem *(cont'd)***
  —The first part of the problem is that "there is no observational evidence for wormholes." (WK 2008j)
  —The second part of the problem is that it is not clear how exactly wormholes (and black holes, for that matter) could allow time travel into the past at all.
  —And the third part of the problem is that if using wormholes (and black holes, for that matter) to travel into the past requires something to go faster than light too, then this raises another issue concerning the two problems as previously indicated in the sub-section on the faster-than-light-speed.

- **Fourth Problem**
  —The fourth problem is that some physicists have argued that, even should time travel into the past be possible, by way of these three methods, the person still would not return to the same past history that he exactly experienced before, but in a different one. (WK 2008i)
  —In other words, "strictly speaking," time travel into the past is not "really" possible, in light of these criticisms, so there is no need to worry that one could travel into the past and kill one's parent, as a way to violate the law of casuality.

Source: From *Sec.4.4.2* of *FPHG*

Table 1.12. The Conceptual Dimensions of Consciousness (and Other Mental States)

- **On Heredity and Time**
  —Heredity and the Environment
  —The Past and the Present

- **On Layers of Mental States and Abnormality**
  —Consciousness, Unconsciousness, and Preconsciousness
  —Normality and Paranormality

- **On Organicity and Motivation**
  —Mechanicity and Organicity
  —Primary Motivations and Multiple Motivations

- **On Other (Mostly Epistemic) Considerations**
  —Synthesis and Analysis
  —Situation and the Subject
  —Process and Outcome
  —Reasoning and Other Modes of Thinking
  —Meta-Conceptual Nominalism and Realism

Notes:: These examples are solely illustrative (not exhaustive), and some of the items can be reclassified somewhere else. Nor are they always mutually exclusive. Since they are generalities, exceptions are expected.
Sources: From *FPHC*. A re-construction, but with my own contribution, originally from G.Lindzey & C.Hall, *Introduction to Theories of Personality* (NY: John Wiley & Sons, 1985).

Table 1.13. The Theoretical Levels of Consciousness (and Other Mental States) (Part I)

- **At the Micro-Physical Theoretical Level**
  —Quantum-Mechanics
  —Electromagnetism

- **At the Chemical Theoretical Level**
  —Biochemistry

- **At the Biological Theoretical Level**
  —Evolutionary Biology
  —Neuroscience
  —Artificial Intelligence

- **At the Psychological Theoretical Level**
  —Psychodynamic Psychology
    - Psychoanalytical Psychology
    - Analytical Psychology
    - Socially Oriented Psychology

  —Experimental (Behavioral) Psychology
    - Operant Reinforcement Theory
    - Stimulus-Response Theory
    - Social Learning Theory

  —Cognitive (Gestalt) Psychology
    - Humanist Psychology
    - Existential Psychology
    - Field Theory

  —Psychometric Psychology
    - Idiographics
    - Constitutional Psychology

  —Social Psychology
    - Symbolic Interactive Theory
    - Social Exchange Theory

(continued on next page)

Table 1.13. The Theoretical Levels of Consciousness (and Other Mental States) (Part II)

- **At the Organizational Theoretical Level**
  —Managerial-Bureaucratic Theory
  —Oligarchic Theory
  —Network Theory

- **At the Institutional Theoretical Level**
  —Functionalist Theory
  —Anomic Theory

- **At the Structural Theoretical Level**
  —Conflict Theory
    - Marxian Theory
    - Critical Theory
    - Weberian Theory

  —Games Theory (in Formal Theory)
  —Feminist Theory
    - Feminist Concerns
    - Feminist Hopes

- **At the Systemic Theoretical Level**
  —Equilibrium Theory
  —System Theory
  —Chaos Theory

- **At the Cosmological Theoretical Level**
  —Superluminal Model
  —The Theory of Floating Consciousness

(continued on next page)

Table 1.13. The Theoretical Levels of Consciousness (and Other Mental States) (Part III)

- **At the Cultural Theoretical Level**
  - —Substantive Theories
    - • Structuralist Theory
    - • Post-Structuralist Theory (in Postmodernism)
  - —Meta-Theories
    - • Epistemic Objectivism vs. Epistemic Historicism
    - • Epistemic Subjectivism vs. Epistemic Non-Subjectivism
      - –Phenomenology
      - –Ethnomethodology
      - –Hermeneutics
    - • Epistemic Relativism vs. Epistemic Absolutism
    - • Epistemic Reductionism vs. Epistemic Emergencism

- **At Other Levels**
  - —Structuration Theory
  - —Reflexive Socioanalysis

Notes: These examples are solely illustrative (not exhaustive), and some of the items can be reclassified somewhere else. Nor are they always mutually exclusive. Since they are generalities, exceptions are expected.
Source: From many different sources as indicated in *FPHC*

Table 1.14. The Thematic Issues of Consciousness (and Other Mental States)

- **The Factor of History**
  —Pre-Modernity
  —Modernity
  —Postmodernity
  —After-Postmodernity

- **The Importance of Needs**
  —Having (e.g., the everyday, the technological)
  —Belonging (e.g., the just and the good)
  —Being (e.g., the true, the holy, the sublime/beautiful, and the good)

Notes: These examples are solely illustrative (not exhaustive). Nor are they always mutually exclusive. Since they are generalities, exceptions are expected. Also, some categories overlap with others; for instance, the good is also related with being-ness, just as the everyday can be connected with belong-ness.
Sources: From *FPHC*, based on *FHC* and *FCD*

Table 1.15. Having, Belonging, and Being in Consciousness

- **Having (e.g., About the Everyday and Technological)**
  —Ex: Physiological needs (e.g., thirst, hunger, sex)
  —Ex: Safety (e.g., stability, freedom from fear and anxiety)

- **Belonging (e.g., About the Good and Just)**
  —Ex: Social interaction
  —Ex: Friendship, acquaintance
  —Ex: Love, family
  —Ex: Self-respect, respect from others

- **Being (e.g., About the True, Holy, Beautiful/Sublime, and Good)**
  —Ex: Understanding of reality about self and world
  —Ex: Spiritual quest for holiness, beauty, sublimity, and goodness

Notes: These examples are solely illustrative (not exhaustive), and some of the items can be reclassified somewhere else. Nor are they always mutually exclusive. Since they are generalities, exceptions are expected.
Sources: From *FPHC*, based on *FHC* and *FCD*

Table 1.16. The Having-Ness of Consciousness (and Other Mental States)

- **At the Micro-Physical Level**
  —Ex: The finer physics of nutrition

- **At the Chemical Level**
  —Ex: Food guide pyramid

- **At the Bio-Psychological Level**
  —Ex: The hierarchy of needs

- **At the Institutional Level**
  —Ex: The production of commodity fetishism

- **At the Organizational Level**
  —Ex: Mass standardization (and customization); bureaucratic rationality

- **At the Structural Level**
  —Ex: The divide between the haves and the have-nots

- **At the Cultural Level**
  —Ex: The bondage of greater expectations

- **At the Systemic Level**
  —Ex: The technological transformation of poverty with transvaluations

- **At the Cosmological Level**
  —Ex: Floating consciousness in deep space without the human physiological needs

Notes: These examples are solely illustrative (not exhaustive), and some of the items can be reclassified somewhere else. Nor are they always mutually exclusive. Since they are generalities, exceptions are expected.
Sources: From *FPHC*. See also *FCD* and *FHC*.

Table 1.17. The Belonging-Ness of Consciousness (and Other Mental States)

- **At the Micro-Physical Level**
  —Ex: Statistical physics and social networks
- **At the Chemical Level**
  —Ex: Olfaction and social relations
- **At the Bio-Psychological Level**
  —Ex: Dual human nature on relationships
- **At the Institutional Level**
  —Ex: Private property and social cleavage
- **At the Organizational Level**
  —Ex: Legal formalism and iron cage
- **At the Structural Level**
  —Ex: Class conflicts and social alienation
- **At the Cultural Level**
  —Ex: Ruthless society and the culture of law
- **At the Systemic Level**
  —Ex: Human extinction and post-human relations
- **At the Cosmological Level**
  —Ex: The transcendence of equality in the cosmos and beyond

Notes: These examples are solely illustrative (not exhaustive), and some of the items can be reclassified somewhere else. Nor are they always mutually exclusive. Since they are generalities, exceptions are expected.
Sources: From *FPHC*. See also *FCD* and *FHC*.

Table 1.18. The Being-Ness of Consciousness (and Other Mental States) (Part I)

- **At the Micro-Physical Level**
  —Ex: Space-time and matter-energy in classical mechanics, quantum mechanics, and the theory of relativity

- **At the Chemical Level**
  —Ex: Drugs and well being

- **At the Biological Level**
  —Ex: Neuroscience and mystical experiences

- **At the Psychological Level**
  —Ex: Reductionism and emergencism

- **At the Institutional Level**
  —Ex: The proliferation of cults and sects, and the corporate-art industry

- **At the Organizational Level**
  —Ex: The organization of ideas in the scheme of things

- **At the Structural Level**
  —Ex: Disciplinary power and control in the order of things

- **At the Systemic Level**
  —Ex: Mutualism and ecological psychology

- **At the Cultural Level**
  —Ex: The recurrence of competing spirits, and floating consciousness as a climax of evolution

(continued on next page)

Table 1.18. The Being-Ness of Consciousness (and Other Mental States) (Part II)

- **At the Cosmological Level**
  —Ex: The search for elsewherewhen, and the evolution of other consciousnesses (e.g., the hyper-spatial consciousness)

Notes: These examples are solely illustrative (not exhaustive), and some of the items can be reclassified somewhere else. Nor are they always mutually exclusive. Since they are generalities, exceptions are expected.
Sources: From *FPHC*. See also *FCD* and *FHC*.

Table 1.19. Cognitive Partiality in Different Mental States

- **The Conscious**
  - —*Biased*
    - Ex: Self-overratingness
    - Ex: Self-modesty
    - Ex: Self-fulfilling prophesy
  - —*Shallow*
    - Ex: Attribution error
    - Ex: Heuristics
    - Ex: Illusory thinking
  - —*Short-term*
    - Ex: Deindividuation
  - —*Materialistic*
    - Ex: Hierarchy of needs

- **The Subconscious**
    - Ex: mental reconstruction
    - Ex: stereotyping

- **The Unconscious**
    - Ex: automatic processing
    - Ex: cognitive intuition

Notes: These examples are solely illustrative (not exhaustive), and some of the items can be reclassified somewhere else. Nor are they always mutually exclusive. Since they are generalities, exceptions are expected. And the comparison is relative, not absolute.
Source: From *BCPC*

Table 1.20. Emotional Non-Neutrality and Behavioral Alteration in Different Mental States

- **The Emotional**
  —*Conscious*
    Ex: Feelings
  —*Subconscious*
    Ex: Moods
  —*Unconscious*
    Ex: Empathy
    Ex: Instincts

- **The Behavioral**
  —*Conscious*
    Ex: Foot-in-the-door
    Ex: Role playing
    Ex: Saying-is-believing
  —*Subconscious*
    Ex: Discrimination
  —*Unconscious*
    Ex: Non-Verbal Communication

Notes: These examples are solely illustrative (not exhaustive), and some of the items can be reclassified somewhere else. Nor are they always mutually exclusive. Since they are generalities, exceptions are expected. And the comparison is relative, not absolute.
Source: From *BCPC*

Table 1.21. The Limits of Intuition in Unconsciousness

- **Cognitive Intuition (or Intuitive Thinking)**
  - —*Features*
    - Ex: mental schemas
    - Ex: expertise
    - Ex: blindsight
  - —*Problems*
    - Ex: inaccurate
    - Ex: non-explanatory

- **Emotional Intuition**
  - —*Features*
    - Ex: empathy
    - Ex: instincts
  - —*Problems*
    - Ex: over-generalized
    - Ex: non-explanatory

- **Behavioral Intuition**
  - —*Features*
    - Ex: prosemics
    - Ex: kinesics
  - —*Problems*
    - Ex: unreliable
    - Ex: non-explanatory

Notes: These examples are solely illustrative (not exhaustive), and some of the items can be reclassified somewhere else. Nor are they always mutually exclusive. Since they are generalities, exceptions are expected. And the comparison is relative, not absolute.
Source: From *BCPC*

Table 1.22. The Wealth/Poverty Dialectics in Different Mental States: The Case of Cognition

- **The Conscious**
  - —*Biased*
    - Ex: If richer in self-serving bias, then poorer in understanding others
  - —*Shallow*
    - Ex: If richer in heuristic thinking, then poorer in comprehending reality
  - —*Short-term*
    - Ex: If richer in short-term tendency, then poorer in long-term planning
  - —*Materialistic*
    - Ex: If richer in physiological obsession, then poorer in spiritual enlightenment

- **The Subconscious**
  - Ex: If richer in mental reconstruction, then poorer in accuracy of recall
  - Ex: If richer in stereotyping, then poorer in understanding groups

- **The Unconscious**
  - Ex: If richer in automatic processing, then poorer in depth of analysis
  - Ex: If richer in cognitive intuition, then poorer in reliability of judgments

Notes: These examples are solely illustrative (not exhaustive), and some of the items can be reclassified somewhere else. Nor are they always mutually exclusive. Since they are generalities, exceptions are expected. And the comparison is relative, not absolute.
Source: From *BCPC*

Table 1.23. The Wealth/Poverty Dialectics in Different Mental States: The Case of Emotion and Behavior

- **Emotion**
  - —*Conscious*
    - Ex: If richer in relying on familiarity and similarity for social attraction, then poorer in having diverse social relations
  - —*Subconscious*
    - Ex: If richer in moodiness, then poorer in sober thinking
  - —*Unconscious*
    - Ex: If richer in empathy, then poorer in task-oriented efficiency
    - Ex: If richer in instincts, then poorer in critical thinking

- **Behavior**
  - —*Conscious*
    - Ex: If richer in foot-in-the-door technique, then poorer in instant Gratification
    - Ex: If richer in role playing, then poorer in identity stability
  - —*Subconscious*
    - Ex: If richer in discriminating, then poorer in social harmony
  - —*Unconscious*
    - Ex: If richer in kinesics and proxemics, then poorer in verbal skills, relatively speaking.

Notes: These examples are solely illustrative (not exhaustive), and some of the items can be reclassified somewhere else. Nor are they always mutually exclusive. Since they are generalities, exceptions are expected. And the comparison is relative, not absolute.
Source: From *BCPC*

Table 1.24. The Theoretical Debate on Nature vs. Nurture (Part I)

- **The Environmental Approach**
  —*Thesis*: It focuses, relatively speaking, more on the environment (culture and society) in explaining the achievement gap among individuals and for that matter, countries or regions, when contrasted with the natural factors.
  —*Discourse*: Examples include Jose Ortega y Gasset ("Man has no nature; what he has is history"), Ashley Montagu ("Man is man because he has no instincts, because everything he is and has become he has learned from his culture, from the man-made part of the environment, from other human beings"), Stephen Jay Gould ("[The] brain [is] capable of a full range of behaviors and predisposed to none"), and Jesse Jackson (who blames white racism for the failure of blacks to close the achievement gap between whites and blacks in America). The works on dependency theory in international political economy (with a Marxian influence) and on the Protestant work ethic (by Max Weber) also point to this environmental direction.

- **The Genetic Approach**
  —*Thesis*: It focuses instead, relatively speaking again, on hereditory factors (e.g., genes and evolution) in explaining the achievement gap among individuals and for that matter, countries or regions, when contrasted with the envrionmental factors.
  —*Discourse*: Examples are Hans Eysenck and William Sheldon (in constitutional psychology), Konrad Lorenz (in his work on innate aggressive human nature), Gary Marcus (on the complexities of human thought by a tiny number of genes) and Robert Plomin, together with Michael Owen and Peter McGuffin (on the genetic basis of complex human behaviors).

(continued on next page)

Table 1.24. The Theoretical Debate on Nature vs. Nurture (Part II)

• **The Compromise Approach**

—*Thesis*: It seeks the middle-of-the-road argument in regard to nature and nurture and regards all differences among individuals and groups as the result of the mixture of both nature and nurture, more or less equally.

—*Discourse*: Examples include C. Murray and R. Herrnstein ("It seems highly likely to us that both genes and environment have something to do with this issue") and Dan Dennett ("Surely 'everyone knows' that the nature-nurture debate was resolved long ago, and neither side wins since everything – is – a – mixture – of – both – and - it's – all – very -complicated...."

• **The Transcendent Approach**

—*Thesis*: It goes beyond both nature and nurture (without, however, committing the compromise fallacy) in showing their closely intertwined interactions in producing the behavioral differences as often seen in individual human endeavors on the micro scale, and for that matter, in country (or regional) endeavors on the macro one—in the context of my five theses, namely, (a) the compromise fallacy, (b) no oppression without self-oppression, (c) no success without failure, (d) the factor of randomness, and (e) the post-human vision, to be elaborated in Chapter Six.

—*Discourse*: Peter Baofu proposed this approach on the basis of his "theory of contrastive advantages" (as an original theoretical contribution to the debate, which was first proposed and analyzed in *The Future of Capitalism and Democracy*). In the end, the human genes will not last, to be eventually superseded by post-human life forms, so the debate between genes and memes has obscured something profoundly important about the future that the world has never known. And the debate is also misleading and faulty in its dichotomy.

Source: From *BNN*. See the book for citations and details.

Table 1.25. Physical Challenges to Hyper-Spatial Consciousness

- **The Understanding of a Higher-Dimensional World of Space-Time**
  —Ex: 4 for traditional aspects of space-time (e.g., length, width, breadth and time) plus 6 more new dimensions in theory of hyper-space, with profound implications for practical applications to new forms of consciousness.

- **The Mastering of Dark Matter and Dark Energy**
  —Ex: "ordinary matter" (e.g., atoms, molecules) as a mere 4.4% of the universe, with 23% made of "cold dark matter" and the rest (about 73%) of mysterious "dark energy," with fundamental significance to questions about the limit of the speed of energy (or info), the availability of energy for use, and the nature of space-time, just to cite some examples.

- **The Exploration of Multiverses**
  —Ex: theoretical speculation of other universes (e.g., "baby universes," "gateways" in black holes, "wave function of the universe," "many worlds," "brane worlds"), with potentially seminary discoveries of different physical laws in relation to matter-energy and space-time, and vital differences to the future of post-human conquest of other universes (for the emergence of new forms of consciousness).

Notes: These examples are solely illustrative (not exhaustive), and some of the items can be reclassified somewhere else. Nor are they always mutually exclusive. Since they are generalities, exceptions are expected. The point here is to give a rough picture of the evolution of consciousness to the hyper-spatial consciousness and others totally unknown to current earthlings. As a note of clarification, it makes no difference to my argument as to whether or not the hyper-spatial consciousness may emerge before, during, and after floating consciousness.
Source: From *Table 4.5* of *FPHC*

Table 1.26. The Theory of Floating Consciousness (Part I)

- **At the Micro-Physical Level**
  —Ex: intelligent life without the human physical-chemical system

- **At the Chemical Level**
  —Ex: space radiation and toxins

- **At the Bio-Psychological Level**
  —Ex: exo-biological evolution in deep space
  —Ex: genetic engineering of new beings

- **At the Institutional Level**
  —Ex: post-capitalism
  —Ex: post-democracy

- **At the Organizational Level**
  —Ex: less legal-formalistic routines

- **At the Structural Level**
  —Ex: alien forms of violence

- **At the Cultural Level**
  —Ex: transcending freedom
  —Ex: transcending equality

(continued on next page)

Table 1.26. The Theory of Floating Consciousness (Part II)

- **At the Cosmological Level**
  —Ex: parallel universes
  —Ex: pocket universes

- **At the Systemic Level**
  —Ex: space habitats (in zero-gravity)

Notes: Each example draws from the works of different scholars in the field. For instance, at the cosmological level, the idea of parallel universes is from the theoretical speculation in quantum cosmology by Stephen Hawking and others, while the one of pocket universes comes from the theoretical work of Allan Guth at MIT. And at the institutional level, I proposed post-capitalism and post-democracy in *FCD* (and later, from *BDPD* and *BCPC*). In addition, the examples are solely illustrative (not exhaustive), and some of the items can be reclassified somewhere else. Nor are they always mutually exclusive. Since they are generalities, exceptions are expected.
Sources: From *FPHC*—and, originally, from *FCD*

Table 1.27. The Potential of Unfolding Unconsciousness

- **Superior Senses**
  - —*Through space*
    - Ex: clair-sentience (in feeling)
    - Ex: clair-voyance (in seeing)
    - Ex: clair-audience (in hearing)
    - Ex: clair-austance (in tasting)
    - Ex: clair-alience (in smelling)
    - Ex: clair-cognizance (in knowing)
  - —*Through time*
    - Ex: pre-cognition (in knowing), pre-sentiment (in feeling)
    - Ex: retro-cognition (in knowing), retro-sentiment (in feeling)

- **Superior Contacts**
  - —*With living minds*
    - Ex: telepathy
  - —*With the dead*
    - Ex: mediumship
  - —*With other-living entities*
    - Ex: telekinesis

Notes: The categories and examples are solely illustrative (not exhaustive), and the comparison is relative (not absolute), nor are they necessarily mutually exclusive. And some can be easily re-classified elsewhere. As generalities, they allow exceptions.
Source: A summary of Ch.5 of *FPHU*.

Table 1.28. The Future Exploration of Unfolding Unconsciousness

- **In the Context of Nature**
  —Ex: the higher-dimensional space-time in hyperspace theory
  —Ex: the particle-wave interactions in quantum mechanics
  —Ex: the study of brain waves in electromagnetic theory

- **In the Context of the Mind**
  —Ex: the unconscious fantasies in psychoanalysis
  —Ex: the mind-altering drugs in biochemistry
  —Ex: the manipulation of neural activity in neurobiology

- **In the Context of Culture**
  —Ex: the correlation between anomalous experience and moral/religious/aesthetic interests in culture studies
  —Ex: the role of myths and language in epistemology

- **In the Context of Society**
  —Ex: different organizational agendas in anomalous research
  —Ex: the scholarly divide on anomalous study
  —Ex: interests and propaganda in institutioanl research
  —Ex: the transformative power of technology in research on social systems
  —Ex: the natural and non-natural selection of other worlds in exo-biology
  —Ex: the role of the post-human conscious mind in qualitative demography

Notes: The examples in each category are solely illustrative (not exhaustive), and the comparison is relative (not absolute), nor are they necessarily mutually exclusive. And some can be easily re-classified elsewhere. As generalities, they allow exceptions.
Source: A summary of Chs.2-5 of *FPHU*, in conjunction with the rest of the book

Table 1.29. Creative Techniques and Traits

- **Creative Techniques**
  - —*Comprehensiveness*
    - Ex: Broadness of Scope
    - Ex: In-Depth Analysis
    - Ex: Multifaceted Taxonomy
    - Ex: Numerous Theories
  - —*Arrangement*
    - Ex: Separation (or Subtraction)
    - Ex: Combination (or Addition)
    - Ex: Replacement
    - Ex: Reversal
    - Ex: Accommodation
    - Ex: Stretching
  - —*Serendipity*
    - Ex: Play
    - Ex: Chance

- **Creative Traits**
  - —*Openness / Risk*
    - Ex: Ability to welcome new ideas or to do new things, with risk
  - —*Discipline*
    - Ex: Ability to sacrifice whatever it takes to succeed
  - —*Resilience / Confidence*
    - Ex: Ability to take failures to mind, not to heart
  - —*Trust*
    - Ex: Ability to cooperate with others for team work, if necessary

Notes: The examples in the categories are solely illustrative (not exhaustive), and the comparison is relative (not absolute), nor are they necessarily mutually exclusive. And some can be easily re-classified elsewhere. As generalities, they allow exceptions.
Source: From *Sec.4.2* of *FPHCT*. For more details, also read the rest of the book.

Table 1.30. The Desirability of Creativity, and Its Dark Sides (Part I)

- **In Relation to Instrumental Rationality**
  —*Desirability*
  - Ex: For the sake of efficiency and effectiveness in this capitalist era of our time.

  —*Undesirability*
  - Ex: What Max Weber (1930) called the "iron cage" of capitalism, where live the "sensualists without heart, specialists without spirit"—or something which the Frankfurt School has forcefully asked us to be freed from.

- **In Relation to Substantive Rationality**
  —*Desirability*
  - Ex: For the sake of God, the King, Motherland, or other comparable ideals.

  —*Undesirability*
  - Ex: Domination and oppression of various forms—something which the moderns had struggled to free themselves from in the first place, since the dawn of modernity.

- **In Relation to Autonomous Rationality**
  —*Desirability*
  - Ex: For the sake of itself, the autonomy of creative endeavor.

  —*Undesirability*
  - Ex: The decadence of its own degeneration (e.g., the donwsides of postmodernism in our postmodern times, or what Nietzsche once called the unbearable "unreality" and "falseness" of the autonomous artist's "innermost existence"—and, in other cases, the painful suffering from different mental illnesses for those highly creative individuals).

(continued on next page)

Table 1.30. The Desirability of Creativity, and Its Dark Sides (Part II)

Notes: The examples in the categories are solely illustrative (not exhaustive), and the comparison is relative (not absolute), nor are they necessarily mutually exclusive. And some can be easily re-classified elsewhere. As generalities, they allow exceptions.
Also, I already went to great lenghts to explain the desirability and its dark sides of all these three rationalities (although there may be others too) in the two volumes of *FHC*—and, for that matter, *FCD*.
Source: From *Sec.4.9* of *FPHCT*. For more details, also read the rest of the book.

Table 1.31. Posthuman-Ism, Post-Humanism, and Trans-Humanism

- **Post-Humanism**
  —The neologism "post-human" used in my books should *not* be confused with another term which looks similar but has a totally different meaning in the literature of postmodernism, namely, "post-humanism"—which constitutes a critique of "humanism" as tradionally understood (especially, though not exclusively, in relation to the idea of progress in science and reason in the Enlightenment project). (WK 2008)
  —My works reject the project of "postmodernism" and propose the future world of what I orignally called "after-postmodernity" in *FHC* and *FCD*, for instance.

- **Trans-Humanism**
  —Also, the neologism "post-human" used in my books should *not* be confused with a similar term which is used to champion the ideology of technology for the fturue co-existence between humans and cyborgs in "trans-humanism." (WK 2008a)
  —Instead, my term "post-human" in relation to "posthuman-ism" also rejects "trans-humanism" (especially, though not exclusively, in relation to the promises of technology) and refers to something else altogether, that is, the future extinction of humans and its post-human successors in deep space and beyond unto multiverses.
  —Chapter Two of my book on ethics (*BEPE*) provides a more comprehensive critique of trans-humanism.

- **Posthuman-Ism**
  —With these two clarifications in mind, the "post-humans" (as envisioned in my books) can take different forms, and I proposed, in my different books, some of them, such as "unfolding unconsciousness," "floating consciousness," "hyper-spatial consciousness," "thinking machines," "thinking robots," "genetically altered superior beings," "cyborgs," and others (as already analyzed in my previous works).
  —They will evolve in conjunction with other visions of mine in relation to nature, society, and culture.

Sources: From *Sec.1.6* of *FPHE* (and also *BEPE*). For more details, also read the rest of the books (and other books of mine).

Table 1.32. Three Great Future Transformations of the Martial Body (Part I)

- **Virtual Battle**
  —Ex: virtual fight
  —Ex: virtual memory

- **The Hyper-Martial Body**
  —*In relation to the physical dimension*
  - Ex: striking (e.g., punching, kicking, trapping, acupressure-striking, and so on in an amazing way)
  - Ex: grapping (e.g., throwing, pinning, joint-locking, and so on in an amazing way)
  - Ex: running and walking (e.g., on different surfaces)
  - Ex: jumping (e.g., through walls, trees, etc.)
  - Ex: flying (e.g., through space)
  - Ex: enduring pain (e.g., absorbing punches, kicks, and so on in an amazing way)
  - Ex: using extremely advanced hi-tech weapons
  - Ex: healing (e.g., with special energy, etc.)
  - Ex: entering (e.g., into different dimensions of space)
  - Ex: sensing (e.g., in extraordinary ways through space and time, etc.)
  - Ex: wearing hi-tech suits for defensive and offensive purposes (e.g., like a hi-tech armor)

  —*In relation to the mental dimension*
  - Ex: the ability to relax the martial body in a way that current humans cannot.
  - Ex: the ability to concentrate the martial mind in a way that current humans are not accustomed to.
  - Ex: the ability to breathe and be timely in preparing the martial body for fighting in a way that current humans are not good at.

- **The Post-Martial Floating Mind**
  —Ex: the mind without the human body
  —Ex: the quest for spirituality beyond martial arts

(continued on next page)

Table 1.32. Three Great Future Transformations of the Martial Body (Part II)

Notes: The examples in the categories are solely illustrative (not exhaustive), and the comparison is relative (not absolute), nor are they necessarily mutually exclusive. And some can be easily re-classified elsewhere. As generalities, they allow exceptions.
Source: From *Sec. 2.5.1* of *FPHMA*. See text for citations.

Table 1.33. Three Great Future Transformations of the Sexual Body

- **Virtual Pleasure**
  —Ex: virtual Sex
  —Ex: virtual Recall

- **The Hyper-Sexual Body**
  —*In relation to the physical dimension*
  - Ex: the ability to perform differenet sexual positions without the various physical limitations that humans currently have.
  - Ex: the ability to produce unusual amount of sexual energy to endure in sexual acts in a way that humans currently cannot.
  - Ex: the ability to engage in multiple sexual experiences comparable to (but different from) non-human sexuality (e.g., hermaphrodite, monoecious, sex-switching, and many other types) that current humans cannot.

  —*In relation to the mental dimension*
  - Ex: the ability to relax the sexual body in a way that current humans are not accustomed to.
  - Ex: the ability to concentrate the sexual mind in a way that current humans are not good at.
  - Ex: the ability to breathe and be patient in preparing the sexual body for sexuality in a way that current humans are not able to.

- **The Post-Sexual Floating Mind**
  —Ex: the mind without the human body
  —Ex: the quest for spirituality beyond sexuality

Notes: The examples in the categories are solely illustrative (not exhaustive), and the comparison is relative (not absolute), nor are they necessarily mutually exclusive. And some can be easily re-classified elsewhere. As generalities, they allow exceptions.
Source: From *Sec. 3.4.2* of *FPHS*. See text for more info.

Table 1.34. The Trinity of Pre-Modernity

- **Pre-Free-Spirited Pre-Modernity (Pre-Modernism) and Its Internal Split**
  —Competing worldviews and values both within and between linear centric (e.g., Islamic, Christian, Judaic, Imperial Roman) and cyclical-centric (e.g., Confucian, Taoist, Hindu, and Buddhist) orientations
  —Compare modernism with pre-modernism here in relation to the seven dimensions of human existence like the true and the holy (e.g., different versions of epistemic dogmas and religious superstitions), the everyday and the technological (e.g.,different versions of non-technophilism and non-consumerism), the beautiful/sublime (e.g., different versions of aesthetic non-autonomy), and the good and the just (e.g., different versions of moral particularism).

- **Pre-Capitalist Pre-Modernity (Pre-Modernization) and Its Own Discontents**
  —Competing versions of societal arrangements (e.g., feudalism, monarchism, and the holy order)

- **Hegemonic Pre-Modernity and Its Countervailing Forces**
  —Different power centers and their enemies (e.g., the Roman Empire and the "barbarian hordes," the "Holy Crusades" and the Muslims, the Middle Kingdom and the invading tribes, different social castes in India, and warring Greek city-states)

Notes: The examples in each category are solely illustrative (not exhaustive) nor necessarily mutually exclusive, and the comparison is relative (not absolute). As generalities, they allow exceptions. Also, it does not matter what the "base" era is in the analysis of any trinity. And in the present context, the "base" era is modernity (for instance, with its "free-spirited modernity" and the other two parts). So, for pre-modernity, the trinity takes the form of, say, "pre-free-spirited pre-modernity," together with the other two parts.
Sources: From Ch.2 of *BCIV* and also the 2 volumes of *FHC*

Table 1.35. The Trinity of Modernity (Part I)

- **Free-Spirited Modernity (Modernism) and Its Internal Split**

  —*On the True and the Holy*
  - The freedom from the dogmas of the past to the better understanding of, and union with, the world and self (Ch.3 of *FHC*)
  - Alternative discourses: about the true (e.g., anti-science discourses) and the holy (non-mainstream theologies) (Ch.3 of *FHC*)

  —*On the Technological and the Everyday*
  - The freedom from life harshness to the higher technophilic, consumeristic lifeform (Ch.2 of *FHC*)
  - Alternative discourses: about the everyday (e.g., transcendental mindsets) and the technological (e.g., Arcadianism) (Ch.2 of *FHC*)

  —*On the Good and the Just*
  - The freedom from the theo-aristocratic tyranny to the moral universality for a just society (Ch.5 of *FHC*)
  - Alternative discourses: about the just (e.g., Communism, Anarchism) and the good (e.g., Nazism/Fascism, and Zarathustrianism) (Chs.5-6 of *FHC*)

  —*On the Beautiful and the Sublime*
  - The freedom from the external distortion of aesthetic pleasure to the boundless infinity of totality in artistic autonomy (Ch.4 of *FHC*)
  - Alternative discourses: about the beautiful/sublime (e.g., kitsch and historical avant-gardism) (Ch.4 of *FHC*)

- **Capitalist Modernity (Modernization) and Its Own Discontents**

  —*During the Industrial Revolution*
  - Ex: Marx on the institution of inequality (Ch.1 of *FHC*)

  —*During the Modern Rational-Instrumental Epoch*
  - Ex: Weber on the politics of soft liberal institutions (Ch.5 of *FHC*)

  —*During the Great Depression*
  - Ex: Keynes on the myth of the free market (Chs.1-3 of *FHC*)

  —*During the Cold War*
  - Ex: Lasch on the narcissistic culture industry (Chs.2-3 of *FHC*)

(continued on next page)

Table 1.35. The Trinity of Modernity (Part II)

- **Hegemonic Modernity and Its Countervailing Forces**
  —*The Legacies of Colonialism and Imperialism*
    - Ex: European colonization of most of the modern world (Ch.1 of *FHC*)
  —*The Struggle for Decolonialization*
    - Ex: The countervailing forces of resentment, rechantment, and regionalism (Chs.1 & 6 of *FHC*)

Notes: The examples in each category are solely illustrative (not exhaustive) nor necessarily mutually exclusive, and the comparison is relative (not absolute). As generalities, they allow exceptions.
Sources: From the 2 volumes of *FHC*—and also from *FCD*

Table 1.36. The Trinity of Post-Modernity

- **Free-Spirited Postmodernity (Postmodernism) and Its Internal Split**
  —*On the True and the Holy*
    - Postmodern performative turn for knowing and its enemies (Ch.3 of *FHC*)
    - Postmodern comparative theology and its opponents (Ch.3 of *FHC*)

  —*On the Technological and the Everyday*
    - Postmodern corporate technological mindset and its adversaries (Ch.2 of *FHC*)
    - Postmodern postmaterialism and its critics (Ch.2 of *FHC*)

  —*On the Good and the Just*
    - Postmodern politics of difference and its foes (Ch.5 of *FHC*)

  —*On the Beautiful and the Sublime*
    - Postmodern deconstruction and its dissenters (Ch.4 of *FHC*)

- **Capitalist Postmodernity (Postmodernization) and Its Own Discontents**
  —*During the Post-Cold War and Beyond*
    - Ex: post-Fordism and its shortcomings (Ch.6 of *FHC*; Chs.6-7 of *FCD*)

- **Hegemonic Postmodernity and Its Countervailing Forces**
  —*The Debate on the Global Village*
    - Ex: uni-civilizationalism vs. multi-civilizationalism (Ch.6 of *FHC*)

  —*The Resistance Movement*
    - Ex: rechantment and the politics of civilizational claims (e.g., Islamic, Confucian and other ethos in relation to the Same) (Ch.6 of *FHC*; Ch.10 of *FCD*)
    - Ex: resentment and the politics of resurgence (e.g., the rising Chinese superpower, the growing EU, and other players in relation to the U.S. and her allies) (Ch.6 of *FHC*; Ch.8 of *FCD*)
    - Ex: regionalism and the politics of inequality (e.g., trans- or international blocs, the North-South divide, NGO's) (Ch.6 of *FHC*; Ch.5 of *FCD*)

Notes: The examples in each category are solely illustrative (not exhaustive) nor necessarily mutually exclusive, and the comparison is relative (not absolute). As generalities, they allow exceptions.
Sources: From *FCD* and the 2 volumes of *FHC*

Table 1.37. The Trinity of After-Postmodernity

- **Free-Spirited After-Postmodernity (After-Postmodernism) and Its Internal Split**
  —The discourse of naked contingency (Ch.10 of *FCD*; Ch.4 of *FPHC*)

- **Post-Capitalist After-Postmodernity (After-Postmodernization) and Its Own Discontents**
  —Different versions of post-capitalism and post-democracy, and their enemies (Ch.10 of *FCD*; Chs.3-4 of *FPHC*)

- **Hegemonic After-Postmodernity and Its Countervailing Forces**
  —The Cyclical Progression of Hegemony in Multiverses (Chs.9-10 of *FCD*; Ch.4 of *FPHC*)

Notes: The examples in each category are solely illustrative (not exhaustive) nor necessarily mutually exclusive, and the comparison is relative (not absolute). As generalities, they allow exceptions.
Sources: From *FCD* and also *FHC*

• Part Two •

Statics

CHAPTER 2
STATICS AND ITS CONTINGENCY

> Nature is a vast tablet, inscribed with signs…; and geology is simply the key by which myriads of these signs, hitherto indecipherable, can be unlocked and perused….
> —Hugh Miller (TSH 2010)

The Durability of Statics

Geology involves both statics and dynamics. As already analyzed in *Sec. 1.3*, statics and dynamics are closely related.

For instance, the example of the "rock cycle" in *Sec. 1.3* describes "the dynamic transitions through geologic time among the three main rock types: sedimentary, metamorphic, and igneous," such that "each type of rock is altered or destroyed when it is forced out of its equilibrium conditions." (WK 2010b)

This then warns us that any study of statics (like the understanding of the types of rocks) requires an examination of the nature of dynamics, as one cannot be completely separated from the other.

With this in mind, this book offers a dialectic treatment of statics and dynamics in that, if statics has its durability, it has its changeability too—just as, if dynamics has its strength, it likewise has its weakness.

The inquiry concernng statics is the subject matter of this chapter (Chapter Two), whereas that of dynamics will be examined in the next chapter (Chapter Three).

In addition, this close relationship between statics and dynamics poses a scholarly challenge here, in that one cannot analyze statics (or dynamics, for that matter) separately.

This then means that the issues concerning statics in this chapter (Chapter Two) and dynamics in the next chapter (Chapter Three) often overlap, as each issue on one can be relevant to the other too.

With this reminder in mind—a good way to analyze statics (in this chapter) is by way of the evaluation of the extent to which it is in fact both possible and desirabe.

This chapter thus takes the challenging task to provide a comprehensive analysis of statics in the context of geology from the four perspectives of the mind, nature, society, and culture (in accordance to my sophisticated methodological holism, as explained in *Sec. 1.9*).

Consequently, this chapter is organized in four main sections, in relation to (2.2) statics and the mind, (2.3) statics and nature, (2.4) statics and society, and (2.5) statics and culture—to be analyzed hereafter, in that order (and summarized in *Table 2.1*).

Statics and the Mind

Statics, when analyzed from the perspective of the mind, can reveal fascinating aspects on its possibility and desirability, and this can be shown by way of two case studies, namely, (2.2.1) gestalt psychology, and the perception of geological structures, and (2.2.2) cemetery rocks, and the geology of fear—to be addressed hereafter, respectively.

Gestalt Psychology, and the Perception of Geological Structures

A good question in geology concerns the nature of perception of geological structures.

For instance, Jorge Carlos Della Favera (2007) and Marco André Malmann Medeiros argued that "the pursuit of geology still rests basically on descriptions. Descriptions are based on perceptions, which are entirely subjective, and depend on the laws of human psychology."

Gestalt Psychology

For Favera (2007) and Medeiros, a good illustration about the impact of human psychology on the study of geological structures has to do with gestalt psychology.

The term "gestalt psychology" refers to "a theory of mind and brain that proposes that the operational principle of the brain is holistic, parallel and analog, with self-organizing tendencies....It opposes to the 'atomistic' principle of operation, a common practice in science, which advocates the analysis of isolated parts of the problem in order to reduce its complexity." (J. Favera 2007)

Thus, in gestalt psychology, "it is not possible to know the whole after their parts but conversely the parts can be known from the whole. The concept 'the whole is more than the summation of parts' is the same to say that 'A+B' is not simply '(A+B)' but a third element 'C' which possesses its own characteristics. It is only from the perception of the totality that the brain can perceive, decode and process an image or a concept." (J. Favera 2007; C. Torrans 1999)

In this sense, "knowledge is a continuous act of organizing and rearranging information aiming a given purpose. Therefore, each new information added to the preexisting knowledge is assimilated and lead to a remodeling of all knowledge. Opposed to the traditional thinking, the Gestalt theory assumes that learning is not a simple accumulation of knowledge but a remodeling of insights, by gradual structuration, leading to a comprehension of problems solving." (J. Favera 2007)

Laws of Gestalt

In Gestalt psychology, there are some laws for "the human perception of forms. They are nothing less than conclusions about the natural brain behavior acting during the perception process." (J. Favera 2007)

For illustration, hereafter are six examples: (J. Favera 2007)

- "Prägnanz Law"—It states that "the forms should be perceived by their simpler character."
- "Similarity Law"—It states that "similar objects tend to be grouped. The similarity may happen in the object color, in texture or element mass sensation."
- "Proximity Law"—It states that "the grouping is made based on the distance the elements are from others."
- "Good Continuity Law"—It states that "it is related to the coincidences of directions or alignment of forms."
- "Closure Law"—It states that "a good shape is completed if it is closed itself, forming a limited figure. Mind adds missing elements to complete a figure."
- "Past Experience Law"—It states that "some forms only could be understood if there is a previous conscience of its existence. For the Christian believers , the phrase 'who see me will never forget' applied to human face...makes sense, but could be also be applied to any complex form perceived from the Gestalt theory."

Of course, there can be other laws of gestalt, so the examples above are illustrative.

The Perception of Sedimentary Structures, and the Laws of Gestalt

For Favera (2007) and Medeiros, these laws of gestalt turn out to be important in understanding the geological discovery of "complex sedimentary structures like the hummocky cross stratification (HCS)" in Brazil in 1979 and later.

The story of this geological discovery began in 1979, when "Petrobras, the Brazilian oil company, was interested in training Brazilian geologists in facies and depositional systems. Therefore the Exploration Manager, Mr. Coutinho, ordered that a group of geologists, including Jorge Della Favera, Miguel Uliana and Ubirajara Mello, should go to the Parnaíba Basin, in the State of Piauí, Brazil, and assessed the local conditions to install a field course on facies and depositional systems." (J. Favera 2007)

Under "the guidance of Mello, the group began studying the fluvial sediments of the Serra Grande Group, near the town of Jaicós. Afterwards they moved to the Pimenteira Formation outcrops, near Picos. There they face a something weird facies association. They thought initially that such an association would be fluvial in nature due to the red color of sandstones." (J. Favera 2007)

However, one member, Jorge Carlos Della Favera, "has had a former contact with the HCS during a SEPM field trip in Texas in 1978 and later on in the South Pyrenees Basin, with the so-called 'Campbell′s truncated-wave ripples' [thanks to the use of the law on the past experience]....With some difficulty the group recognized the HCS in the first Pimenteira outcrop and has been shown as a common feature in the rest of outcrops near Picos. As soon as they moved upwards in the stratigraphic column, new HCS recognitions were made. After this field trip, the group...concluded that HCS was present in all the lithostratigraphic units in the Parnaíba Basin, a surprising finding to them." (J. Favera 2007)

Because "such a structure has not been reported previously in Brazil, Della Favera...announced the discovery in the Brazilian Geological Society Meeting in 1980. During more than ten years, several groups of Brazilian geologists, including University professors, attended that courses which were conducted by Jorge Della Favera and Rodi Ávila Medeiros. In this period, the students′ behavior before the HCS recognition was always the same: they only recognized the structure after a simple drawing by themselves of its basic properties" (thanks to the use of the Prägnanz law). (J. Favera 2007)

Then, "in 1980, Della Favera was charged to make a geological guide to the Sergipe-Alagoas Basin. He was amazed by seeing uncountable HCS in practically every unit studied. As the idea at that time that HCS would develop only in marine settings it looked weird that this structure would occur also in non-marine units. When he recognized the HCS pattern on a tiled wall in a restaurant, he thought he was becoming out of his mind. However, it can be realized now that at that moment he had assimilated completely the HCS Gestalt" (thanks to the laws of gestalt) (J. Favera 2007)

Finally, "Della Favera…realized…the reason that the former researchers—mainly the late Rodi Ávila Medeiros, who was a keen sedimentologist—failed in not describing HCS, because they did not have a past experience to see it as a unit defined by the whole" (without the use of the law on past experience). (J. Favera 2007)

Some Critical Questions

However, the claim by Favera and Medeiros about the perception of sedimentary structures on the basis of the laws of gestalt (with the geological discovery of HCS in Brazil as a case study) is not without problems. Consider a few criticisms below, for illustration.

Firstly, a major criticism of the claim by Favera and Medeiros is that the appeal to the past experience by Favera (from his previous trip in Texas) as the reason to successfully recognize the HCS in Brazil begs the question somehow, since it did not explain how Favera came to learn about the HCS in his previous contact with the HCS during a SEPM field trip in Texas in 1978 in the first place.

Just because someone does not recognize something because he does not know about it does not necessarily have anything to do with gestalt psychology per se. The situation varies from case to case.

And secondly, another major criticism of the claim by Favera and Medeiros is that the whole account was made up after the fact, such that they then proposed another explanation to suggest that the students "only recognized the structure after a simple drawing by themselves of its basic properties" (thanks to the use of the Prägnanz law).

But it is not clear if this was really the reason that the students finally recognized the HCS. Was there something missing, that is, something else which was involved in the subsequent recognition of the structures by the students?

A Lesson to Learn

However, these criticisms of the claim by Favera and Medeiros should not be misinterpreted as a dismissal of its usefulness.

Rather, the criticisms are shown here for us to learn something from the two opposing sides of the debate, such that the possibility and desirability of statics (from the perspective of the mind with gestalt psychology in the discovery of HCS as a case study here) are not to the extent that its respective defenders would like us to believe.

More importantly, this analysis can teach us something valuable about the ontological principles in existential dialectics, and good examples include the absoluteness-relativeness principle, the partiality-totality principle, the predictability-unpredictability principle, the explicability-inexplicability principle, the preciseness-vagueness principle, the simpleness-complicatedness principle, the openness-hiddenness principle, the denseness-emptiness principle, the slowness-quickness principle, the expansion-contraction principle, the theory-praxis principle, the convention-novelty principle, the evolution-transformation principle, the symmetry-asymmetry principle, the regression-progression principle, and the same-difference principle.

For instance, in relation to the absoluteness-relativeness principle, if there is absolute viewpoint (e.g., the absolute viewpoint by Favera and Medeiros that "the former researchers—mainly the late Rodi Ávila Medeiros, who was a keen sedimentologist—failed in not describing HCS, because they did not have a past experience to see it as a unit defined by the whole"), there is relative counterpart (e.g., what is true for Favera and Medeiros in regard to the failure of former researchers to recognize HCS because of the lack of past experience is not necessarily so for the critics, because this explanation is too simplistic). And the reverse direction also holds true.

In relation to the partiality-totality principle, if there is partiality (e.g., the partial view of the simple parts of the HCS by the students, such that they did not recognize the HCS), there is totaltiy (e.g., the holistic view of the HCS as a whole by the students later, thanks to the use of the Prägnanz law in gestalt psychology, such that the whole is not the sum of its parts). And the reverse direction also holds true.

In relation to the predictability-unpredictability principle, if there is predictability (e.g, the predictable tendency of an individual, according to gestalt psychology, to "perceive, decode and process an image or a concept," "only from the perception of the totality"), there is unpredictability (e.g., the more difficult task to predict which specific law of gestalt, or which specific combination of them, an individual will use or not use in regard to an image or a concept on a particular occasion). And the reverse direction also holds true.

In relation to the explicability-inexplicability principle, if there is explicability (e.g., the explanation by Favera of his ability to successfully recognize the HCS in Brazil because of the appeal to his past experience from his previous trip in Texas), there is inexplicability (e.g., the lack of sufficient explanation by Favera about this appeal to his past experience, since it did not explain how Favera came to learn about the HCS in his previous contact with the HCS during a SEPM field trip in Texas in 1978 in the first place). And the reverse direction also holds true.

In relation to the preciseness-vagueness principle, if there is preciseness (e.g., the precise identification of the 6 laws of gestalt by Favera and Medeiros for their application to the discovery of the HCS in Brazil), there is vagueness (e.g,. the vagueness in the identification, since it is not really clear how the individual concepts in the laws like "similarity," "proximity," etc. are supposed to mean, since they can mean different things to different individuals and are applied in different ways on different occassions). And the reverse direction also holds true.

In relation to the simpleness-complicatedness principle, if there is simpleness (e.g., the relatively simple analysis by Favera and Medeiros of the discovery of the HCS in Brazil on the basis of the laws of gestalt), there is complicatedness (e.g., the relatively more complicated analysis of the discovery of the HCS in Brazil by way of showing some problems in their claim). And the reverse direction also holds true.

In relation to the openness-hiddenness principle, if there is openness (e.g., the open exploration of the relationship between gestalt psychology and the discovery of the HCS in Brazil by Favera and Medeiros), there is hiddenness (e.g., the hidden bias in the analysis by Favera and Medeiros because of their obsession with gestalt psychology in understanding the reasons behind the discovery of the HCS in Brazil, while ignoring other possible causes). And the reverse direction also holds true.

In relation to the denseness-emptiness principle, if there is denseness (e.g., the relativley denser concentration of the HCS in the Parnaíba Basin as discovered by Favera and his team), there is emptiness (e.g., the relatively less dense, or more empty, concentration of the HCS outside the Parnaíba Basin that Favera and his team did not succeed in discovering much). And the reverse direction also holds true.

In relation to the slowness-quickness principle, if there is slowness (e.g., the relatively slower readiness of the students to recognize the HCS in Brazil at the beginning, when they did not benefit from the help of the laws of gestalt like the Prägnanz law), there is quickness (e.g., the relatively quicker readiness of Favera to recognize the HCS in Brazil because of his previous trip in Texas, thanks to the use of the law on the past experience). And the reverse direction also holds true.

In relation to the expansion-contraction principle, if there is expansion (e.g., the relative expansion of the influence of gestalt psychology in relation to the work of Favera and his team, after they discovered the linkage between the laws of gestalt and the perception of geological structures), there is contraction (e.g., the relative decline of the influence of the "atomistic" practice in science in relation to the work of Favera and his team, after they discovered the linkage between the laws of gestalt and the perception of geological structures). And the reverse direction also holds true.

In relation to the theory-praxis principle, if there is theory (e.g., the theoretical construction of gestalt psychology like the laws of gestalt), there is praxis (e.g, the practical application of the laws of gestalt for the discovery of the HCS in Brazil by Favera and his team). And the reverse direction also holds true.

In relation to the convention-novelty principle, if there is convention (e.g., the conventional wisdom in regard to "the 'atomistic'...practice in science, which advocates the analysis of isolated parts of the problem in order to reduce its complexity"), there is novelty (e.g,. the alternative novel challenge to this conventional wisdom about the "atomistic" practice in science, by way of a new practice proposed by Favera and Medeiros along the line of gestalt psychology). And the reverse direction also holds true.

In relation to the evolution-transformation principle, if there is evolution (e.g., the natural evolution of humans to make observations about nature in a primitive way, in the state of nature), there is transformation (e.g,. the technical transformation of human ability to observe nature by way of the invention of geology, with the discovery of HCS and others). And the reverse direction also holds true.

In relation to the symmetry-asymmetry principle, if there is symmetry (e.g., the co-existence of different practices in science, be they "atomistic," "holistic," etc.), there is asymmetry (e.g., the acceptance of the "atomistic" practice in science in the older days—but the popularity of the "holistic" practice in science nowadays, like the use of gestalt psychology). And the reverse direction also holds true.

In relation to the regression-progression principle, if there is regression (e.g., the regression made in the application of gestalt psychology for geological fieldwork, as shown in the problems pointed out by the critics), there is progression (e.g., the progress made in the application of gestalt psychology for geological fieldwork, as shown in the discovery of the HCS in Brazil). And the reverse direction also holds true.

And in relation to the same-difference principle, if there is similarity in outcome (e.g., the contribution to the molding of beliefs and behaviors, regardless of whether this be done by way of the "atomistic" practice in science or by way of the "holistic" counterpart in gestalt psychology), there is difference in outcome (e.g., the contribution to the molding of beliefs and behaviors by way of the "atomistic" practice in science for the focus on individual parts of an object or a concept—but the contribution to the molding of beliefs and behaviors by way of the "holistic" counterpart in gestalt psychology for the focus on the whole instead).

Cemetery Rocks, and the Geology of Fear

Another fascinating way to examine statics (or structure) in geology from the perspective of the mind has to do with the issue concerning cemetery rocks in relation to the geology of fear.

Metamorphic Rocks for Graves

The use of rocks for graves is not new. For instance, back in 1764, a cemetery was built using metamorphic rocks at the Market Street in Brighton, Massachusetts, and now, after more than 200 years of existence, it is known as "the historic Market Street Burying Ground." (M. Lyman 2009)

What is interesting about the history of these rocks is that "[t]he dominant gravestone material of Colonial America was slate, a fine-grained metamorphic rock also used to make roof tiles, chalkboards, and pool tables. Thanks to the parallel alignment of slate's mineral matrix, the rock breaks readily into smooth plates, and because it began its geologic life as shale—a sedimentary rock usually composed of clay minerals—it is also relatively soft and easy to carve." (M. Lyman 2009)

But the trade-off of using metamorphic rocks with these qualities (for soft and easy carving) is that they "also make the slate especially susceptible to elemental weathering. Centuries of rain, ice, and snow have cracked, faded, pitted and streaked the old slate markers, making the sunken stones look ghoulishly worn. The result is slate stone cemeteries that easily out-creep their modern, granite gravestone counterparts." (M. Lyman 2009)

The historic Market Street Burying Ground is an excellent example of this trade-off, "as are two particularly eerie slate gravestones overlooking the busy street. They mark the final resting places of Abijah Learned and his wife Abigail. The markers are pallid as zombie flesh and a black discoloration creeps in from the stones' weathered edges. Abijah's stone is engraved with the solemn inscription, *Behold the spirit of the just ascend to God on high, and while their bodies sleep in dust their souls shall never die*. Like many of the cemetery's gravestones, his stands nearly four feet tall and is less than two inches thick, dimensions that render slate stones vulnerable to weather and time. The wear has left the markers tilting sickly in the soft soil, or, in the case of Abigail's gravestone, broken jaggedly in half." (M. Lyman 2009)

The Geology of Fear

One consequence does not escape the attention of many visitors to the graves like those at the historic Market Street Burying Ground and others nearby, and it has to do with its contribution to "the geology of fear," especially at the psychological level. (M. Lyman 2009)

On the surface, "rocks influence the look of a cemetery, but psychologists suggest that a graveyard's feel is more of a mental matter. According to Neil Lutsky, Kenan Professor of Psychology at Carlton College, the dread associated with old gravestones stems from cultural stereotypes and people's inherent fear of death. Humans, after all, are unique among animals in our knowledge of death's inevitability and this creates powerful psychological insecurities." (M. Lyman 2009)

So, in the older cemeteries like the historic Market Street Burying Ground and others nearby, Lutsky suggested that these "older cemeteries correspond in appearance to prototypes for spooky, broken down, ancient graves as represented in television and movies. Perhaps their age reminds us of the constancy of death." (M. Lyman 2009)

Social psychologists even have "a popular hypothesis about why humans fear death," and it is called "Terror Management Theory," which "asserts that we erect defenses against reminders of our own mortality"; so Lutsky added that "[c]emeteries serve as such threatening reminders." (M. Lyman 2009)

Another social psychologist, Jeff Greenberg at the University of Arizona, argued that, "for some people, weathering equals woe"—and Greenberg thus explained: "The older [stones] are more decayed, they reflect people who are forgotten. And people are afraid of being forgotten. Watching horror movies and ghost stories, people connect the older [gravestones] with nameless ghosts. There is a sense of obscurity." (M. Lyman 2009)

Some Disagreements

However, not all social psychologists accept this view about the geology of fear in older cemeteries, or in short, the "older is creepier" notion.

A good counter-argument against the "older is creepier" notion is that, as Lutsky admittted, some people "actually find...modern ones [not the older ones] more troubling," because "the rows are too close, the sameness too threatening." (M. Lyman 2009)

In fact, Lutsky recognized that, for some people, "there is something disturbing in the precision of modern [not older] cemeteries: the perfectly aligned grave rows, the flawless carvings, the laser-carved verses, and the impeccably manicured grass." (M. Lyman 2009)

In addition, for other people, "[p]erfection to this degree feels false, as synthesized as the plastic flowers that lie upon the markers. These modern graveyards may not be classically spooky but they are certainly dreary in their dullness." (M. Lyman 2009)

Belonging to Neither Side

Lest misunderstanding occurs, these alternative views against the "older is creepier" notion should not be misconstrued as a rejection of the claim about the geology of fear.

Instead, the alternative views are introduced here to show the two opposing sides of the debate, such that the possibility and desirability of statics (from the perspective of the mind with the geology of fear as a case study here) are not to the extent that its respective defenders would like us to believe.

After all, some geologists may regard metamorphic rocks as beautiful, but for some visitors to the historic Market Street Burying Ground, the rocks look frightening, to say the least.

In any event, this analysis of the cemetery rocks and the geology of fear can teach us something refreshing about the ontological principles in existential dialectics, and good examples include the absoluteness-relativeness principle, the partiality-totality principle, the predictability-unpredictability principle, the explicability-inexplicability principle, the preciseness-vagueness principle, the simpleness-complicatedness principle, the openness-hiddenness principle, the denseness-emptiness principle, the change-constancy principle, the order-chaos principle, the slowness-quickness principle, the expansion-contraction principle, the theory-praxis principle, the convention-novelty principle, the evolution-transformation principle, the symmetry-asymmetry principle, the softness-hardness principle, the regression-progression principle, and the same-difference principle.

For instance, in relation to the absoluteness-relativeness principle, if there is absolute viewpoint (e.g., the absolute view by those who believe in the "older is creepier" notion), there is relative counterpart (e.g., what is true for those who regard older cemeteries as more troubling is not necessarily so for those who regard newer or modern cemeteries as more troubling instead). And the reverse direction also holds true.

In relation to the partiality-totality principle, if there is partialtiy (e.g., the partial view of a metamorphic rock in an old cemetery in America), there is holistic counterpart (e.g., the holistic view of a metamorphic rock coupled with some creepy images, in an old American cemetery, such that the whole is not the sum of its parts). And the reverse direction also holds true.

In relation to the predictability-unpredictability principle, if there is predictability (e.g., the predictable tendency of slates to be "susceptible to elemental weathering," like those in the historic Market Street Burying Ground), there is unpredictability (e.g,. the more difficult task to predict whether or to what extent older cemeteries with the use of metamorphic rocks like slates are more troubling than modern cemeteries with the use of hard-to-crack granites, for some particular visitors on a given occasion). And the reverse direction also holds true.

In relation to the explicability-inexplicability principle, if there is explicability (e.g., the explanation of the creepy look of older cemeteries on the basis of the use of slates, which have been "cracked, faded, pitted and streaked"), there is inexplicability (e.g., the lack of sufficient explanation, if understood only within the "older is creepier" notion, of why, for some other individuals, what is more troubling is not older cemeteries but the modern ones). And the reverse direction also holds true.

In relation to the preciseness-vagueness principle, if there is preciseness (e.g., the precise identification of the fear of death as the reason for some visitors to find old cemeteries creepy), there is vagueness (e.g., the vagueness in this identification of the fear of death, since it is not clear what this fear of death really means to different individuals, as it may refer to the inevitability of death, the awfulness of death, the relief in death, etc.). And the reverse direction also holds true.

In relation to the simpleness-complicatedness principle, if there is simpleness (e.g., the relatively simplistic account of old cemeteries on the basis of the "older is creepier" notion), there is complicatedness (e.g., the relatively more complicated analysis of the "older is creepier" notion, by also considering alternative reactions by other visitors, who may find modern, not older, cemeteries more troubling instead). And the reverse direction also holds true.

In relation to the openness-hiddenness principle, if there is openness (e.g., the open exploration of the relationship between geological rocks and their use for graves), there is hiddenness (e.g., the hidden bias in the "older is creepier" notion, because of its obsession with older cemeteries, without giving sufficient attention to new cemeteries). And the reverse direction also holds true.

In relation to the denseness-emptiness principle, if there is denseness (e.g., the relatively denser concentration of the use of slates in older cemeteries like the historic Market Street Burying Ground), there is emptiness (e.g., the relatively less dense, or more empty, concentration of the use of slates in modern cemeteries in America). And the reverse direction also holds true.

In relation to the change-constancy principle, if there is change (e.g., the ever new ways of building cemeteries, like the use of slates, granites, etc.), there is constancy (e.g., the constancy of an uneasy feeling among visitors to cemeteries, regardless of which kiind of rocks will be used to build them). And the reverse direction also holds true.

In relation to the order-chaos principle, if there is order (e.g., the relativley orderly appearance of cemeteries with the use of slates, when they are newly built), there is chaos (e.g., the relatively chaotic appearance of cemeteries with the use of slates, after some centuries when the slates are "cracked, faded, pitted and streaked"). And the reverse direction also holds true.

In relation to the slowness-quickness principle, if there is slowness (e.g., the relatively slower speed of granites to be "cracked, faded, pitted and streaked" by weathering), there is quickness (e.g., the relatively quicker speed of slates to be "cracked, faded, pitted and streaked" by weathering). And the reverse direction also holds true.

In relation to the expansion-contraction principle, if there is expansion (e.g.,. the relative expansion of the use of granites in modern American cemeteries), there is contraction (e.g., the relative decline of the use of slates in modern Amerian cemeteries). And the reverse direction also holds true.

In relation to the theory-praxis principle, if there is theory (e.g., the theoretical construction of the nature of rocks in geology), there is praxis (e.g., the practical application of the understanding of rocks in geology for the use of building cemeteries). And the reverse direction also holds true.

In relation to the convention-novelty principle, if there is convention (e.g., the conventional wisdom about using slates to build graves in colonial America), there is novelty (e.g., the alternative novel challenge to this conventional wisdom, by way of the new method of using granites to build graves in modern America). And the reverse direction also holds true.

In relation to the evolution-transformation principle, if there is evolution (e.g, the natural evolution of humans to handle the death of individuals in a primitive way in the state of nature), there is transformation (e.g., the technical transformation of humans to handle the death of individuals by the invention of modern cemeteries, with the use of granites). And the reverse direction also holds true.

In relation to the symmetry-asymmetry principle, if there is symmetry (e.g., both slates and granites can be used to build graves), there is asymmetry (e.g., granites are less susceptible than slates to decay due to weathering). And the reverse direction also holds true.

In relation to the softness-hardness principle, if there is softness (e.g., slate is "relatively soft and easy to carve"), there is hardness (e.g.,. "granite is so much harder than slate"). (M. Lyman 2009) And the reverse direction also holds true.

In relation to the regression-progression principle, if there is regression (e.g., the regression made in the use of slates to build cemeteries, like the problems of decayed slates and the geology of fear among visitors), there is progression (e.g., the progress made in knowing how to build more durable cemeteries). And the reverse direction also holds true.

And in relation to the same-difference principle, if there is similarity in outcome (e.g., the contribution of cemeteries to uneasy feelings among visitors, regardlesss of whether they are built with the use of slates, granites, or else), there is difference in outcome (e.g., the contribution of cemeteries to uneasy feelings among visitors by way of the use of slates, with its "older is creepier" consequence—but the contribution of cemeteries to uneasy feelings among visitors by way of the use of granites, with its "dreary-in-dullness" consequence). And the reverse direction also holds true.

Statics and Nature

Statics, when examined from the perspective of nature, can shed further light on its possibility and desirability, and two good case studies here are (a) the debate on tectonic plates and (b) the contention on rock classification, to be addressed below, in that order.

A reminder is needed here, in that these two case studies about the statics of geology overlap with the dynamics of geology too (as the intricate relationship between statics and dynamics was already described in *Sec. 1.3*).

The Debate on Tectonic Plates

The term "plate tectonics" is "from the Late Latin *tectonicus*, from the Greek: *τεκτονικός*" to mean "pertaining to building"—and refers to "a scientific theory which describes the large scale motions of Earth's lithosphere. The theory builds on the older concepts of continental drift, developed during the first decades of the 20th century by Alfred Wegener, and seafloor spreading, developed in the 1960s." (WK 2010g)

Main Features

Tectonic plates have different features to be understood. Consider a few of them below, for illustration.

Firstly, one major feature of tectonic plates has to do with "the outer layers of the Earth" which "are divided into lithosphere and asthenosphere. This is based on differences in mechanical properties and in the method for the transfer of heat. Mechanically, the lithosphere is cooler and more rigid, while the asthenosphere is hotter and flows more easily. In terms of heat transfer, the lithosphere loses heat by conduction whereas the asthenosphere also transfers heat by convection." (WK 2010g)

Thus, this distinction allows us to understand "the key principle of plate tectonics," in that "the lithosphere exists as separate and distinct tectonic plates, which ride on the fluid-like (visco-elastic solid) asthenosphere. Plate motions range up to a typical 10–40 mm/a [per year] (Mid-Atlantic Ridge; about as fast as fingernails grow), to about 160 mm/a [per year] (Nazca Plate; about as fast as hair grows)." (WK 2010g; S. Huang 1997; P. Hancock 2000)

Secondly, another main feature of tectonic plates is that they "consist of lithospheric mantle overlain by either of two types of crustal material [on the outside]: oceanic crust…and continental crust." (WK 2010g)

In relation to "oceanic crust"—an "average oceanic lithosphere is typically 100 km thick; its thickness is a function of its age: as time passes, it conductively cools and becomes thicker. Because it is formed at mid-ocean ridges and spreads outwards, its thickness is therefore a function of its distance from the mid-ocean ridge where it was formed. For a typical distance oceanic lithosphere must travel before being subducted, the thickness varies ~6 km thick at mid-ocean ridges to greater than 100 km at subduction zones; for shorter or longer distances, the subduction zone (and therefore also the mean) thickness becomes smaller or larger, respectively." (WK 2010g; D.Turcotte 2002)

In relation to "continental crust"—an average "continental lithosphere is typically ~200 km thick, though this also varies considerably between basins, mountain ranges, and stable cratonic interiors of continents. The two types of crust also differ in thickness, with continental crust being considerably thicker than oceanic (35 km vs. 6 km)." (WK 2010g; D.Turcotte 2002)

However, although "tectonic plates can include continental crust or oceanic crust,...many plates contain both. For example, the African Plate includes the continent and parts of the floor of the Atlantic and Indian Oceans. The distinction between oceanic crust and continental crust is based on their modes of formation. Oceanic crust is formed at sea-floor spreading centers, and continental crust is formed through arc volcanism and accretion of terranes through tectonic processes; though some of these terranes may contain ophiolite sequences, which are pieces of oceanic crust, these are considered part of the continent when they exit the standard cycle of formation and spreading centers and subduction beneath continents." (WK 2010g)

And thirdly, another interesting feature of tectonic plates concerns "plate boundaries," where they meet. Geologists classify three major types of plate boundaries, as shown below (and summarized in *Table 2.2*). (WK 2010g)

- "Transform Boundaries"—This is "where plates slide or, perhaps more accurately, grind past each other along transform faults. The relative motion of the two plates is either sinistral (left side toward the observer) or dextral (right side toward the observer). The San Andreas Fault in California is an example of a transform boundary exhibiting dextral motion."
- "Divergent Boundaries"—This is "where two plates slide apart from each other. Mid-ocean ridges (e.g., Mid-Atlantic Ridge) and active zones of rifting (such as Africa's Great Rift Valley) are both examples of divergent boundaries."
- "Convergent Boundaries"—This is "where two plates slide towards each other commonly forming either a subduction zone (if one plate moves underneath the other) or a continental collision (if the two plates contain continental crust). Deep marine trenches are typically associated with subduction zones. The subducting slab contains many hydrous minerals, which release their water on heating; this water

then causes the mantle to melt, producing volcanism. Examples of this are the Andes mountain range in South America and the Japanese island arc."

Of course, there are other types, like the "plate boundary zones," which are "where the effects of the interactions are unclear and the broad belt boundaries are not well defined." (WK 2010g; USGS 1999; PT 2010)

This feature of tectonic plates about "plate boundaries" is highy important, because "plate boundaries are commonly associated with geological events such as earthquakes and the creation of topographic features such as mountains, volcanoes, mid-ocean ridges, and oceanic trenches. The majority of the world's active volcanoes occur along plate boundaries, with the Pacific Plate's Ring of Fire being most active and most widely known." (WK 2010g)

Main Types

Tectonic plates can be classified into two main types, namely, (a) "primary plates" and (b) "secondary plates."

In relation to the "primary plates," there are seven of them, as shown below (and summarized in *Table 2.2*): (WK 2010)

- "North American Plate"
- "South American Plate"
- "African Plate"
- "Eurasian Plate"
- "Indo-Australian Plate"
- "Pacific Plate"
- "Antarctic Plate"

And in relation to the "secondary plates," there are also seven of them, as shown below (and summarized in *Table 2.2*). (WK 2010g)

- "Juan de Fuca Plate"
- "Cocos Plate"
- "Caribbean Plate"
- "Nazca Plate"
- "Scotia Plate"
- "Arabian Plate"
- "Philippine Sea Plate"

A follow-up question, of course, concerns the causes of plate motions.

Main Causes

A good question to follow up the discussion of tectonic plates concerns their causes, that is, What are the driving forces of plate movements?

Geologists identify three major causes of tectonic plate movements, namely, (a) "friction," (b) "gravitation," and (c) "tidal forces"—to be discussed in what follows, in that order (and summarized in *Table 2.2*).

(a) Friction

The first major cause of tectonic plate movements has to do with the "friction" between "the relative density of oceanic lithosphere and the relative weakness of the asthenosphere." (WK 2010g)

For instance, "[w]hen it forms at mid-ocean ridges, the oceanic lithosphere is initially less dense than the underlying asthenosphere, but it becomes denser with age, as it conductively cools and thickens. The greater density of old lithosphere relative to the underlying asthenosphere allows it to sink into the deep mantle at subduction zones, providing most of the driving force for plate motions. The weakness of the asthenosphere allows the tectonic plates to move easily towards a subduction zone." (WK 2010g; P. Mendia-Landa 2007)

(b) Gravitation

The second major cause of tectonic plate movements has to do with the "gravitation" of tectonic plates, or more precisely, "gravitational sliding," when "plate motion is driven by the higher elevation of plates at ocean ridges." (WK 2010g)

For instance, "as oceanic lithosphere is formed at spreading ridges from hot mantle material, it gradually cools and thickens with age (and thus distance from the ridge). Cool oceanic lithosphere is significantly denser than the hot mantle material from which it is derived and so with increasing thickness it gradually subsides into the mantle to compensate the greater load. The result is a slight lateral incline with distance from the ridge axis." (WK 2010g)

(c) Tidal Forces

And the third major cause of tectonic plate movements has to do with "tidal forces." (WK 2010g)

In fact, "Alfred Wegener had proposed tidal forces as a mechanism for continental drift. However, the forces were considered far to small to cause continental motion as the concept then was of continents plowing through oceanic crust. However, Harry H. Hess had shown by 1962 that oceanic crust was in motion with the continents." (WK 2010g; UCMP 2010; G. Moore 1973)

Later, in 1973, "George W. Moore of the USGS presented evidence that tidal forces or tidal lag was a driving force for plate tectonics. In a more recent 2006 study, scientists argued that the westward component of every plate's motion is due to Earth's rotation and the tidal friction of the moon. As the Earth spins eastward beneath the moon, the moon's gravity ever so slightly pulls the Earth's surface layer back westward. It has also been suggested that this observation may also explain why Venus and Mars have no plate tectonics since Venus has no moon, and Mars' moons are too small to have significant tidal effects on Mars." (WK 2010g; B. Scoppola 2006; R. Lovett 2006)

Geologists in Debate

Yet, geologists often disagree about the claims about tectonic plates. Consider a few criticisms below, for illustration (as summarized in *Table 2.2*).

Firstly, a major criticism about the literature on tectonic plates is that the focus on the three major types of plate boundaries often does not give sufficient attention to the "plate boundary zones," which are "where the effects of the interactions are unclear and the broad belt boundaries are not well defined." (WK 2010g; USGS 1999; PT 2010)

Secondly, another major criticism about the literature on tectonic plates is that it is not clear why there are exactly "seven" primary plates and "seven" secondary plates. Why is this magic number of "seven" for both types? Is there something aesthetic in the classification (which is a bit arbitrary)?

Thirdly, still another major criticism about the literature on tectonic plates is that "[t]he sources of plate motion are a matter of intensive research and discussion among earth scientists," such that there can be other forces (unlike the three aforecited) to be accounted for.

And fourthly, the last major criticism about the literature on tectonic plates is that geologists disagree about the "relative significance of each mechanism" (or cause).

For instance, although it is generally accepted that "the actual vector of a plate's motion must necessarily be a function of all the forces acting upon the plate," what is contentious concerns "the problem regarding what degree each process contributes to the motion of each tectonic plate." (WK 2010g)

Some geologists suggested that "lithospheric plates attached to downgoing (subducting) plates move much faster than plates not attached to subducting plates. The Pacific plate, for instance, is essentially surrounded by zones of subduction (the so-called Ring of Fire) and moves much faster than the plates of the Atlantic basin, which are attached (perhaps one could say 'welded') to adjacent continents instead of subducting plates. It is thus thought that forces associated with the downgoing plate (slab pull and slab suction) are the driving forces which determine the motion of plates, except for those plates which are not being subducted." (WK 2010g; C. Conrad 2002)

Yet, the debate continues, because "the driving forces of plate motion continue to be active subjects of on-going research within geophysics." (WK 2010g)

A Didactic Exercise

But this debate among geologists about tectonic plates does not suggest that the theory of plate tectonics is totally nonsensical.

On the contrary, the debate among geologists as introduced above is to show us the two opposing sides on the nature of tectonic plates, such that the possibility and desirability of statics (from the perspective of the mind with the debate on tectonic plates as a case study here) are not to the extent that its respective defenders would like us to believe.

In addition, this analysis of tectonic plates can teach us something refreshing about the ontological principles in existential dialectics, and good examples include the absoluteness-relativeness principle, the predictability-unpredictability principle, the explicability-inexplicability principle, the preciseness-vagueness principle, the simpleness-complicatedness principle, the openness-hiddenness principle, the denseness-emptiness principle, the slowness-quickness principle, the expansion-contraction principle, the convention-novelty principle, the evolution-transformation principle, the symmetry-asymmetry principle, the softness-hardness principle, the regression-progression principle, and the same-difference principle.

For instance, in relation to the absoluteness-relativeness principle, if there is absolute viewpoint (e.g., the absolute view by some geologists that the "excess density of the oceanic lithosphere sinking in subduction zones is the most powerful source of plate motion"), there is relative counterpart (e.g., what is true for some geologists in regard to the "excess density of the oceanic lithosphere sinking in subduction zones" as "the most powerful source of plate motion" is not necessarily so for others, because "it is still a matter of some debate," as geologists still disagree about the "relative significance of each mechanism"). (WK 2010g) And the reverse direction also holds true.

In relation to the predictability-unpredictability principle, if there is predictability (e.g, the predictable tendency of some major driving forces of plate motions, like "friction," "gravitation," and "tidal forces"), there is unpredictability (e.g., the more difficult task to predict "what degree each process contributes to the motion of each tectonic plate"). And the reverse direction also holds true.

In relation to the explicability-inexplicability principle, if there is explicability (e.g., the explanation, within the theory of plate tectonics, of plate motions on the basis of the three main driving forces), there is inexplicability (e.g., the lack of convincing explanation, within the theory of plate tectonics, of why there cannot be any other forces which should be accounted for—not just friction, gravitation, and tidal forces,). And the reverse direction also holds true.

In relation to the preciseness-vagueness principle, if there is preciseness (e.g., the precise measurement of an "average oceanic lithosphere" to be "100 km thick"), there is vagueness (e.g., the vagueness in the measurement of the average oceanic lithosphere, because it is not clear how thicker it becomes each year, how much the rate varies from one specific location to another, etc.). And the reverse direction also holds true.

In relation to the simpleness-complicatedness principle, if there is simpleness (e.g., the relatively simple analysis of plate tectonics on the basis of its features and the driving forces), there is complicatedness (e.g., the relatively more complicated analysis of plate tectonics by considering some of the problems still unresolved, especially for the critics). And the reverse direction also holds true.

In relation to the openness-hiddenness principle, if there is openness (e.g., the open exploration of three main types of plate boundaries), there is hiddenness (e.g., the hidden bias of the exploration of the three main types of plate boundaries, because it often does not give sufficient attention to the "plate boundary zones," which are "where the effects of the interactions are unclear and the broad belt boundaries are not well defined"). And the reverse direction also holds true.

In relation to the denseness-emptiness principle, if there is denseness (e.g., the relatively denser nature of "cool oceanic lithosphere," which "is significantly denser than the hot mantle material from which it is derived"), there is emptiness (e.g., the relatively less dense nature of "hot mantle material" when compared with "cool oceanic lithosphere"). And the reverse direction also holds true.

In relation to the slowness-quickness principle, if there is slowness (e.g, the relatively slower speed of plate motions in the Mid-Atlantic

Ridge, with "a typical 10–40 mm/a [per year]...about as fast as fingernails grow"), there is quickness (e.g., the relatively faster speed of plate motions in the Nazca Plate, with "about 160 mm/a [per year]...about as fast as hair grows"). And the reverse direction also holds true.

In relation to the expansion-contraction principle, if there is expansion (e.g., the relatively more expansive nature of plate motions in "divergent boundaries," when contrasted with "convergent boundaries"), there is contraction (e.g., the relatively more contractive nature of plate motions in "convergent boundaries," when contrasted with "divergent boundaries"). And the reverse direction also holds true.

In relation to the convention-novelty principle, if there is convention (e.g., the conventional wisdom before the theory of plate tectonics, especially "in the late 19th and early 20th centuries," when "geologists assumed that the Earth's major features were fixed, and that most geologic features such as mountain ranges could be explained by vertical crustal movement, through geosynclinal theory"), there is novelty (e.g., the alternative novel challenge to the conventional wisdom of "geosynclinal theory," by the proposal of the new theory of plate tectonics, for plate motions). (WK 2010g) And the reverse direction also holds true.

In relation to the evolution-transformation principle, if there is evolution (e.g.,the natural evolution of humans to make observations about nature in a non-scientific way in the state of nature), there is transformation (e.g, the technical transformation of human ability to understand nature by the invention of geology like the theory of plate tectonics). And the reverse direction also holds true.

In relation to the symmetry-asymmetry principle, if there is symmetry (e.g., the existence of different geological theories over time, like geosynclinal theory and the theory of plate tectonics), there is asymmetry (e.g., the popularity of geosynclinal theory "in the late 19th and early 20th centuries"—but the popularity of the theory of plate tectonics nowadays). And the reverse direction also holds true.

In relation to the softness-hardness principle, if there is softness (e.g., the sliding of two plates along each other without collision in "transform boundaries), there is hardness (e.g., the collision between two continental plates in "convergent boundaries"). And the reverse direction also holds true.

In relation to the regression-progression principle, if there is regression (e.g., the regression made in the theory of plate tectonics, as shown in the criticisms by the critics), there is progression (e.g., the progress made in the theory of plate tectonics for a better understanding of plate motions). And the reverse direction also holds true.

And in relation to the same-difference principle, if there is similarity in outcome (e.g., the contribution to the understanding of geolotical features on Earth, like existence of "mountain ranges," regardless of whether this is understood by way of the geosynclinal theory or the theory of plate tectonics), there is difference in outcome (e.g., the contribution to the understanding of geolotical features on Earth, like existence of "mountain ranges," by way of the geosynclinal theory, in that they are so because of "vertical crustal movement"—but the contribution to the understanding of geolotical features on Earth, like existence of "mountain ranges," by way of the theory of plate tectonics, in that they are the result of plate motions). And the reverse direction also holds true.

The Contention on Rock Classification

Another interesting way to examine statics (or structure) in geology from the perspective of nature concerns the contention on rock classification in petrology (or the study of rocks).

In geology, the term "rock" (or "stone") refers to "a naturally occurring solid aggregate of minerals and/or mineraloids. The Earth's outer solid layer, the lithosphere, is made of rock....The scientific study of rocks is called petrology, and petrology is an essential component of geology." (WK 2010h)

Rock Classification

In general, "rocks are classified by mineral and chemical composition, by the texture of the constituent particles and by the processes that formed them." (WK 2010h)

With these criteria, geologists classify rocks into three types, namely, (a) "igneous," (b) "sedimentary," and (c) "metamorphic" rocks, to be described below, in that order.

(a) Igneous Rocks

The first type of rocks is known as "igneous rocks," which "are formed when molten magma cools and are divided into two main categories: plutonic [or intrusive] rocks and volcanic [or extrusive] rocks. Plutonic or intrusive rocks result when magma cools and crystallizes slowly within the Earth's crust (example: granite), while volcanic or extrusive rocks result from magma reaching the surface either as lava or fragmental ejecta (examples: pumice and basalt)." (WK 2010h; H. Blatt 1996)

Thus, "igneous rocks can be seen at mid ocean ridges, areas of island arc volcanism or in intra-plate hotspots." (WK 2010i)

(b) Sedimentary Rocks

The second type of rocks is known as "sedimentary rocks," which "are formed through the gradual accumulation of sediment: for example, sand on a beach or mud on a river bed. As the sediment is buried it is compacted as more and more material is deposited on top. Eventually the sediment will become so dense that it is essentially rock. This process is known as lithification." (WK 2010i)

In general, "[s]edimentary rocks form at or near the Earth's surface. Mud rocks comprise 65% (mudstone, shale and siltstone); sandstones 20 to 25% and carbonate rocks 10 to 15% (limestone and dolostone)." (WK 2010h; H. Blatt 1996)

(c) Metamorphic Rocks

And the third type of rocks is known as "metamorphic rocks," which "are rocks which once existed as igneous or sedimentary rocks but have been subjected to varying degrees of pressure and heat within the Earth's crust. The processes involved will change the composition and fabric of the rock and their original nature is often hard to distinguish." (WK 2010i)

In other words, metamorphic rocks "are formed by subjecting any rock type (including previously formed metamorphic rock) to different temperature and pressure conditions than those in which the original rock was formed. These temperatures and pressures are always higher than those at the Earth's surface and must be sufficiently high so as to change the original minerals into other mineral types or else into other forms of the same minerals (e.g., by recrystallisation)." (WK 2010h; H. Blatt 1996)

In general, "metamorphic rocks are typically found in areas of mountain building." (WK 2010i)

Some Problems with Rock Classification

Of course, these three types can consist of different sub-types, such that each type can be "subdivided into many groups," depending on the criteria given by a scholar. (WK 2010h) For instance, igneous rocks can be subdivided into "andesites," "anorthosites," etc.; sedimentary rocks can be subdivided into "anthracites," "argillites," etc.; and metamorphic rocks can be subdivided "amphibolites," "blueschists," etc. (WK 2010j)

Even with this qualification, rock classification, simple and neat as it is, comes with a price, as there are problems associated with it. Consider a few examples below, for illustration.

Firstly, one major problem with rock classification is that "[t]here are…no hard and fast boundaries between allied rocks. By increase or decrease in the proportions of their constituent minerals they pass by every gradation into one another, the distinctive structures also of one kind of rock may often be traced gradually merging into those of another. Hence the definitions adopted in establishing rock nomenclature merely correspond to selected points (more or less arbitrary) in a continuously graduated series." (WK 2010h)

Secondly, still another problem with rock classification is that the complicated history of the formation of rocks has yet to be completely understood. For instance, although "every stage in the origin of clays, sands and gravels can be seen in process around us," the problem still remains today, in that, "where these have been converted into coherent shales, sandstone and conglomerates, and still more where they have experienced some degree of metamorphism, there are many obscure points about their history upon which experiment may yet throw light." (WK 2010i)

And thirdly, one more problem with rock classification is that rock classification becomes more complicated when the three types of rocks often interact with each other because of the dynamic processes of "the rock cycle." (WK 2010b)

The term "rock cycle" refers to "a fundamental concept in geology that describes the dynamic transitions through geologic time among the three main rock types: sedimentary, metamorphic, and igneous….[E]ach type of rock is altered or destroyed when it is forced out of its equilibrium conditions." (WK 2010b)

For instance, "the sedimentary rock can…be subsequently turned into a metamorphic rock due to heat and pressure, and the metamorphic rock can be weathered, eroded, deposited, and lithified, becoming a sedimentary rock. Sedimentary rock may also be re-eroded and redeposited, and metamorphic rock may also undergo additional metamorphism. All three types of rocks may be re-melted; when this happens, a new magma is formed, from which an igneous rock may once again crystallize." (WK 2010)

So, in this way, "the rock cycle is an illustration that explains how the 3 rock types are related to each other and how processes change from one type to another over time." (WK 2010b)

An Overall Perspective

Yet, these problems about rock classification do not mean that it should be thown out of the window.

Instead, the discussion of the problems has the scholarly virtue to show us the two opposing sides in the debate, such that the possibility and desirability of statics (from the perspective of nature with the contention on rock classification as a case study here) are not to the extent that its respective defenders would like us to believe.

Furthermore, this analysis of the contention on rock classification can teach us something important about the ontological principles in existential dialectics, and good examples include the formalness-informalness principle, the absoluteness-relativeness principle, the predictability-unpredictability principle, the explicability-inexplicability principle, the finiteness-transfiniteness principle, the preciseness-vagueness principle, the simpleness-complicatedness principle, the openness-hiddenness principle, the denseness-emptiness principle, the slowness-quickness principle, the expansion-contraction principle, the convention-novelty principle, the evolution-transformation principle, the symmetry-asymmetry principle, the softness-hardness principle, the regression-progression principle, and the same-difference principle.

For instance, in relation to the formalness-informalness principle, if there is formalness (e.g., the formal logical requirement of completeness and soundness in a system of ideas), there is informalness (e.g., the nonformal existence of incompleteness in the system of rock classification, as there remain "many obscure points about their history" to be understood). And the reverse direction also holds true.

In relation to the absoluteness-relativeness principle, if there is absoluteness (e.g., the absolute view by petrologists that rocks can be classified in terms of three types), there is relativeness (e.g., what is true for petrologists in regard to the classification of rocks into three types is not necessarily so for the critics, who are more cautious, because of "many obscure points about their history upon which experiment may yet throw light"). And the reverse direction also holds true.

In relation to the predictability-unpredictability principle, if there is predictability (e.g., the predictable tendency of petrologists to classify rocks into three types), there is unpredictability (e.g., the more difficult task to predict the nature of "allied rocks" in a given case, because "[t]here are…no hard and fast boundaries between allied rocks"). And the reverse direction also holds true.

In relation to the explicability-inexplicability principle, if there is explicability (e.g., the explanation, by petrologists, of rock classification on the basis of "mineral and chemical composition,…the texture of the constituent particles and…the processes that formed them"), there is inexplicability (e.g., the lack of sufficient explanation of why rock classification should not include other criteria except the ones of "mineral and chemical composition,...the texture of the constituent particles and...the processes that formed them," given the fact that there remain "many obscure points about their history" to be understood). And the reverse direction also holds true.

In relation to the finiteness-transfiniteness principle, if there is finiteness (e.g., the finite number of the types of rocks in Earth, that is, the three types), there is transfiniteness (e.g., the transfinite number of all the indiviudal rocks in Earth). And the reverse direction also holds true.

In relation to the preciseness-vagueness principle, if there is preciseness (e.g., the precise identification of three types of rocks in rock classification), there is vagueness (e.g., the vagueness in the rock classification, since it is not clear which "selected points [more or less arbitrary]" a given rock can belong to, "in a continuously graduated series"). And the reverse direction also holds true.

In relation to the simpleness-complicatedness principle, if there is simpleness (e.g., the relatively simple analysis of rock classification into three types), there is complicatedness (e.g., the relatively more complicated analysis of rock classification by considering some difficult problems associated with the classification). And the reverse direction also holds true.

In relation to the openness-hiddenness principle, if there is openness (e.g., the open exploration of rocks in terms of their classification into three different types), there is hiddenness (e.g., the hidden bias of rock classification, since its obsession with three types does not give sufficient attention to "many obscure points about their history upon which experiment may yet throw light"). And the reverse direction also holds true.

In relation to the denseness-emptiness principle, if there is denseness (e.g., the relatiely denser nature of the sediment which is buried, as " more and more material is deposited on top," before "the sediment will become so dense that it is essentially rock"), there is emptiness (e.g., the relatively less dense, or more empty, nature of the recent material which is deposited on top of the sediment buried far beneath). And the reverse direction also holds true.

In relation to the slowness-quickness principle, if there is slowness (e.g., the relatively slower speed of magma, in the case of intrusive rock formation, when it "cools and crystallizes slowly within the Earth's crust"), there is quickness (e.g., the relatively faster speed of magma, in the case of extrusive rock formation, when it "reach[es]...the surface either as lava or fragmental ejecta"). And the reverse direction also holds true.

In relation to the expansion-contraction principle, if there is expansion (e.g., the relative expansion of magma, in the case of extrusive rock formation, when it "reach[es]...the surface either as lava or fragmental ejecta"), there is contraction (e.g,.. the relative contraction of a sediment, when it is buried and becomes "compacted as more and more material is deposited on top"). And the reverse direction also holds true.

In relation to the convention-novelty principle, if there is convention (e.g., the conventional wisdom of studying rocks by fieldwork), there is novelty (e.g., the alternative novel challenge to this conventional wisdom of doing geology by the new method of "experimental petrology" in our times which "employs high-pressure, high-temperature apparatus to investigate the geochemistry and phase relations of," say, "rocks of the lower crust and upper mantle that rarely survive the journey to the surface in pristine condition"). (WK 2010d) And the reverse direction also holds true.

In relation to the evolution-transformation principle, if there is evolution (e.g., the natural evolution of humans to observe nature in a primitive way in the state of nature), there is transformation (e.g., the technical transformation of human ability to observe nature by the invention of "high-pressure, high-temperature apparatus" for the study of, say, "the geochemistry and phase relations" of rocks). (WK 2010d) And the reverse direction also holds true.

In relation to the symmetry-asymmetry principle, if there is symmetry (e.g., the existence of different ways to do geology, be they about the use of mineralogy, petrography, geochemistry, etc.), there is asymmetry (e.g., the popularity of using "mineralogy" and "petrography," for instance, in the older days to study "the composition and texture of rocks"—but the acceptance of using "geochemistry" and "geophysics," for instance, nowadays to study the "geochemical trends and cycles and the use of thermodynamic data and experiments to better understand the origins of rocks"). (WK 2010d) And the reverse direction also holds true.

In relation to the softness-hardness principle, if there is softness (e.g., "the gradual accumulation of sediment," such that different layers of materials lie on top of each other peacefully, as "more and more material is deposited on top"), there is hardness (e.g., the use of "sufficiently high" pressure and heat to burn and change "the composition and fabric of the rock"). And the reverse direction also holds true.

In relation to the regression-progression principle, if there is regression (e.g., the regression made in rock classification, as shown by the problems associated with it), there is progression (e.g., the progress made in rock classification, for the better understanding of different types of rocks). And the reverse direction also holds true.

And in relation to the same-difference principle, if there is similarity in outcome (e.g., the contribution to the molding of beliefs about the world in a certain way, regardless of whether this be achieved by way of a neat rock classification or a more critical one), there is difference in outcome (e.g., the contribution to the molding of beliefs about the world by way of a neat rock classification into three types—but the contribution to the molding of beliefs about the world by way of a more critical analysis with more challenging questions without complete answers). And the reverse direction also holds true.

Statics and Society

Statics, when examined from the perspective of society, can illuminate its possibility and desirability in a distinctive way, and this can be shown by way of two case studies, namely, (2.4.1) economic geology, and the social use of resources, and (2.4.2) hydrogeology, and the technological limits in society—to be addressed hereafter, in that order.

Economic Geology, and the Social Use of Resources

The use of geology can be best understood in the context of society, in that "earth materials...can be used for economic and/or industrial purposes," which falls into the jurisdiction of "economic geology." (WK 2010k)

The Cross-Disciplinary Nature of Economic Geology

Economic geology thus focuses on these useful earth materials, which include, for instance, "precious and base metals, nonmetallic minerals, construction-grade stone, petroleum minerals, coal, and water," etc. (WK 2010k)

Because of this practical usefulness of these earth materials, economic geology is essentially cross-disciplinary, as it not only makes use of "the techniques employed by other earth science disciplines (such as geochemistry, mineralogy, geophysics, and structural geology)" but also "is of prime interest to investment bankers, stock analysts and other professions such as engineers, environmental scientists, and conservationists because of the far-reaching impact that extractive industries have on society, the economy, and the environment." (WK 2010k)

Mineral Resource Classification

The earth materials useful for economic and/or industrial purposes are numerous. For instance, according to one classification, these earth materials can be classified into two major types, as shown below (and summarized in *Table 2.3*). (WK 2010o)

- "Biotic"—These materials (or resources in the current context) "are obtained from the biosphere, such as forests and their products, animals, birds and their products, fish and other marine organisms. Mineral fuels such as coal and petroleum are also included in this category because they formed from decayed organic matter."
- "Abiotic"—These materials (or resources in the current context) "include non-living things. Examples include land, water, air and ores such as gold, iron, copper, silver etc."

For illustration, let's elaborate the last category in the second type of resources, namely, "ores," in what follows.

The term "ore" refers to "a type of rock that contains minerals with important elements including metals. The ores are extracted through mining; these are then refined to extract the valuable element(s)." (WK 2010p)

In turn, the term "mineral" refers to "a naturally occurring solid chemical substance that is formed through geological processes and that has a characteristic chemical composition, a highly ordered atomic structure, and specific physical properties. By comparison, a rock is an aggregate of minerals and/or mineraloids and does not have a specific chemical composition." (WK 2010m)

In general, minerals can be classified in three major types, as shown below (and summarized in *Table 2.3*). (WK 2010l)

- "Mineral Occurrences"—These materials have the "prospects of geological interest but not necessarily of economic interest."
- "Mineral Resources"—These materials "are potentially valuable,...for which reasonable prospects exist for eventual economic extraction."
- "Mineral [or Ore] Reserves"—These materials are "valuable and legally and economically and technically feasible to extract." And the term "ore deposit" thus refers to "an 'ore reserve', and may or may not have additional 'resources.'"

In the third type of "mineral reserves," the classification can be subdivided into "probable reserves" or "proven reserves." (WK 2010l)

In any event, "the conversion of resources into reserves requires the application of various modifying factors," as shown below: (WK 2010l)

- "Mining and Geological Factors"—These refer to the "knowledge of the geology of the deposit sufficient that it is predictable and verifiable; extraction and mine plans based on ore models; quantification of geotechnical risk," etc.
- "Metallurgical Factors"—These refers to the "scrutiny of assay data to ensure accuracy of the information supplied by the laboratory.... Essentially, once a deposit is elevated to reserve status, it is an economic entity and an asset upon which loans and equity can be drawn...generally to pay for its extraction at (hopefully) a profit."
- "Eonomic Factors"
- "Environmental Factors"
- "Marketing Factors"
- "Legal Factors"
- "Governmental Factors"
- "Social Factors"

Surely, there can be other factors, but the ones aforecited suffice for the illustration at hand.

Once "mineral resources" are regarded as "mineral reserves" (or "ore reserves"), there are three considerations which determine the abundance of "ore deposits," as shown below (and summarized in *Table 2.3*). (WK 2010n)

- "Source"—This refers to the requirement that "metal must come from somewhere, and be liberated by some process."
- "Transport"—This refers to the requirement that something is needed to "first...move the metal bearing fluids or solid minerals into the right position," and "to the act of physically moving the metal, as well as chemical or physical phenomenon which encourage movement."
- "Trapping"—This refers to the requirment "to concentrate the metal via some physical, chemical or geological mechanism into a concentration which forms mineable ore."

In this way, "[t]he biggest deposits are formed when the source is large, the transport mechanism is efficient, and the trap is active and ready at the right time." (WK 2010n)

Some Problems with the Mineral Resource Classification

However, there are some problems with mineral resource classification aforediscussed. Consider a few of them below, for illustration (as summarized in *Table 2.3*) .

Firstly, a major problem with mineral resource classification is that the classification, 'because it is an economic function, is governed by statutes, regulations and industry best practice norms." (WK 2010l) So, there is no consensus on the classification, although some well-known (often cited) ones are "the Canadian CIM classification," "the Australasian Joint Ore Reserves Committee Code (JORC Code)," and "the South African Code for the Reporting of Mineral Resources and Mineral Reserves (SAMREC)." (WK 2010l)

Secondly, another major problem with mineral resource classification is that not all of the three types are equally important. For instance, it is often said that "mineral occurrences are generally the least important and least economic. They include all known occurrences of minerals of economic interest, including obviously uneconomic outcrops and manifestations." (WK 2010l)

And thirdly, still another major problem with mineral resource classification is that the identification of mineral resources on Earth has its dark side, in that it facilitates, for example, the European colonization of Africa (and elsewhere) for resources in their heydays of industrialization in modern times.

In the two volumes of *The Future of Human Civilization* (2000), I went to great lengths to show this as what I originally called "hegemonic modernity" in the "trinity of modernity" (besides "capitalist modernity" and "free-spirited modernity").

A Refreshing Revelation

However, these problems with the mineral resource classification should not give the wrong impression that it should be rejected outright.

Rather, the discussion of the problems with the mineral resource classification can reveal the two opposing sides in the debate, such that the possibility and desirability of statics (from the perspective of society with economic geology as a case study here) are not to the extent that its respective defenders would like us to believe.

In addition, this analysis of economic geology and the social use of resources can show us something refreshing about the ontological principles in existential dialectics, and good examples include the absoluteness-relativeness principle, the predictability-unpredictability principle, the explicability-inexplicability principle, the finiteness-transfiniteness principle, the preciseness-vagueness principle, the simpleness-complicatedness principle, the openness-hiddenness principle, the denseness-emptiness principle, the slowness-quickness principle, the expansion-contraction principle, the theory-praxis principle, the convention-novelty principle, the evolution-transformation principle, the symmetry-asymmetry principle, the softness-hardness principle, the regression-progression principle, and the same-difference principle.

For instance, in relation to the absoluteness-relativeness principle, if there is absoluteness (e.g., the absolute view of each type of mineral resource classification as the right one), there is relativeness (e.g., what is right for one mineral resource classification like "the Canadian CIM classification" is not necessarily so for another mineral resource classification like "the South African Code for the Reporting of Mineral Resources and Mineral Reserves"). And the reverse direction also holds true.

In relation to the predictability-unpredictability principle, if there is predictability (e.g., the predictable tendency of mineral resources to fall into one of the three types in the mineral resource classification), there is unpredictability (e.g., the more difficult task to predict exactly what future mineral resource classification will be like, since classifications change over time and there are also different types of classification available in each historical era). And the reverse direction also holds true.

In relation to the explicability-inexplicability principle, if there is explicability (e.g., the explanation, by economic geologists, of the importance of earth materials because of their usefulness for economic and/industrial purposes), there is inexplicability (e.g., the lack of sufficient explanation by economic geologists of why this exploitation of earth materials for economic and/or industrial purposes is necessarily desirable, since, for the critics like the environmentalists and the anti-colonialists, the exploitation has its dark sides, which must be redressed). And the reverse direction also holds true.

In relation to the finiteness-transfiniteness principle, if there is finiteness (e.g., the finite number of the types of natural resources in Earth, that is, the two types known as biotic and abiotic resources), there is transfiniteness (e.g., the transfinite number of all the indiviudal units of biotic and abiotic resources in Earth, be they measured in terms of gallons, kilograms, etc.). And the reverse direction also holds true.

In relation to the preciseness-vagueness principle, if there is preciseness (e.g., the precise identification of the three types of Earth materials in the mineral resource classification), there is vagueness (e.g., the vagueness in the classification of the three types, since there is no consensus on the classification and there are different versions like "the Canadian CIM classification," "the Australasian Joint Ore Reserves Committee Code," and "the South African Code for the Reporting of Mineral Resources and Mineral Reserves"). And the reverse direction also holds true.

In relation to the simpleness-complicatedness principle, if there is simpleness (e.g., the relatively simple analysis of mineral resource classifcation into three types), there is complicatedness (e.g., the relatively more complicated analysis of mineral resource classification by also considering the problems associated with it, as pointed out by the critics). And the reverse direction also holds true.

In relation to the openness-hiddenness principle, if there is openness (e.g., the open exploration of mineral resource classification, in accordance to a particular version like "the Canadian CIM classification," "the Australasian Joint Ore Reserves Committee Code," or "the South African Code for the Reporting of Mineral Resources and Mineral Reserves"), there is hiddenness (e.g., the hidden bias in each version of mineral resource classification, since there is no consensus on one classification as the best for all). And the reverse direction also holds true.

In relation to the denseness-emptiness principle, if there is denseness (e.g., the relatively denser concentration of minerals with economic value for extraction in "mineral reserves"), there is also emptiness (e.g., the relatively less dense, or more empty, concentration of minerals with economic value for extraction in "mineral occurrences"). And the reverse direction also holds true.

In relation to the slowness-quickness principle, if there is slowness (e.g., the relatively slower speed of minerals to be transported to the right place for mining, if there is no well developed transport system), there is quickness (e.g., the relatively faster speed of minerals to be transported to the right place for mining, if there are well developed transport systems). And the reverse direction also holds true.

In relation to the expansion-contraction principle, if there is expansion (e.g., the relatively more developed ability to exploit the resources in nature for economic and/or industrial purposes), there is contraction (e.g., the relatively less developed ability to exist harmoniously with nature without dominating or exploiting it, as advocated by the critics against industrialization in the back-to-nature movement). And the reverse direction also holds true.

In relation to the theory-praxis principle, if there is theory (e.g., the theoretical construction of geology for the understanding of Earth materials), there is praxis (e.g., the practical application of geology for economic and/or industrial purposes, as in economic geology). And the reverse direction also holds true.

In relation to the convention-novelty principle, if there is convention (e.g., the conventional wisdom in modern times about the need to use Earth materials for economic and/or industrial uses), there is novelty (e.g., the alternative novel challenge to this modern industrialist conventional wisdom about Earth materials, by the anti-industrial and back-to-nature critics for a more harmonious co-existence with nature, without the instrumental attempt to exploit or dominate it). And the reverse direction also holds true.

In relation to the evolution-transformation principle, if there is evolution (e.g., the natural evolution of humans to use Earth resources for existence at the subsistence level in the state of nature), there is transformation (e.g., the technical transformation of human ability to use Earth resources by the invention of machines and new systems of knowledge like economic geology for large-scale exploitation of nature). And the reverse direction also holds true.

In relation to the symmetry-asymmetry principle, if there is symmetry (e.g., the co-existence of different views about human relationship with nature, be they about harmonious co-existence or exploitative stewardship), there is asymmetry (e.g., the popularity of harmonious co-existence in the older days of antiquity—but the acceptance of exploitative stewardship towards nature in the modern days of industrialization). And the reverse direction also holds true.

In relation to the softness-hardness principle, if there is softness (e.g., the call for harmonious co-existence with nature), there is hardness (e.g., the call for exploitative domination over nature). And the reverse direction also holds true.

In relation to the regression-progression principle, if there is regression (e.g., the regression made in the use of Earth materials in economic geology, as shown in environmental and colonial problems pointed out by the critics), there is progress (e.g, the progress made in the use of Earth materials for the material enhancement of human life). And the reverse direction also holds true.

And in relation to the same-difference principle, if there is similarity in outcome (e.g., the contribution to the molding and control of beliefs and values, regardless of whether this be achieved by way of an instrumental rationality towards nature for exploitative domination or by way of a substantive rationality towards nature for harmonious co-existence), there is similarity in outcome (e.g., the contribution to the molding and control of beliefs and values by way of an instrumental rationality towards nature for exploitative domination—but the contribution to the molding and control of beliefs and values by way of a substantive rationality towards nature for harmonious co-existence). And the reverse direction also holds true.

Hydrogeology, and the Technological Limits in Society

Another (related) way to evaluate the possibility and desirability of statics from the perspective of society has to do with hydrogeology and the technological limits in society at a given historical era.

The term "hydrogeology" is from "hydro-" (to mean "water" and "geology" (to mean "the study of the Earth"), so it is "the area of geology that deals with the distribution and movement of groundwater in the soil and rocks of the Earth's crust, (commonly in aquifers)." (WK 2010r)

However, the term "hydrogeology" is often used "interchangeably" with a related term "geohydrology"; however, there is a "minor distinction," in that "a geohydrologist" is a hydrologist who applies "geology" to his work (in hydrology), whereas "a hydrogeologist" is a geologist who applyies "hydrology" to his work (in geology). (WK 2010r) This is true, even though both hydrology and geology are two different geosciences (as introduced in *Sec. 1.2*).

The Interdisciplinary Nature of Hydrogeology

In practice, hydrogeology is interdisciplinary, precisely because it makes use of different fields related to the soil, water, climate, rocks, law, politics, etc.

In other words, hydrogeology has "to account fully for the chemical, physical, biological and even legal interactions between soil, water, nature and society. The study of the interaction between groundwater movement and geology can be quite complex." (WK 2010r)

For instance, "the very shallow flow of water in the subsurface (the upper 3 m or 10 ft) is pertinent to the fields of soil science, agriculture and civil engineering, as well as to hydrogeology. The general flow of fluids (water, hydrocarbons, geothermal fluids, etc.) in deeper formations is also a concern of geologists, geophysicists and petroleum geologists." (WK 2010r)

So, this "interplay of the different facets of a multi-component system often requires knowledge in several diverse fields at both the experimental and theoretical levels." (WK 2010r)

Main Concerns of Hydrogeology

In spite of this difficult challenge to account for the complexity of hydrogeology, the main task of a hydrogeologist is "the prediction of future behavior of an aquifer system, based on analysis of past and present observations. Some hypothetical, but characteristic questions asked" in hydrogeology can be shown below, for illustration. (WK 2010r)

- Ex: "Can the aquifer support another subdivision?"
- Ex: "Will the river dry up if the farmer doubles his irrigation?"
- Ex: "Did the chemicals from the dry cleaning facility travel through the aquifer to my well and make me sick?"
- Ex: "Will the plume of effluent leaving my neighbor's septic system flow to my drinking water well?"

Surely, these questions are not exhaustive but illustrative for the purpose at hand.

Main Physical Properties of Aquifers

To answer questions like the ones aforecited, a hydrogeologist has to investiage some physical properties of a given aquifer in question. Hereafter are six main physical properties of aquifers for illustration. (WK 2010r)

(a) Hydraulic Head

The first physical property of an aquifer concerns the "hydraulic head (h)," which refers to the "water level," and its change is "the driving force which causes water to move from one place to another. It is composed of pressure head (ψ) and elevation head (z). The head gradient is the change in hydraulic head per length of flowpath, and appears in Darcy's law as being proportional to the discharge." (WK 2010r)

For instance, "in an aquifer, it can be calculated from the depth to water in a piezometric well (a specialized water well), and given information of the piezometer's elevation and screen depth. Hydraulic head can similarly be measured in a column of water using a standpipe piezometer by measuring the height of the water surface in the tube relative to a common datum." (WK 2010s)

In other cases, "in wells tapping unconfined aquifers the water level in a well is used as a proxy for hydraulic head, assuming there is no vertical gradient of pressure." (WK 2010r)

(b) Porosity

The second physical property of an aquifer concerns "porosity (*n*)," which has "a fraction between 0 and 1 indicating the amount of pore space between unconsolidated soil particles or within a fractured rock. Typically, the majority of groundwater (and anything dissolved in it) moves through the porosity available to flow (sometimes called effective porosity)." (WK 2010r)

Water tends to flow more easily when the pores are connected, and the term "permeability is an expression of the connectedness of the pores. For instance, an unfractured rock unit may have a high porosity (it has lots of holes between its constituent grains), but a low permeability (none of the pores are connected)." (WK 2010r)

However, "porosity" by itself "does not directly affect the distribution of hydraulic head in an aquifer, but it has a very strong effect on the migration of dissolved contaminants, since it affects groundwater flow velocities through an inversely proportional relationship." (WK 2010r)

(c) Water Content

The third physical property of an aquifer concerns "water content (θ)," which refers to "the fraction of the total rock which is filled with liquid water. This is also a fraction between 0 and 1, but it must also be less than or equal to the total porosity." (WK 2010r)

The importance of water content should not be unestimated, because, for insance, "in vadose [unsaturated, not yet full of moisture] zone hydrology,...the hydraulic conductivity is a strongly nonlinear function of water content." (WK 2010r)

(d) Hydraulic Conductivity

The fourth physical property of an aquifer concerns "hydraulic conductivity (*K*) and transmissivity (*T*)", which "are indirect aquifer properties (they cannot be measured directly). *T* is the *K* integrated over the vertical thickness (*b*) of the aquifer ($T=Kb$ when *K* is constant over the entire thickness)." (WK 2010r)

Thus, "these properties are measures of an aquifer's ability to transmit water." (WK 2010r)

(e) Specific Storage and Specific Yield

The fifh physical property of an aquifer concerns "specific storage (*Ss*)" and "its depth-integrated equivalent, storativity ($S=Ssb$)," which are also "indirect aquifer properties (they cannot be measured directly); they indicate the amount of groundwater released from storage due to a unit depressurization of a confined aquifer. They are fractions between 0 and 1." (WK 2010r)

Like porosity, "specific yield (Sy) is also a ratio between 0 and 1" but must be "Sy ≤ porosity," that is, "the value for specific yield is less than the value for porosity because some water will remain in the medium even after drainage due to molecular forces. Often the porosity or effective porosity is used as an upper bound to the specific yield." (WK 2010r)

(f) Contaminant Transport Properties

And the last physical property of an aquifer concerns "contaminant transport properties," because it is important to consider "how the moving groundwater water will move dissolved contaminants around....The contaminants can be man-made (e.g., petroleum products, nitrate or Chromium) or naturally occurring (e.g., arsenic, salinity). Besides needing to understand where the groundwater is flowing, based on the other hydrologic properties discussed above, there are additional aquifer properties which affect how dissolved contaminants move with groundwater." (WK 2010r)

Three specific examples are in order here, namely, (f1) "hydrodynamic dispersion," (f2) "molecular diffusion," and (f3) "retardation by absorption," to be summarized below. (WK 2010r)

(f1) In regard to "hydrodynamic dispersivity"—this refers to "how much contaminants stray away from the path of the groundwater which is carrying it. Some of the contaminants will be 'behind' or 'ahead' the mean groundwater, giving rise to a longitudinal dispersivity..., and some will be 'to the sides of' the pure advective groundwater flow, leading to a transverse dispersivity." (WK 2010r)

Dispersion occurs, because "each water 'particle,' passing beyond a soil particle, must choose where to go, whether left or right or up or down, so that the water 'particles' (and their solute) are gradually spread in all directions around the mean path." (WK 2010r)

(f2) In regard to "molecular diffusion"—this refers to "the random thermal movement of molecules and small particles in gases and liquids. It is an important phenomenon for small distances (it is essential for the achievement of thermodinamic equilibria), but, as the time necessary to cover a distance by diffusion is proportional to the square of the distance itself, it is ineffective for spreading a solute over macroscopic distances." (WK 2010r)

(f3) In regard to "retardation by adsorption"—this refers to "the motion of the contaminant to deviate from the average groundwater motion. It is analogous to the retardation factor of chromatography. Unlike diffusion and dispersion, which simply spread the contaminant, the retardation factor changes its global average velocity, so that it can be much slower than that of water. This is due to a chemico-physical effect:

the adsorption to the soil, which holds the contaminant back and does not allow it to progress until the quantity corresponding to the chemical adsorption equilibrium has been adsorbed." (WK 2010r)

Analytical and Numerical Methods

However, it is one thing to propose some physical properties of aquifers for water management, but it is something else to know their empirical validity in the first place.

In hydrogeology, there are different methods used to calculate and understand the physical properties of aquifers. But the important point to remember here is that, "no matter which method we use…, we need both initial conditions (heads at time (t) = 0) and boundary conditions (representing either the physical boundaries of the domain, or an approximation of the domain beyond that point). Often the initial conditions are supplied to a transient simulation, by a corresponding steady-state simulation (where the time derivative in the groundwater flow equation is set equal to 0)." (WK 2010r)

In general, there are two methods often used in hydrogeology, namely, (a) "analytical methods" and (b) "numerical methods," to be summarized hereafter.

(a) Analytical Methods

In analytical methods, hydrogeologists "use the structure of mathematics to arrive at a simple, elegant solution, but the required derivation for all but the simplest domain geometries can be quite complex (involving non-standard coordinates, conformal mapping, etc.)." (WK 2010r)

In general, "analytic solutions typically are also simply an equation that can give a quick answer based on a few basic parameters. The Theis equation is a very simple (yet still very useful) analytic solution to the groundwater flow equation, typically used to analyze the results of an aquifer test or slug test." (WK 2010r)

In this sense, analytical methods (unlike numerical methods) provide solutions "under a simplified set of conditions exactly." (WK 2010r)

(b) Numerical Methods

In numerical methods, hydrogelogists have a "quite large" set of different methods, "obviously being of use to most fields of engineering and science in general. Numerical methods have been around much longer than computers have." (WK 2010r)

For instance, "in the 1920s [Lewis] Richardson developed some of the finite difference schemes still in use today, but they were calculated by hand, using paper and pencil, by human 'calculators,' but they have become very important through the availability of fast and cheap personal computers." (WK 2010r)

To simplify, "[t]here are two broad categories of numerical methods: gridded or discretized methods and non-gridded or mesh-free methods." (WK 2010r)

For instance, in regard to gridded methods, "in the common finite difference method and finite element method (FEM) the domain is completely gridded ('cut' into a grid or mesh of small elements)." (WK 2010r)

But in regard to non-gridded methods, "[t]he analytic element method (AEM) and the boundary integral equation method (BIEM…, sometimes also called BEM, or Boundary Element Method) are only discretized at boundaries or along flow elements (line sinks, area sources, etc.)," while "the majority of the domain is mesh-free." (WK 2010r)

In either way, numerical methods (unlike analytical methods) provide solutions "under more general conditions to an approximation." (WK 2010r)

Problems in Hydrogeological Works

Yet, there are problems in hydrogeological works, which reflect, in a way, the technological limits in society at a given historical era. Consider a few problems below, for illustration.

Firstly, a major problem in hydrogeological works is that their very nature is simply very "complex," with uncertain outcomes.

For instance, because of its "interdisciplinary" complexity, hydrogeology works "can be difficult to account fully for the chemical, physical, biological and even legal interactions between soil, water, nature and society. The study of the interaction between groundwater movement and geology can be quite complex," for instance, with uncertain outcomes.

Secondly, another major problem in hydrogeological works is that the lack of information can limit the solution to a given problem.

For instance, "dispersivity is actually a factor which represents our lack of information about the system we are simulating. There are many small details about the aquifer which are being averaged when using a macroscopic approach (e.g., tiny beds of gravel and clay in sand aquifers)," and "they manifest themselves as an apparent dispersivity. Because of this,…the dispersivity found for transport through 1 m^3 of aquifer is different than that for transport through 1 cm^3 of the same aquifer material." (WK 2010r)

Thirdly, still another major problem in hydrogeological works is that not everyone is happy with the two methods available, namely, analytical and numerical methods.

For instance, some hydrogeologists try to make good use of both methods, even though they are not popular or widely used in the profession. One good example is "the mesh-free methods like the Analytic Element Method (AEM) and the Boundary Element Method (BEM), which are closer to analytic solutions, but they do approximate the groundwater flow equation in some way," while they also use gridded methods. (WK 2010r) Unfortunately, these hybrid methods "have not seen as widespread use outside academic and research groups." (WK 2010r)

And fourthy, one additional major problem in hydrogeological works is that each method comes with a price to pay.

For instance, one drawback of analytical methods is that they tend to be simplistic, because they operate "under a simplified set of conditions exactly." (WK 2010r)

On the other hand, one drawback of numerical methods is that they tend to be not accurate, because they operate "under more general conditions to an approximation." (WK 2010r)

And one drawback of hybrid methods is that they tend to share the drawbacks of both methods, albeit in different degrees.

Beyond the Fray

However, these problems in the hydrogeological works does not mean that hydrogeology is a worthless field.

Instead, the discussion of the problems in hydrogeological works in general is to reveal to us the two opposing sides in the debate, such that the possibility and desirability of statics (from the perspective of society with hydrogeology as a case study here) are not to the extent that its respective defenders would like us to believe.

Furthermore, this analysis of hydrogeology and the technological limits in society at a given historical era can teach us something valuable about the ontological principles in existential dialectics, and good examples include the formalness-informalness principle, the absoluteness-relativeness principle, the partiality-totality principle, the predictability-unpredictability principle, the explicability-inexplicability principle, the finiteness-transfiniteness principle, the preciseness-vagueness principle, the simpleness-complicatedness principle, the openness-hiddenness principle, the denseness-emptiness principle, the slowness-quickness principle, the expansion-contraction principle, the theory-praxis principle, the convention-novelty principle, the evolution-transformation principle, the symmetry-asymmetry principle, the softness-hardness principle, the regression-progression principle, and the same-difference principle.

For instance, in relation to the formalness-informalness principle, if there is formalness (e.g., the formal logical requirement of a system of ideas to be sound), there is informalness (e.g., the nonformal existence of some sort of inaccuracy in hydrogeological works by numerical methods, because of their assumption "under more general conditions to an approximation," especially in computer simulation). And the reverse direction also holds true.

In relation to the absoluteness-relativeness principle, if there is absoluteness (e.g., the absolute view of analytical methods, because of their mathematical operation "under a simplified set of conditions exactly"), there is relativeness (e.g., what is the right focus for analytical methods in regard to exactness is not necessarily so for numerical methods in regard to approximation). And the reverse direction also holds true.

In relation to the partiality-totality principle, if there is partiality (e.g., the partial view of each of the possible factors in hydrogeology, be it chemical, physical, biological, legal, political, etc.), there is totality (e.g., the holistic view of all the possible factors in hydrogeology as "the different facets of a multi-component system," such that the interactions of them all are highly complex, with uncertain outcomes, so that the whole is not the sum of its parts). And the reverse direction also holds true.

In relation to the predictability-unpredictability principle, if there is predictability (e.g., the predictable tendency of a hydrogeologist to "predict…future behavior of an aquifer system, based on analysis of past and present observations"), there is unpredictability (e.g., the more difficult task for a hydrogeologist to predict exactly the outcomes of an aquifer system, because of the multiple factors involved, whose interactions are highly complex). And the reverse direction also holds true.

In relation to the explicability-inexplicability principle, if there is explicability (e.g., the explanation, if viewed within analytical methods, of the need for "a simplified set of conditions," because of the mathematical requirement of "exactness"), there is inexplicability (e.g., the lack of convincing explanation, if viewed within analytical methods, of why this assumption for "a simplified set of conditions" is necessarily needed, since there can be other ways of doing hydrogeology, like the numerical methods "under more general conditions to an approximation"). And the reverse direction also holds true.

In relation to the finiteness-transfiniteness principle, if there is finiteness (e.g., the finite number of ways that "contaminants move with groundwater," like the three ways via "hydrodynamic dispersion," "molecular diffusion," and "retardation by absorption"), there is transfiniteness (e.g, the transfinite number of contaminants like "molecules" and "small particles in gases and liquids" that move within groundwater). And the reverse direction also holds true.

In relation to the preciseness-vagueness principle, if there is preciseness (e.g., the precise identification of the six physical properties of an aquifer system), there is vagueness (e.g., the vagueness in understanding the properties and the solutions in a given aquifer system, because, in numerical methods, for instance, the outcomes are "under more general conditions to an approximation"). And the reverse direction also holds true.

In relation to the simpleness-complicatedness principle, if there is simpleness (e.g., the relatively simplistic account for an aquifer system in analytical methods, because they operate "under a simplified set of conditions"), there is complicatedness (e.g., the relatively more complicated account for an aquifer system by also considering the different problems associated with analytic methods). And the reverse direction also holds true.

In relation to the openness-hiddenness principle, if there is openness (e.g., the open exploration of analytical methods for the solutions to an aquifer system), there is hiddenness (e.g., the hidden bias in the analytical methods because of their simplistic assumption for "a simplified set of conditions exactly"). And the reverse direction also holds true.

In relation to the denseness-emptiness principle, if there is denseness (e.g., the relatively denser concentration of contaminants like "molecules and small particles in gases and liquids" within groundwater, when they are held back by the "retardation by absorption" to the soil), there is emptiness (e.g., the relatively less dense, or more empty, concentration of contaminants like "molecules and small particles in gases and liquids" within groundwater, when they are not held back by the "retardation by absorption" to the soil). And the reverse direction also holds true.

In relation to the slowness-quickness principle, if there is slowness (e.g., the relative slowing-down of the speed of a contaminant in the case of the "retardation by absorption" to the soil within groundwater), there is quickness (e.g., the relative quicker speed of a contaminant in the cases of of "diffusion and dispersion, which simply help spreading the contaminant" within groundwater). And the reverse direction also holds true.

In relation to the expansion-contraction principle, if there is expansion (e.g., the relatively more developed ability in those specializing in analytic methods to exploit mathematics for "simple, elegant" solutions), there is contraction (e.g,. the relatively less developed ability in those specializing in analytic methods to develop more realistic models "under more general conditions to an approximation"). And the reverse direction also holds true.

In relation to the theory-praxis principle, if there is theory (e.g., the theoretical works of computer science), there is praxis (e.g., the use of computers for the practical application of hydrogeology, like the use of numerical methods for the solutions of an aquifer system, with the help of computer simulation and calculation). And the reverse direction also holds true.

In relation to the convention-novelty principle, if there is convention (e.g., the conventional wisdom about doing calculation by hand in the older days, like the example concerning Richardson in the 1920s), there is novelty (e.g., the alternative novel challenge to this conventional wisdom of doing calculation by hand, with the use of computers instead in our times). And the reverse direction also holds true.

In reltion to the evolution-transformation principle, if there is evolution (e.g., the natural evolution of humans to count in a simple way in the state of nature), there is transformtion (e.g., the technical transformation of human ability to count by way of the invention of supercomputers nowadays). And the reverse direction also holds true.

In relation to the symmetry-asymmetry principle, if there is symmetry (e.g., the co-existence of different methods in hydrogeology, like numerical and analytical methods), there is asymmetry (e.g., the popularity of numerical and analytical methods in our times—but the acceptance of hybrid methods combining both methods within some "academic and research groups" in our times, which means that these hybrid methods "have not seen as widespread use outside academic and research groups"). And the reverse direction also holds true.

In relation to the softness-hardness principle, if there is softness (e.g., diffusion and dispersion to help the spread of contaminants within goundwater), there is hardness (e.g., the retardation by absorption to oppose, or at least to slown down, the spread of contaminants within goundwater). And the reverse direction also holds true.

In relation to the regression-progression principle, if there is regression (e.g., the regression made in hydrogeology, as shown in the problems pointed out by the critics), there is progression (e.g., the progress made in hydrogeology, for the better understanding of aquifer systems). And the reverse direction also holds true.

And in relation to the same-difference principle, if there is similarity in outcome (e.g., the contribution to the molding of beliefs and values, regardless of whether this be achieved by way of using analytical methods or numerical methods), there is difference in outcome (e.g., the contribution to the molding of beliefs and values by way of using analytical methods for a more simplistic but exact way of understanding reality—but the contribution to the molding of beliefs and values by way of using numerical methods for a more general but approximate way of understanting reality). And the reverse direction also holds true.

Statics and Culture

Statics, when examined from the perspective of culture, can cast an illuminating light on its possibility and desirability, and this can be shown by way of two case studies, namely, (2.5.1) fieldwork, and the methodological culture in geology, and (2.5.2) geology, paleontology, and the classificatory tradition—to be addressed hereafter, in that order.

Fieldwork, and the Methodological Culture in Geology

The discussion of analytical and numerical methods (in the previous section on hydrogeology) raises the larger question concerning the methodological culture of investigating reality in geology.

Besides analytical methods (with the use of mathematics) and numerical methods (especially with the use of computer simulation), geologists have used "fieldwork" and "laboratory work" to "decipher Earth history." (WK 2010)

In *Sec.1.4*, different versions of fieldwork and laboratory work in geology were already introduced and summarized, so there is no need to repeat them here.

The task now is to compare and contrast the different methods from the historical perspective of doing geology in relation to its cultural tradition.

The Methodological Culture of Doing Fieldwork in Geology

For much of the history of geology, there has been a methodological tradition for prefering fieldwork—until recently. This is true, even when geologists conducts some experiments in nature, as it is often said that "[g]eology has a long history of field experiments, since the time of Avicenna," (WK 2010u)

One only needs to take a look at the different versions of fieldwork as introduced in *Sec. 1.4* to understand this fieldwork preference in geology, and good examples are "geological mapping," "surveying of topographic features," "subsurface mapping," "collection of samples," "excavation of fossil material," "measuement of characteristics of glaciers," etc. (WK 2010)

The term "field work" refers to "the collection of raw data," as "contrasted with laboratory or experimental research which is conducted in a quasi-controlled environment." (WK 2010t)

In general, "the term is mainly used in the natural and social sciences studies, such as in anthropology, folkloristics, archaeology, biology, ecology, environmental science, ethnomusicology, geography, geology, geophysics, linguistics, paleontology, and sociology....It is more technically known to scientific methodologists as field research." (WK 2010t)

The Rise of Competing Traditions in Our Times

However, things have changed quite a bit in our times, when other methods of doing geology are also accepted.

For instance, geologists nowadays also make good use of analytical methods (with the use of mathematics) and numerical methods (especially with the use of computer simulation), as already described in the previous section.

In addition, there is also the competing tradition of doing laboratory work in geology. In *Sec. 1.4*, I already introduced some noticeable examples of laboratory work which were done in the fields of "petrology," "stratigraphy," and "structural geology," just to cite three instances.

The term "laboratory" in laboratory work refers to "a facility that provides controlled conditions in which scientific research, experiments, and measurement may be performed. The title of laboratory is also used for certain other facilities where the processes or equipment used are similar to those in scientific laboratories" (like "the computer lab," "the medical lab," etc.). (WK 2010v)

The Trade-off of Each Type of Methods

A good question to ask then concerns whether fieldwork is better than lab work, for example—or vice versa. The answer is complicated enough, in that each has its advantages and disadvantages.

For instance, in relation to fieldwork, there are some major disadvantages, as shown below. (WK 2010u) Although these examples are quite general, they apply to fieldworks both within geology and outside it.

- "Fairness of Randomization"—For instance, "in 'negative income tax' experiments communities may lobby for their community to get a cash transfer so the assignment is not purely random."
- "Contamination of the randomization"
- "Difficulty of Replicability"—For instance, "field experiments often require special access or permission, or technical detail" (e.g., "the instructions for precisely how to replicate a field experiment are rarely if ever available in economics").
- "Limits on Ability to Obtain Informed Consent of Participants"
- "Field Testing is Always Less Controlled than Lab Testing"—For instance, "this increases the variability of the types and magnitudes of stress in field testing. The resulting data, therefore, are more varied: larger standard deviation, less precision and accuracy, etc."

Of course, there can be other problems too, but these examples here suffice for the illustration at hand.

By the same logic, in regard to lab work, there are some major disadvantages too, as shown below.

- "Lab Safety"—For instance, "laboratory hazards are as varied as the subjects of study in laboratories, and might include poisons; infectious agents; flammable, explosive, or radioactive materials; moving machinery; extreme temperatures; strong magnetic fields or high voltage." (WK 2010v)
- Unreality of Control—Lab conditions "can be controlled with more precision and certainty," to the point that they are not realistic. (WK 2010w)
- "Approximation"—For instance, computer lab work (like computer simulation in numerical methods) tends to be not accurate, because they operate "under more general conditions to an approximation."

Again, there can be other problems too, but these examples here suffice for the illustration at hand.

From a Broader Horizon

In any event, the important point to remember here is not that one should reject field methods in geology in favor of lab methods, or vice-versa.

Instead, the discussion of the trade-off of each type of methods in geology is to draw our attention to the cultural tradition which shapes the methodological preference of a given type in a historical era, such that the possibility and desirability of statics (from the perspective of culture with field methods vs. lab methods in geology as a case study here) are not to the extent that its respective defenders would like us to believe.

In addition, this analysis of fieldwork and the methodological culture in geology can teach us something insightful about the ontological principles in existential dialectics, and good examples include the formalness-informalness principle, the absoluteness-relativeness principle, the predictability-unpredictability principle, the explicability-inexplicability principle, the preciseness-vagueness principle, the simpleness-complicatedness principle, the openness-hiddenness principle, the denseness-emptiness principle, the change-constancy principle, the slowness-quickness principle, the expansion-contraction principle, the theory-praxis principle, the convention-novelty principle, the evolution-transformation principle, the symmetry-asymmetry principle, the regression-progression principle, and the same-difference principle.

For instance, in relation to the formalness-informalness principle, if there is formalness (e.g., the formal logical requirement of a system of ideas to be complete and sound), there is informalness (e.g., the nonformal existence of some degree of incompleteness and unsoundness in the methods used in geology, like the inherent assumptions in regard to the unreality of control in lab methods and the contamination and unfairness of randomization in field methods). And the reverse direction also holds true.

In relation to the absoluteness-relativeness principle, if there is absoluteness (e.g., the absolute view of each type of methods in geology as the right one), there is relativeness (e.g., what is right for fieldwork is not necesssarily so for lab work, since each has its own merits and problems). And the reverse direction also holds true.

In relation to the predictability-unpredictability principle, if there is predictability (e.g., the predictable tendency of a geologist doing field work to go out to the field for "the collection of raw data"), there is unpredictability (e.g., the more difficult task to predict exactly the validity of a given fieldwork on a particular occasion, because of the inherent problems associated with field methods). And the reverse direction also holds true.

In relation to the explicability-inexplicability principle, if there is explicability (e.g., the explanation of the preference for fieldwork, by those specializing in this method as shown by the examples earlier in Chapter One, on the basis that "[g]eology has a long history of field experiments, since the time of Avicenna," as a kind of methodological tradition), there is inexplicabilty (e.g., the lack of sufficient explanation, if looked within this methodological tradition, of why fieldwork should be the preferred method, in light of historical hindsight nowadays that there are other methods available in our times). And the reverse direction also holds true.

In relation to the preciseness-vagueness principle, if there is preciseness (e.g., the precise control of the conditions in a lab experiment), there is vagueness (e.g., the vagueness about the validity of the outcomes of a lab experiment, if its conditions are so precisely controlled and the results become not realistic, such that their validity is not clear). And the reverse direction also holds true.

In relation to the simpleness-complicatedness principle, if there is simpleness (e.g., the relatively simple treatment of a method by its own proponents), there is complicatedness (e.g., the relatively complicated treatment of a method by its critics, with the relentless consideration of fundamental problems associated with the method). And the reverse direction also holds true.

In relation to the openness-hiddenness principle, if there is openness (e.g., the open exploration of the usefulness of fieldwork in geology, for example, in regard to "geological mapping," "surveying of topographic features," etc.), there is hiddenness (e.g., the hidden bias in fieldwork, as shown by the disadvantages pointed out by the critics). And the reverse direction also holds true.

In relation to the denseness-emptiness principle, if there is denseness (e.g., the relatively denser concentration of people doing fieldwork in geology in the older days), there is emptiness (e.g., the relatively less dense, or more empty, concentration of people doing computer simulation in geology in the older days). And the reverse direction also holds true.

In relation to the change-constancy principle, if there is change (e.g., the ever new ways to do geology in its history, be they about field methods, lab methods, analytical methods, etc.), there is constancy (e.g., the ever constant existence of some methods governing the field of geology at a given point in time). And the reverse direction also holds true.

In relation to the slowness-quickness principle, if there is slowness (e.g., the relatively slower readiness of a geologist doing computer lab work to go out to the field to collect raw data), there is quickness (e.g., the relatively quicker readiness of a geologist doing the excavation of fossil material to go out to the field to collect raw data). And the reverse direction also holds true.

In relation to the expansion-contraction principle, if there is expansion (e.g., the relatively more developed ability of a geologist doing computer lab work to engage in computer simulation), there is contraction (e.g., the relatively less developed ability of a geologist doing computer lab work to engage in well drilling and logging). And the reverse direction also holds true.

In relation to the theory-praxis principle, if there is theory (e.g,. the theoretical construction of electronics, chemistry, etc.), there is praxis (e.g., the practical application of electronics, chemistry, etc. for the field of petrology with the use of the "electron microprobe," for instance). And the reverse direction also holds true.

In relation to the convention-novelty principle, if there is convention (e.g., the conventional wisdom of doing fieldwork in much of the history of geology until recently), there is novelty (e.g,. the alternative novel challenge to the conventional wsidom of doing fieldwork in geology, by the new methods like computer lab work, analytical methods, etc.). And the reverse direction also holds true.

In relation to the evolution-transformation principle, if there is evolution (e.g., the natural evolution of humans to observe rocks with their senses in the state of nature), there is transformation (e.g., the technical transformation of human ability to observe rocks by the invention of "optical microscopy" and "electron microprobe," for instance). And the reverse direction also holds true.

In relation to the symmetry-asymmetry principle, if there is symmetry (e.g., the co-existence of different methods in the history of geology), there is asymmetry (e.g., the popularity of fieldwork in the older days of doing geology, but the acceptance of different competing methods in doing geology nowadays). And the reverse direction also holds true.

In relation to the regression-progression principle, if there is regression (e.g., the regression made in field methods, as shown by the problems associated with them), there is progression (e.g, the progress made in field methods for the collection of raw data in doing geology). And the reverse direction also holds true.

And in relation to the same-difference principle, if there is similarity in outcome (e.g., the contribution to the molding and regulation of beliefs and values, regardless of whether this be done by way of field methods or by way of formal methods), there is difference in outcome (e.g., the contribution to the molding and regulation of beliefs and values by way of field methods for a more empiricist mindset to understand reality—but the contribution to the molding and regulation of beliefs and values by way of formal methods for a more analytical and numerical mindset to understand reality). And the reverse direction also holds true.

Geology, Paleontology, and the Classificatory Tradition

Another way to evaluate the possibility and desirability of statics from the perspective of culture has to do with geology, paleontology, and the classificatory tradition.

The Relationship between Geology and Paleontology

Geology and paleontology (another geoscience for "the study of ancient life") have something closely in common, namely, the interest in the study of ancient organisms in the biosphere. (WK 2010x; R. Cowen 2000)

Although geologists focus, for instance, primarily on soil (in pedology) and rocks (in petrology and lithology), they also have an interest in the biosphere, because of the existence of petroleum and coal as decayed organic materials, which geologists want to uncover for economic and/or industrial uses—and, moreover, because of "the paleontological evidence [of fossils] important for estimating the ages of rocks." (WK 2010x)

Thus lies the linkage between geology and paleontology, and therefore it is no wonder that, for many scholars studying the subject matter, "paleontology lies on the border between biology and geology." (WK 2010x)

In fact, "for historical reasons, paleontology is part of the geology departments of many universities, because in the 19th and early 20th centuries geology departments found paleontological evidence important for estimating the ages of rocks while biology departments showed little interest." (WK 2010x; R. Laudan 1992)

The Mainstream Classification of Ancient Organisms

An interresting illustration of the close relationship between geology and paleontology concerns the study of ancient organisms, like their classification.

Based on the evidences provided by "body fossils" (like "wood, bones, and shells"), "trace fossils" (like "tracks," "burrows," "fossil feces," and "marks left by feeding") and "geochemical observations" (like "eucaryotic cells" and "carbon isotope ratios")—paleontologists are able to come up with a classfication of ancient organisms. (WK 2010x; UCMP 2010a; M. Fedonkin 2007)

Nowadays, "it is...generally accepted that classification should reflect the Darwinian principle of common descent. Since the 1960s a trend called cladistic taxonomy (or cladistics or cladism) has emerged, arranging taxa in an evolutionary tree. If a taxon includes all the descendants of some ancestral form, it is called monophyletic, as opposed to paraphyletic. Other groups are called polyphyletic." (WK 2010z)

Cladistic taxonomy based on an evolutionary tree yields "phylogenetic nomenclature (PN) or phylogenetic taxonomy," where the word "phylogenetic" refers to the "genealogic trees" of species to "map…out the proposed divisions and common ancestors of all living species." (WK 2010aa & 2010bb)

In other words, phylogenetic taxonomy is "an alternative to rank-based nomenclature, applying definitions from cladistics (or phylogenetic systematics). Its two defining features are the use of phylogenetic definitions of biological taxon names, and the lack of obligatory ranks. It is currently not regulated, but the PhyloCode (International Code of Phylogenetic Nomenclature) is intended to regulate it once implemented." (WK 2010aa)

For instance, in an "evolutionary [or phylogentic] tree of life," the classification suggests that there are three "domains" (i.e., "bacteria," "archaea," and "eukaryotes"), with the same ancestral origin. (WK 2010bb)

Even so, the mainstream view about "the phylogeny of life" has its different versions, "ranging from 'numerical taxonomy' (phenetics) over 'evolutionary taxonomy' (gradistics) to 'phylogenetic systematics' (cladistics)," for instance—as summarized in *Table 2.4*. (WK 2010aa)

The Classificatory Dispute

However, not everyone is happy with the current adoption of the classification of ancient organisms according to an "evolutionary tree" based on the Darwinian principle of common descent (even though it has different versions like phenetics, gradistics, and cladistics).

Consider a few alternatives below, for illustration, namely, (a) the "Linnaean taxonomy," (b) the "Creationist" view, and (c) the "molecular systematics"—as summarized in *Table 2.4*.

(a) The Linnaean Taxonomy

Firstly, one classificatory alternative is the "Linnaean taxonomy" proposed by Carolus Linnaeus, who wrote *Systema Naturae* (1735), where "nature was divided into three kingdoms: mineral, vegetable and animal. Linnaeus used five ranks: class, order, genus, species, and variety." (WK 2010z)

Over time, however, "the number of commonly used ranks has increased from the five to seven Linnaeus used (variety, species, genus, order, class; kingdom, arguably empire) to about twenty-one (by the addition of family, phylum/division, domain, and the prefixes super- and sub- and occasionally infra-; 'empire' is not used, and 'variety' is only used in botany"). (WK 2010aa)

In addition, Linnaeus "abandoned long descriptive names of classes and orders and two-word generic names (e. g., *Bursa pastoris*) still used by his immediate predecessors (Rivinus and Pitton de Tournefort) and replaced them with single-word names, provided genera with detailed diagnoses (*characteres naturales*), and reduced numerous varieties to their species, thus saving botany from the chaos of new forms produced by horticulturalists." (WK 2010z)

(b) The Creationist View

Secondly, another classificatory alternative is the "creationist" view which is "hostile" to any classification which "is based upon evolution," like the Darwinian evolutionary principle of common descent. (WK 2010z)

In fact, some creationists accept "a literal interpretation of Genesis creation, accepting it as an accurate historical account of creation of the universe in its present form over the course of six twenty-four hour days." (WK 2010cc)

For instance, "Duane Gish [1995] of the Institute for Creation Research defined 'special creation' as being creation using supernatural processes," in that "[w]e do not know how the Creator created, [or] what processes He used, for He used processes which are not now operating anywhere in the natural universe. This is why we refer to creation as special creation. We cannot discover by scientific investigation anything about the creative processes used by the Creator." (WK 2010cc)

Others like "Dennis Jensen [1999] of the American Scientific Affiliation," argue that "special creation means that complex living things did not descend from simpler ones but were created independently." (WK 2010cc) And "James B. Stenson [1984] writes that…there is a 'special creation' of each separate species in six 24 hour days, starting a few thousand years ago." (WK 2010cc)

(c) The Molecular Systematics

And thirdly, still another classificatory alternative is the "molecular systematics," which is both similar to and different from the mainstream evolutionary "phylogenetic taxonomy." (WK 2010dd)

On the one hand, molecular systematics is similar to the mainstream evolutionary "phylogenetic taxonomy," because it also "assumes that classification must correspond to phylogenetic descent, and that "all valid taxa must be monophyletic." (WK 2010dd)

On the other hand, molecular systematics is different from the mainstream evolutionary "phylogenetic taxonomy," because it replaces the mainstream classification "based on visible and ecologically relevant characteristics by one based on genetic details" in relation to their molecular structure. (WK 2010dd)

This difference is important, because molecular systematics provides a different way to understand the evolutionary tree of life. For instance, "[e]very living organism contains DNA, RNA, and proteins. Closely related organisms generally have a high degree of agreement in the molecular structure of these substances, while the molecules of organisms distantly related usually show a pattern of dissimilarity. Conserved sequences, such as mitochondrial DNA, are expected to accumulate mutations over time, and assuming a constant rate of mutation provide a molecular clock for dating divergence. Molecular phylogeny uses such data to build a 'relationship tree' that shows the probable evolution of various organisms. Not until recent decades, however, has it been possible to isolate and identify these molecular structures." (WK 2010dd)

In Retrospect

Of course, the introduction of these classificatory altenrnatives does not mean an endorsement of them in rejection of the mainstream versions of the evolutionary tree of life.

Rather, the introduction to these alternatives is to show us the opposing side in the debate, such that the possibility and desirability of statics (from the perspective of culture with the discussion on the mainstream versions of the evolutionary tree of life as a case study here) are not to the extent that its respective defenders would like us to believe.

In addition, this analysis of geology, paleontology, and the classificatory tradition can reveal something valuable about the ontological principles in existential dialectics, and good examples include the formalness-informalness principle, the absoluteness-relativeness principle, the predictability-unpredictability principle, the explicability-inexplicability principle, the preciseness-vagueness principle, the simpleness-complicatedness principle, the openness-hiddenness principle, the denseness-emptiness principle, the change-constancy principle, the order-chaos principle, the slowness-quickness principle, the expansion-contraction principle, the theory-praxis principle, the convention-novelty principle, the evolution-transformation principle, the symmetry-asymmetry principle, the regression-progression principle, and the same-difference principle.

For instance, in relation to the formalness-informalness principle, if there is formalness (e.g., the formal requirement of completeness in a system of ideas), there is informalness (e.g., the nonformal existence of incompleteness in the creationist view, since Duane Gish himself admitted that "[w]e do not know how the Creator created....This is why we refer to creation as special creation. We cannot discover by scientific investigation anything about the creative processes used by the Creator"). And the reverse direction also holds true.

In relation to the absoluteness-relativeness principle, if there is absoluteness (e.g., the absolute view of creationism in regard to "special creation"), there is relativeness (e.g., what is true for the creationists like Gish in relation to "special creation" is not so for Darwin in regard to "evolution" which is "blind," without the creator). And the reverse direction also holds true.

In relation to the predictability-unpredictability principle, if there is predictability (e.g., the predictable tendency of ranking in a Linnean classification), there is unpredictaiblity (e.g., the more difficult task to predict the specifc ranks any taxa belong to, since "the number of generally recognized ranks is limited," which "means that many taxa must go unnamed because no ranks are available for them"). (WK 2010aa) And the reverse direction also holds true.

In relation to the explicability-inexplicability principle, if there is explicability (e.g., the explanation, within molecular systematics, of its classification as a better alternative to the mainstream evolutionary "phylogenetic taxonomy," on the basis of "genetic details" in relation to their molecular structure"), there is unexplicability (e.g., the lack of sufficient explanation, within molecular systematics, of why it is a better alternative, since it does not always work, as shown in the case about animals, "because in mammals mitochondria are inherited only from the mother," which means that "inheritance in the paternal line might not be detected"). (WK 2010dd) And the reverse direction also holds true.

In relation to the preciseness-vagueness principle, if there is preciseness (e.g., the precise identification by Carolus Linnaeus, of five and later seven ranks in his classification), there is vagueness (e.g., the vagueness in the five or seven ranks by Linnaeus, since it is not clear of why there must be five or seven, and the number was later increased to 21 after his death). And the reverse direction also holds true.

In relation to the simpleness-complicatedness principle, if there is simpleness (e.g., the relatively simple classification of ancient organisms on the basis of the evolutionary principle of common descent), there is complicatedness (e.g., the relatively complicated analysis of the classification by also considering alternative classifications which question the mainstream evolutionary versions). And the reverse direction also holds true.

In relation to the openness-hiddenness principle, if there is openness (e.g., the open exploration of another classification of ancient organisms from the creationist viewpoint), there is hiddenness (e.g., the hidden bias in the creationist view, because it cannot be proved, as Duane Gish himself said that "[w]e do not know how the Creator created," and also because it is baised against its nemesis, the evolutionary theory of common descent). And the reverse direction also holds true.

In relation to the denseness-emptiness principle, if there is denseness (e.g., the relatively denser concentration of "mitochondria" in mammals along the "maternal" line to be detected by molecular systematics, because "in mammals mitochondria are inherited only from the mother"), there is emptiness (e.g., the relatively less dense, or more empty, concentration of "mitochondria" in mammals along the "paternal" line to be detected by molecular systematics, because "in mammals mitochondria are inherited only from the mother"). And the reverse direction also holds true.

In relation to the change-constancy principle, if there is change (e.g., the ever new ways of classification of ancient organisms in the history of paleontology, like "phenetics," "gradistics," "cladistics," "Linnean taxonomy," "creationism," "molecular systematics," etc.), there is constancy (e.g., the ever constant existence of some classifications of ancient organisms in the history of paleontology). And the reverse direction also holds true.

In relation to the order-chaos principle, if there is order (e.g., the relatively orderly arrangement of organisms in a classificatory scheme like the Linnean taxonomy), there is chaos (e.g., the relatively chaotic view of organisms before the Linnean taxonomy, when the field of "botany" suffered "from the chaos of new forms produced by horticulturalists"). (WK 2010z) And the reverse direction also holds true.

In relation to the slowness-quickness principle, if there is slowness (e.g., the relatively slower readiness of a creationist to classify organisms by common descent), there is quickness (e.g., the relatively quicker readiness of a specialist in phylogenetic taxonomy to classify organisms by common descent). And the reverse direction also holds true.

In relation to the expansion-contraction principle, if there is expansion (e.g., the relatively more developed ability of a creationist to classify things on the basis of religious revelation), there is contraction (e.g., the relatively less developed ability of a creationist to classify things on the basis of scientific investigation). And the reverse direction also holds true.

In relation to the theory-praxis principle, if there is theory (e.g., the theoretical construction of molecular biology), there is praxis (e.g., the practical application of molecular biology to molecular systematics for the classification of organisms). And the reverse direction also holds true.

In relation to the convention-novelty principle, if there is convention (e.g., the conventional wisdom about classifying organisms by Carolus Linnaeus in the 18th century, on the basis of kingdoms and ranks, for insance), there is novelty (e.g., the novel challenge to this conventional wisdom by the new approach known as the evolutionary classification by common descent). And the reverse direction also holds true.

In relation to the evolution-transformation principle, if there is evolution (e.g., the natural evolution of humans in the state of nature to observe animals and other things in nature, in a rather primitive way), there is transformation (e.g.,. the technical transformation of human abiltity to understand animals and things in nautre, by the invention of molecular systematics). And the reverse direction also holds true.

In relation to the symmetry-asymmetry principle, if there is symmetry (e.g., the co-existence of different ways of classifying organisms in the history of paleontology), there is asymmetry (e.g., the popularity of the creationist view of special creation in the older days, but the rise of the evolutionary versions of classifying organisms in our time). And the reverse direction also holds true.

In relation to the regression-progression principle, if there is regression (e.g., the regression made in molecular systematics, as shown in the example about the inability to detect mitochondria in mammals along the "paternal" line), there is progression (e.g., the progress made in molecular systemtics, to help us understand the classification of organisms in relation to genetic details). And the reverse direction also holds true.

And in relation to the same-difference principle, if there is similarity in outcome (e.g., the contribution to the molding and regulation of beliefs and values, regardless of whether this is achieved by way of creationist classification or by way of evolutionary classification), there is difference in outcome (e.g., the contribution to the molding and regulation of beliefs and values by way of creationist classification for a religious view of the world and its spiritual view of life—but the contribution to the molding and regulation of beliefs and values by way of evolutionary classification for a scientific view of the world and its instrumental way of life). And the reverse direction also holds true.

The Changeability of Statics

This comprehensive analysis of the possibility and desirability of statics, from the perspectives of the mind, nature, society, and culture, is no idle academic exercise, since it is to show that statics is both possible and desirable, but not to the extent that the spokespersons from each side would like us to believe.

In the end, statics can have its durability, just as it can have its changeability too, as another side of the same mirror.

But this is only one part of a larger picture, since the other part is dynamics. Since this chapter deals with statics, the next chapter will therefore account for the issue of dynamics.

These two-fold understanding of statics and dynamics is important, because, as already pointed out in *Sec. 1.3*, both statics (structure) and dynamics (process) are intricately related.

With this in mind, let's now turn to Chapter Three for the study of dynamics and its restriction—for which we now turn to chapter three.

Table 2.1. Statics and Its Contingency

- **Statics and the Mind**
  —Ex: gestalt psychology, and the perception of geological structures
  —Ex: cemetery rocks, and the geology of fear

- **Statics and Nature**
  —Ex: the debate on tectonic plates
  —Ex: the contention on rock classification

- **Statics and Society**
  —Ex: economic geology, and the social use of resources
  —Ex: hydrogeology, and the technological limits in society

- **Statics and Culture**
  —Ex: fieldwork, and the methodological culture in geology
  —Ex: geology, paleontology, and the classificatory tradition

Notes: The examples in the categories are solely illustrative (not exhaustive), and the comparison is relative (not absolute), nor are they necessarily mutually exclusive. And some can be easily re-classified elsewhere. As generalities, they allow exceptions.
Source: A summary of Ch.2 of *FPHGEOL*

Table 2.2. The Discourse on Tectonic Plates

• **Types of Plate Boundaries**
—Ex: "transform boundaries"
—Ex: "divergent boundaries"
—Ex: "convergent boundaries"

• **Types of Tectonic Plates**
—Ex: "primary plates"
—Ex: "secondary plates"

• **Causes of Tectonic Plate Movements**
—Ex: "friction"
—Ex: "gravitation"
—Ex: "tidal forces"

• **Criticisms about the Discourse on Tectonic Plates**
—Firstly, the focus on the three major types of plate boundaries often does not give sufficient attention to the "plate boundary zones," which are "where the effects of the interactions are unclear and the broad belt boundaries are not well defined." (WK 2010g; USGS 1999; PT 2010)
—Secondly, it is not clear why there are exactly "seven" primary plates and "seven" secondary plates. Why is this magic number of "seven" for both types?
—Thirdly, "[t]he sources of plate motion are a matter of intensive research and discussion among earth scientists," such that there can be other forces (unlike the three aforecited) to be accounted for.
—And fourthly, geologists disagree about the "relative significance of each mechanism" (or cause).

Notes: The examples in the categories are solely illustrative (not exhaustive), and the comparison is relative (not absolute), nor are they necessarily mutually exclusive. And some can be easily re-classified elsewhere. As generalities, they allow exceptions.
Source: From *Sec. 2.3.1* of *FPHGEOL*. See book for citations, like WK (2010g), and so on.

Table 2.3. The Contention on Mineral Resource Classification

- **Types of Resources**
  —Ex: "biotic"
  —Ex: "abiotic"

- **Types of Minerals**
  —Ex: "mineral occurrences"
  —Ex: "mineral resources"
  —Ex: "mineral reserves"

- **Factors for the Determination of Deposit Abundance**
  —Ex: "source"
  —Ex: "transport"
  —Ex: "trapping"

- **Problems in the Classification**
  —Firstly, the classification, 'because it is an economic function, is governed by statutes, regulations and industry best practice norms." (WK 2010l) So, there is no consensus on the classification.
  —Secondly, not all of the three types are equally important. For instance, it is often said that "mineral occurrences are generally the least important and least economic. They include all known occurrences of minerals of economic interest, including obviously uneconomic outcrops and manifestations." (WK 2010l)
  —And thirdly, the identification of mineral resources on Earth has its dark side, in that it facilitates, for example, the European colonization of Africa (and elsewhere) for resources in their heydays of industrialization in modern times. In the two volumes of *The Future of Human Civilization* (2000), I went to great lengths to show this as what I originally called "hegemonic modernity" in the "trinity of modernity" (besides "capitalist modernity" and "free-spirited modernity").

Notes: The examples in the categories are solely illustrative (not exhaustive), and the comparison is relative (not absolute), nor are they necessarily mutually exclusive. And some can be easily re-classified elsewhere. As generalities, they allow exceptions.
Source: From *Sec. 2.4.1* of *FPHGEOL*. See book for citations, like WK (2010l, 2010n, & 2010o), and others.

Table 2.4. The Classificatory Dispute in Paleontology (Part I)

- **The Evolutionary Phylogentic Taxonomy**

—Nowadays, "it is...generally accepted that classification should reflect the Darwinian principle of common descent. Since the 1960s a trend called cladistic taxonomy (or cladistics or cladism) has emerged, arranging taxa in an evolutionary tree. If a taxon includes all the descendants of some ancestral form, it is called monophyletic, as opposed to paraphyletic. Other groups are called polyphyletic." (WK 2010z)

—In other words, phylogenetic taxonomy is "an alternative to rank-based nomenclature, applying definitions from cladistics (or phylogenetic systematics). Its two defining features are the use of phylogenetic definitions of biological taxon names, and the lack of obligatory ranks." (WK 2010aa) For instance, in an "evolutionary [or phylogentic] tree of life," the classification suggests that there are three "domains" (i.e., "bacteria," "archaea," and "eukaryotes"), with the same ancestral origin. (WK 2010bb)

—Even so, the mainstream view about "the phylogeny of life" has its different versions, "ranging from 'numerical taxonomy' (phenetics) over 'evolutionary taxonomy' (gradistics) to 'phylogenetic systematics' (cladistics)," for instance. (WK 2010aa)

- **The Linnaean Taxonomy**

—One classificatory alternative is the "Linnaean taxonomy" proposed by Carolus Linnaeus, who wrote *Systema Naturae* (1735), where "nature was divided into three kingdoms: mineral, vegetable and animal. Linnaeus used five ranks: class, order, genus, species, and variety." (WK 2010z)

- **The Creationist View**

—Another classificatory alternative is the "creationist" view which is "hostile" to any classification which "is based upon evolution," like the Darwinian evolutionary principle of common descent. (WK 2010z)

(continued on next page)

Table 2.4. The Classificatory Dispute in Paleontology (Part II)

- **The Molecular Systematics**
  —Another classificatory alternative is the "molecular systematics," which is both similar to and different from the mainstream evolutionary "phylogenetic taxonomy." (WK 2010dd)
  —On the one hand, molecular systematics is similar to the mainstream evolutionary "phylogenetic taxonomy," because it also "assumes that classification must correspond to phylogenetic descent, and that "all valid taxa must be monophyletic." (WK 2010dd)
  —On the other hand, molecular systematics is different from the mainstream evolutionary "phylogenetic taxonomy," because it replaces the mainstream classification "based on visible and ecologically relevant characteristics by one based on genetic details" in relation to their molecular structure. (WK 2010dd)

Notes: The examples in the categories are solely illustrative (not exhaustive), and the comparison is relative (not absolute), nor are they necessarily mutually exclusive. And some can be easily re-classified elsewhere. As generalities, they allow exceptions.
Source: From *Sec. 2.5.2* of *FPHGEOL*. See book for citations.

• PART THREE •

Dynamics

CHAPTER 3
DYNAMICS AND ITS RESTRICTION

The present is the key to the past.
—Charles Lyell (WK 2010)

The Strength of Dynamics

Geology has both statics (structure) and dynamics (process). As already summarized in *Sec. 1.3*, statics and dynamics are closely related.

The inquiry concernng statics was already addressed in Chapter Two, so this chapter is to focus on the nature of dynamics.

Again, as already cautioned in Chapter Two, because of the close relationship between statics and dynamics, one cannot really analyze dynamics (or statics, for that matter) separately.

This then means that the issues concerning dynamics in this chapter (Chapter Three) and statics in Chapter Two often overlap.

With this caution in mind—a good way to analyze dynamics and its restriction is by way of the evaluation of the extent to which dynamics is in fact both possible and desirabe.

This chapter thus provides a comprehensive analysis of dynamics in the context of geology from the four perspectives of the mind, nature, society, and culture (in accordance to my sophisticated methodological holism, as explained in *Sec. 1.9*).

Therefore, this chapter is organized in four main sections, in relation to (3.2) dynamics and the mind, (3.3) dynamics and nature, (3.4) dynamics and society, and (3.5) dynamics and culture—to be examined hereafter, respectively (and summarized in *Table 3.1*).

Dynamics and the Mind

Dynamics, when analyzed from the perspective of the mind, can shed some illuminating light on its possibility and desirability, and this can be shown by way of two case studies, namely, (3.2.1) geology, and the controversy about feminine intuition, and (3.2.2) geology, and the rediscovery of the mind—to be addressed hereafter, in that order.

Geology, and the Controversy about Feminine Intuition

A fascinating question to start the inquiry of dynamics from the perspective of the mind concerns the practice of geology from the feminine vantage point of view, especially in relation to the claim about feminine intuition.

The Didactics of a Personal Story

Sarah Andrews (2010) in "What Women Bring to Geology" precisely made this controversial point between the differences in the mind which allow women to do geology in a distinctive way different from that of men.

Andrews (2010) began the essay by claiming that "[w]omen bring unique talents to geology." But why? A short answer is, Feminine intuition.

But a long answer begins with her personal story: "Once upon a time in the oil patch, I was asked to help interpret some drill core that was laid out in Amoco's research facility in Tulsa. I was just a kid from one of the regional offices, but I happened to have experience with sand dunes, ergo eolian sandstones. Two men who had much bigger, juicier postgraduate degrees and decades more experience than I, and who had been puzzling over the core for some weeks, walked behind me as I strolled past the layout 'reading' the depositional history of the rocks aloud to them. I do mean that I strolled: I had learned from working with eolianites to first build an overview rather than immediately try to decode each separate lamination. Mine was an intuitive approach, built on a kinesthetic sense of what the rock meant at the scale of its depositional rhythms, and how that might affect an oilfield. As I reached the last box of core, I turned and looked at the men. I was greeted by two very dark countenances. One man glared at me, unwilling to name his emotion. The other snapped, 'How do you know all that?'" (Andrews 2010)

Her answer, this time a longer one, has to do with feminine intuition, in the context of the work of Jungian psychologist Irene Claremont de Castellejo [1980], which Andrews (2010) cited for a more theoretical discussion of the gender differences in doing geology, in that, "regardless of one's physical gender—the receipt of inspiration and information is feminine, and that capacity to process that input into a product is masculine. Think of a fully visible landscape washed by a soft, scintillating glow versus a tightly-focused flashlight beam illuminating one spot in an otherwise darkened room. To put this in fully sexual terms, think receptive versus penetrating." (S. Andrews 2010)

The Nature of Feminine Intuition

For Andrews (2010), women simply "approach science intuitively," and her "definition of intuition is open communication between the conscious and unconscious minds, and that communication is opened most widely through the soft interface of diffuse awareness. The more widely and intentionally we open that door, the better we can compare, sort, and categorize data, and predict results."

Firstly, intuition exists for both men and women, as Andrews (2010) herself thus acknowledged: "All geologists, male and female, are rich in feminine diffuse awareness. It is within the soft, receptive ground of diffuse awareness that we transcend the boundary that stretches between conscious and unconscious thought, and engage the power of intuition, that place where the 'aha!' of consciousness and our lightning-speed access to the stored observations of the unconscious unite. At this nexus we receive and integrate connections, embrace ambiguity, and enter the state of four-dimensional flow in which we know what we know. Christine Turner (2000) points out that this relaxed, receptive state of mental functioning is most pronounced when geologists do fieldwork."

Secondly, Andrews (2010) also argued that both the feminine diffuseness and the masculine focusedness are needed for a better result of doing deology: "When speaking to geoscience audiences about these concepts, I sweep my hands around the sides of my head to indicate feminine diffuse awareness, and point to the center of my forehead to represent focused masculine. The women in the room smile and nod their heads knowingly. We know this modality well. The enriched zone of the feminine is our home ground. If we struggle at all, it is to arise from the flow of observation long enough to put metes and bounds to it, call it good, and write up our results. It is in conscious integration of the feminine and masculine that we become whole, striking a balance between taking in and producing outward."

Thirdly, Andrews (2010) claimed the biological basis of the "woman geologist's native capacity for diffuse awareness," which "is not simply an artifact of personality, but also the direct consequence of her chromosomal—and hence hormonal—makeup. Being female can present intermittent physical symptoms best treated with analgesics or a hot bath, but estrogen and progesterone also boost our diffuse and focused states. In the years between menarche and menopause, I learned to greet the new moon for the depth of intuitive/receptive capacity I found there. In the full moon I emerged recharged and refreshed to present my findings. I cycled between inward-looking receptivity and outward, knock-heads assertiveness. I planned my work around the intensity of the extremes and used them to great advantage."

And fourthly, Andews (2010) also acknowledged the difficult time that women often face in a predominatly masculine culture in geology, where the discourse of the feminine diffusiveness (or intuition) is not popular: "Which is not to suggest that any of this is easy. As women immersed in a male dominated professional culture, not to mention the patriarchal values of our culture as a whole, we struggle to strike a balance between feminine and masculine cultural norms. Shall we bear children, or concentrate on career, or have both?...If the demands of motherhood haven't taught us to multitask, parallel process, and be flexible, nothing ever will. We take our lumps and come back again and again, building a new home and life in each place fate deposits us."

Main Problems with the Claim about Feminine Intuition

The claim by Andrews (2010) about feminine intuition in geology is not wihtout its controversy, of course. Consider a few criticisms below, for illustration.

Firstly, one major criticism of the claim by Andrews is that it is not clear to what extent she is offering an insight about the gender differences on intuition or instead she contributes to the cultural perpetuation of the stereotypes concerning males and females.

For instance, some feminists like Marilyn Suiter liked her work, because she thought that it helps to "overthrow…the paradigm that women aren't as good at science as men." (S. Andrews 2010) But the critics, her work can instead help the perpetuation of the stereotype of why women are not as good at science as men, because the "capacity to process…input into a product is masculine," which means that women are not naturally good as men in analysis, as the majority of the major results in geology have been sorted out and analyzed by men in well-written forms.

Secondly, another major criticim of the claim by Andrews is that it is not clear if she has succeeded in proving anything, since she gave a lot of her personal stories to back up her claim (in her essay), but very little solid scientific evidences (if at all) to defend it. *A fortiori*, her appeal to Jungian psychology is not helping either, because the work by Jung has the notorious repulation as unscientific or pseudo-scientific for the critics. Even worse, it is not even clear in the first place if, or to what extent, all these personal stories by here are really true or something that she made up or exaggerated.

Thirdly, still another major criticim of the claim by Andrews is that, if it is not easy at all, as she also acknowledged, for women to overcome the "male dominated professional culture, not to mention the patriarchal values of our culture as a whole," then this "struggle to strike a balance between feminine and masculine cultural norms" before deciding to pursue a career in science for women cannot go very far, which then means that there is a limit of how far women can be successful in science at all—contrary to her claim (or wish) to have overthrown the paradigm that women are not as good at science as men.

Fourthly, one additional major criticim of the claim by Andrews is that the appeal to intuition is a tricky business, since what is intuitive to some may not be so obvious to another. But then, this is also the end of the conversation, because an intuitive claim cannot prove itself.

And fifthy, the last major criticim of the claim by Andrews is that it is also not clear how much her claim about intuition in a given case is really intuitive at all, as it can be something else that she learned from her past experiences.

For instance, in the personal story that she told at the beginning of the essay, she acknowledged the benefit of her past experiences for reading the history of the rocks: "I happened to have experience with sand dunes, ergo eolian sandstones….I had learned from working with eolianites to first build an overview rather than immediately try to decode each separate lamination." (S. Andrews 2010) Thus, this seems that her past experiences, not her intuition, helped her read the rocks in this case.

Siding with Neither

However, these criticisms of the claim by Andrews about feminine intuition in geology should not be misunderstood as a downright dismissal of its merit.

Rather, they are shown here to teach us the two opposing sides of the debate, such that the possibility and desirability of dynamics (from the perspective of the mind with feminine intuition in geology as a case study here) are not to the extent that its respective defenders would like us to believe.

In addition, this analysis of feminine intuition in geology can teach us something valuable about the ontological principles in existential dialectics, and good examples include the formalness-informalness principle, the absoluteness-relativeness principle, the partiality-totality principle, the predictability-unpredictability principle, the explicability-inexplicability principle, the preciseness-vagueness principle, the simpleness-complicatedness principle, the openness-hiddenness principle, the denseness-emptiness principle, the slowness-quickness principle, the expansion-contraction principle, the theory-praxis principle, the convention-novelty principle, the evolution-transformation principle, the symmetry-asymmetry principle, the regression-progression principle, and the same-difference principle.

For instance, in relation to the formalness-informalness principle, if there is formalness (e.g., the formal logical requirement of completeness and soundness in a system of ideas), there is infornalness (e.g., the nonformal existence of incompleteness in the claim by Andrews about intuition, since intuition cannot be proved). And the reverse direction also holds true.

In relation to the absoluteness-relativeness principle, if there is absoluteness (e.g., the absolute claim by Andrews about feminine intuition, without seriously questioning its validity), there is relativeness (e.g., what is right for Andrews in regard to intuition is not necessarily so for the critics, who seriously question its validity). And the reverse direction also holds true.

In relation to the partiality-totality principle, if there is partiality (e.g., the partial view of a geologist with a masculine focusedness to decode the rocks by reading each separately first), there is totality (e.g., the holistic view of a geologist with a feminine diffusiveness to decode the rocks by "first build[ing] an overview rather than immediately try[ing] to decode each separate lamination"). And the reverse direction also holds true.

In relation to the predictability-unpredictability principle, if there is predictability (e.g., the predictable tendency of a geologist with a feminine diffusiness to study rocks by intuition than by analysis), there is unpredictability (e.g., the more difficult task to predict to what extent exactly a particular intuition on a given occasion will yield accurate result). And the reverse direction also holds true.

In relation to the explicability-inexplicability principle, if there is explanation (e.g., the explanation by Andrews that women bring unique talents to geology on the basis of her claim about feminine intuition), there is inexplicability (e.g., the lack of sufficient explanation by Andrews of why this so-called unique talent of women is not really an embarassing handicap at all, since, for the critics, her work can instead explain the opposite conclusion, namely, that women are not as good at science as men, because the "capacity to process…input into a product is masculine," which means that women are not naturally good as men in analysis). And the reverse direction also holds true.

In relation to the preciseness-vagueness principle, if there is preciseness (e.g., the precise identification of intuition as the unique talent that women bring to geology, according to Andrews), there is vagueness (e.g., the vagueness in the identification about intuition, since what is intuitive for one person may not be so for another, so it can mean different things to different individuals). And the reverse direction also holds true.

In relation to the simpleness-complicatedness principle, if there is simpleness (e.g., the relatively simple account for intuition by Andrews), there is complicatedness (e.g., the relatively more complicated account for intuition by also considering the critical problems pointed out by the critics). And the reverse direction also holds true.

In relation to the openness-hiddenness principle, if there is openness (e.g., the open exploration of the role of feminine intuition in geology), there is hiddenness (e.g., the hidden bias of this claim about femine intuition in geology, because of its feminist ideological motivation to "overthrow…the paradigm that women aren't as good at science as men"). And the reverse direction also holds true.

In relation to the denseness-emptiness principle, if there is denseness (e.g., the relatively denser concentration of feminists who accept the claim by Andrews that women brings unique talents in geology by feminine intuition), there is emptiness (e.g., the relatively less dense, or more empty, concentration of the critics who accept the claim by Andrews that women brings unique talents in geology by feminine intuition). And the reverse direction also holds true.

In relation to the slowness-quickness principle, if there is slowness (e.g., the relatively slower speed of trying to understand something by someone who uses analytical proof, not by intuition), there is quickness (e.g., the relatively quicker speed of trying to understand something by someone who uses intuition). And the reverse direction also holds true.

In relation to the expansion-contraction principle, if there is expansion (e.g., the relatively more developed ability of a woman who is endowed with the feminine intuition to see things "diffusively"), there is contraction (e.g., the relatively less developed ability of a woman who is endowed with the feminine intuition to "analyze" things with focusedness).

In relation to the theory-praxis principle, if there is theory (e.g., the Jungian theory of psychology), there is praxis (e.g., the practical application of the Jungian psychology for the study of the feminine intuition in geology by Andrews and her followers). And the reverse direction also holds true.

In relation to the convention-novelty principle, if there is convention (e.g., the conventional wisdom that "women aren't as good at science as men"), there is novelty (e.g., the alternative challenge to this conventional wisdom by Andrews that "women bring unique talents to geology"). And the reverse direction also holds true.

In relation to the evolution-transformation principle, if there is evolution (e.g., the natural evolution of human males and females in the state of nature to relate with each other in a relaively primitive way), there is transformation (e.g, the technical transformation of human relations between the two genders by way of the invention of different ideologies like feminism, sexism, etc.). And the reverse direction also holds true.

In relation to the symmetry-asymmetry principle, if there is symmetry (e.g., the co-existence of different views about the two genders in the history of geology), there is asymmetry (e.g., the popularity of the view that "women aren't as good at science as men" in the older days—but the popularity of the view that "women bring unique talents to geology" among the feminists in our time). And the reverse direction also holds true.

In relation to the regression-progression principle, if there is regression (e.g., the regression made in the claim by Andrews, as shown in the problems pointed out by the critics), there is progression (e.g., the progress made in the claim by Andrews, in helping us understand the role of intuition in doing geology). And the reverse direction also holds true.

And in relation to the same-difference principle, if there is similarity in outcome (e.g., the contribution to the molding and control of beliefs, values, and behavior, regardless of whether this be achieved by way of feminism or by way of the opposite view), there is difference in outcome (e.g., the contribution to the molding and control of beliefs, values, and behavior by way of feminism for a more equalitarian way of life between the two genders—but the contribution to the molding and control of beliefs, values, and behavior by way of the opposite view of gender diffrences for a more inequalitarian way of life between the two genders). And the reverse direction also holds true.

Geology, and the Rediscovery of the Mind

Another interesting way to study dynamics from the perspective of the mind concerns geology and the recovery of the mind.

Harold Morowitz (1980) in his article "Rediscovering the Mind" claimed the inceasing awareness of scientists in the hard sciences (including geology) nowadays about the importance of taking into consideration the role of the mind in doing science.

His essential point here is the need to avoid the kind of redunctionism which had plagued the hard sciences like physics, chemistry, biology, and geology in much of the modern era.

The Danger of Scientific Reductionism

For Morowitz, the story begins with the pervasiveness of scientific reductionism in hard sciences like physics, chemistry, biology, and geology in the modern era.

For instance, "[t]he study of life at all levels, from social to molecular behavior, has in modern times relied on reductionism as the chief explanatory concept. This approach to knowledge tries to comprehend one level of scientific phenomena in terms of concepts at a lower and presumably more fundamental level. In chemistry, large-scale reactions are accounted for by examining the behavior of molecules. Similarly, physiologists study the activity of living cells in terms of processes carried out by organelles and other subcellular entities. And in geology, the formations and properties of minerals are described using the features of the constituent crystals. The essence of these cases is seeking explanation in underlying structures and activities." (H. Morowitz 1980)

This reductionism, for Morowitz (1980) had also influenced the works of the best and the brightest scientists in the modern era: "Reductionism…is exemplified by the viewpoint in Carl Sagan's best-selling book *The Dragons of Eden*. He writes: 'My fundamental premise about the brain is that its workings—what we sometimes call 'mind'—are a consequence of its anatomy and physiology and nothing more.' As a further demonstration of this trend of thought, we note that Sagan's glossary does not contain the words mind, consciousness, perception, awareness, or thought, but rather deals with entries such as synapse, lobotomy, proteins, and electrodes."

The history of reductionism has a long history indeed, "beginning with the early Darwinians and their contemporaries working in physiological psychology. Before the nineteenth century, the mind-body duality, which was central to Descartes' philosophy, had tended to place the human mind outside the domain of biology. Then [there was] the stress that the human mind outside the domain of biology. Then [there was] the stress that the evolutionists placed on our 'apeness' made us subject to biological study by methods appropriate to nonhuman primates and, by extension, to other animals. The Pavlovian school reinforced that theme, and it became a cornerstone of many behavioral theories. While no general agreement has emerged among psychologists as to how far reductionism should be carried, most will readily concede that our actions have hormonal, neurological, and physiological components. Although Sagan's premise lies within a general tradition in psychology, it is radical in aiming at complete explanation in terms of the underlying level." (H. Morowitz 1980)

Even in psychology, "various schools…were attempting to reduce their science to biology, other life scientists were also looking for more basic levels of explanation. Their outlook can be seen in the writings of a popular spokesman of molecular biology, Francis Crick. In his book, *Of Molecules and Men*, a contemporary attack on vitalism—the doctrine that biology needs to be explained in terms of life forces lying outside the domain of physics—Crick States: 'The ultimate aim of the modern movement in biology is in fact to explain all biology in terms of physics and chemistry.' He goes on to say that by physics and chemistry he refers to the atomic level, where our knowledge is secure. By use of the italicized all he expresses the position of radical reductionism that has been the dominant viewpoint among an entire generation of biochemists and molecular biologists." (H. Morowitz 1980)

The Anti-Reductionist Turn

For Morowitz, things started to change in the 20th century, especially with the anti-reductionist turn.

The modern reductionist outlook for a neat (orderly) world, for Morowitz (1980), "was unceremoniously upset," first "with the introduction of the theory of relativity by Albert Einstein in 1905….The new theory postulated that observers in different systems moving with respect to each other would perceive the world differently. The observer thus became involved in establishing physical reality. The Scientist was losing the spectator's role and becoming an active participant in the system under study."

Later, "[w]ith the development of quantum mechanics, the role of the observer became an even more central part of physical theory, an essential component in defining an event. The mind of the observer emerged as a necessary element in the structure of the theory. The implications of the developing paradigm greatly surprised early quantum physicists and led them to study epistemology and the philosophy of science" (H. Morowitz 1980)

For instance, "Werner Heisenberg, one of the founders of the new physics,…[i]n *Philosophical Problems of Quantum Physics*,…wrote of physicists having to renounce thoughts of an objective time scale common to all observers, and of events in time and space that are independent of our ability to observe them. Heisenberg stressed that the laws of nature no longer dealt with elementary particles, but with our knowledge of these particles—that is, with the contents of our minds." (H. Morowitz 1980)

Others, like "Erwin Schrodinger, the man who formulated the fundamental equation of quantum mechanics," made a similar point, as he "wrote an extraordinary little book in 1958 called *Mind and Matter*. In this series of essays, he moved from the results of the new physics to a rather mystical view of the universe that he identified with the 'perennial philosophy' of Aldous Huxley. Schrodinger was the first of the quantum theoreticians to express sympathy with the Upanishads and Eastern philosophical thought. A growing body of literature now embodies this perspective, including two popular works, *The Tao of Physics* by Fritjof Capra and *The Dancing Wu Li Masters* by Gary Zukav." (H. Morowitz 1980)

What is central here concerns the relationship between the mind who studies nature and the natural event to be studied: "In more formal terms, a complex system can only be described by using a probability distribution that relates the possible outcomes of and experiment. In order to decide among the various alternatives, a measurement is required. This measurement is what constitutes an event, as distinguished from the probability, which is a mathematical abstraction. However, the only simple and consistent description physicists were able to assign to a measurement involved an observer's becoming aware of the result. Thus the physical event and the content of the human mind were inseparable. This linkage forced many researchers to seriously consider consciousness as an integral part of the structure of physics. Such interpretations moved science toward the idealist as contrasted with the realist conception of philosophy." (H. Morowitz 1980)

In the article "Remarks on the Mind-Body Question," Nobel laureate Eugene Wigner stressed that "most physical scientists have returned to the recognition that thought—meaning the mind—is primary. He goes on to state: 'It was not possible to formulate the laws of quantum mechanics in a fully consistent way without reference to the consciousness.' And he concludes by noting how remarkable it is that the scientific study of the world led to the content of consciousness as an ultimate reality." (H. Morowitz 1980)

In fact, this mind-event interaction can also be described in another way, by "[t]he introduction of information theory and its application to thermodynamics," in that, for instance, "entropy, a basic concept of that science, is a measure of the observer's ignorance of the atomic details of the system. When we measure the pressure, volume, and temperature of an object, we have a residual lack of knowledge of the exact position and velocity of the component atoms and molecules. The numerical value of the amount of information we are missing is proportional to the entropy. In earlier thermodynamics, entropy had represented, in an engineering sense, the energy of the system unavailable to perform external work. In the modern view, the human mind enters once again, and entropy relates not just to the state of the system but to our knowledge of that state." (H. Morowitz 1980)

So, in the end, Horowitz (1980) argued that "[w]e now understand the troublesome features in a forceful commitment to uncritical reductionism as a solution to the problem of the mind. We have discussed the weaknesses of that position. In addition to being weak, it is a dangerous view, since the way we respond to our fellow human beings is dependent on the way we conceptualize them in our theoretical formulations....The human psyche is part of the observed data of science."

In the case of geology, for Horowitz (1980), the lesson to learn here to avoid the impoverishing reductionist tradition in the older days when, "in geology, the formations and properties of minerals are described using the features of the constituent crystals. The essence of these cases is seeking explanation in underlying structures and activities"—that is, the danger of reductionsim.

Some Critical Thoughts

But the anti-reducitonist view by Horowitz (1980) is not without problems. Consider a few criticisms below, for illustration.

Firstly, one criticism of the anti-reducitonist view by Horowitz is that it runs the risk of commiting the opposite academic sin, namely, what I already discussed in my previous books and summarized in *Sec. 1.9* as "reverse-reductionism," in that it is more fashionable nowadays to welcome diverse views as if they were equally valid. In the current context, one is tempted to tolerate different views of the mind to be integrated into the study of natural events (with little qualifications), for instance. This can be another mistake, albeit in the opposite direction this time.

Secondly, another criticism of the anti-reducitonist view by Horowitz is that he made a general analysis of the reductionist tendency in modern hard sciences and the challenge to it by the anti-reductionist turn, to the piont that he had very little to say about its relationship to the field of geology, except mentioning in passing in different sentences that it also affected the works by geologists.

Thirdly, still another criticism of the anti-reducitonist view by Horowitz is that he did not offer any analysis of the differences between the humanistic fields where the study of the mind is more pervasive, and the hard sciences, where the study of the mind can be justifiedly less so, such that the extent to which the relevance of the role of the mind varies from one field to the next, even though the factor of the mind can be relevant to both.

And fourthly, one additional criticism of the anti-reducitonist view by Horowitz is that he also implicitly assumes that the rediscovery of the mind for the hard sciences is necessarily desirable. But he did not give a more balanced analysis of why or under what conditions the introduction to thefactor of the mind can be counterproductive for the hard sciences too.

Neither Reductionism Nor Reverse-Reductionism

Of course, these criticisms of the anti-reductionist views by Horowitz should not be misinterpreted as an unconditional support of reductionism either.

Instead, they are shown here to teach us the two opposing sides of the debate (so as to side neither with reductionism nor with reverse-reductonism), such that the possibility and desirability of dynamics (from the perspective of the mind with the rediscovery of the mind as a case study here) are not to the extent that its respective defenders would like us to believe.

Furthermore, this analysis of geology and the rediscovery of the mind can teach us something insightful about the ontological principles in existential dialectics, and good examples include the formalness-informalness principle, the absoluteness-relativeness principle, the partiality-totality principle, the predictability-unpredictability principle, the explicability-inexplicability principle, the preciseness-vagueness principle, the simpleness-complicatedness principle, the openness-hiddenness principle, the denseness-emptiness principle, the order-chaos principle, the slowness-quickness principle, the expansion-contraction principle, the theory-praxis principle, the convention-novelty principle, the evolution-transformation principle, the softness-hardness principle, the symmetry-asymmetry principle, the regression-progression principle, and the same-difference principle.

For instance, in relation to the formalness-informalness principle, if there is formalness (e.g., the formal logical requirment of completedness in a system of ideas), there is informalness (e.g., the nonformal existence of the contamination, due to the role of the observer, in measuring a natural event in quantum mechanics, with negative implications for the unrealistic search for an objective Archmedian point). And the reverse direction also holds true.

In relation to the absoluteness-relativeness principle, if there is absoluteness (e.g., the absolute anti-reductinist view of Morowitz in regard to his uncritical integration of the mind factor into the study of the hard sciences), there is relativeness (e.g., what is right for Morowitz in regard to his uncritical advocation for the integration of the mind factor into the study of the hard sciences is not necessarily so for those critics in the hard sciences who are less enthusiastic about the integration, since the hard sciences, for the critics, are much different from the humanities, although they have some commonalities). And the reverse direction also holds true.

In relation to the partiality-totality principle, if there is partiality (e.g., the partial view of those in the hard sciences in the older days, when the role of the observer was not taken into account when measuring natural events), there is totality (e.g., the totalistic view of someone like Morowitz who wants to include all factors into the equation, like the mind-event integration). And the reverse direction also holds true.

In relation to the predictability-unpredictability principle, if there is predictability (e.g., the predictable tendency of an anti-reductionist like Morowitz to include the mind in studying nature), there is unpredictability (e.g., the more difficult task to predict exactly the degree of relevance of this mind-event integration for a given study in the hard sciences, which are different from the humanities, even if the role of the mind is relevant to both). And the reverse direction also holds true.

In relation to the explicability-inexplicability principle, if there is explicability (e.g., the explanation, by Morowitz, of the importance of the mind in studying nature on the basis of the contamination of the role of the observer in measuring natural events as shown in quantum mechanics), there is inexplicability (e.g., the lack of sufficient explanation, by Morowitz, of why the role of the observer is equally or more or less relevant to both the humanities and the hard sciences, while the two are quite different from each other, even though they have something in common). And the reverse direction also holds true.

In relation to the preciseness-vagueness principle, if there is preciseness (e.g,. the precise identification of the role of the mind by Morowitz for a better way to do geology and other sciences in general), there is vagueness (e.g., the vagueness in what is exactly meant by "the mind," since it can refer to different things to different individuals). And the reverse direction also holds true.

In relation to the simpleness-complicatedness principle, if there is simpleness (e.g., the relatively simple analysis of the problem of reductionism by Morowitz), there is complicatedness (e.g., the relatively more complicated analysis of the problem of reductionism by also considering the problems associated with the view by Morowitz himself). And the reverse direction also holds true.

In relation to the openness-hiddenness principle, if there is openness (e.g., the open exploration of the role of the observer in debunking reductionism by Morowitz), there is hiddenness (e.g., the hidden bias in the analysis of Morowitz, because of its one-dimensional enthusiasm against reductionism, without due consideration of the problems associated with reverse-reductionism). And the reverse direction also holds true.

In relation to the denseness-emptiness principle, if there is denseness (e.g., the relatively denser concentration of reductionist scientists in the hard sciences to not include the role of the observer in studying nature), there is emptiness (e.g., the relatively less dense, or more empty, concentration of anti-reductionist scientists like Morowitz to not include the role of the observer in studying nature). And the reverse direction also holds true.

In relation to the order-chaos principle, if there is order (e.g., the relatively orderly, neat world in Newtonian mechanics), there is chaos (e.g., the relatively chaotic, uncertain world in quantum mechanics). And the reverse direction also holds true.

In relation to the slowness-quickness principle, if there is slowness (e.g., the relatively slower readiness of a hard-core scientist to take full consideration of the role of the mind in studying natural events), there is quickness (e., the relatively quicker readiness of a humanist like a Taoist to consider the role of the mind in studying nature). And the reverse direction also holds true.

In relation to the expansion-contraction principle, if there is expansion (e.g., the relatively more developed ability of a Taoist to be in harmony with nature by practising the mind-nature integration), there is contraction (e.g., the relatively less developed ability of a Taoist to intrude into nature by dominating it as a hard-core scientists would like to do with no scruples). And the reverse direction also holds true.

In relation to the theory-praxis principle, if there is theory (e.g., the theory of quantum mechanics), there is praxis (e.g., the practical application of quantum mechanics, by Morowizt and other antireductionists, for the mind-nature integration in the hard sciences). And the reverse direction also holds true.

In relation to the convention-novelty principle, if there is convention (e.g., the conventinal wisdom about reductionism in the older days in the hard sciences), there is novelty (e.g., the alternative novel challenge to this conventional wisdom about reductionism, by the contemporary anti-reductionist movement by Morowitz and others in the hard sciences). And the reverse direction also holds true.

In relation to the evolution-transformation principle, if there is evolution (e.g., the natural evolution of humans to study nature in a primitive way in the state of nature), there is transformation (e.g., the technical transformation of human ability to study nature by the invention of quantum mechanics, information theory, etc. in the 20th century). And the reverse direction also holds true.

In relation to the softness-hardness principle, if there is softness (e.g. the inclusion of the role of the mind in studying nature by anti-reductionist scientists), there is hardness (e.g., the exclusion of the role of the mind in studying nature by reductionist scientists). And the reverse direction also holds true.

In relation to the symmetry-asymmetry principle, if there is symmetry (e.g., the co-existence of different views about the mind and nature in the history of ideas), there is asymmetry (e.g., the popularity of Newtonian physics without the consideration of the role of the mind in the older days—but the popularity of quantum mechanism with the consideration of the role of the mind nowadays). And the reverse direction also holds true.

In relation to the regression-progression principle, if there is regression (e.g., the regression made in the anti-reductionist view by Morowitz, as shown in the problems associated with it), there is progression (e.g., the progress made in the anti-reductionist view by Morowitz, to help us better understand the role of the mind in understanding nature). And the reverse direction also holds true.

And in relation to the same-difference principle, if there is similarity in outcome (e.g., the contribution to the molding and regulation of beliefs and values, regardless of whether this is done by way of the tao of physics or by way of Newtonian physics), there is difference in outcome (e.g., the contribution to the molding and regulation of beliefs and values by way of the tao of physics for a more harmonious relationship with nature—but the contribution to the molding and regulation of beliefs and values by way of Newtonian physics, for a more dominating relationship with nature). And the reverse direction also holds true.

Dynamics and Nature

Dynamics, when examined from the perspective of nature, can reveal something refreshing enough about its possibility and desirability, and this can be shown by way of two case studics, namely, (3.3.1) volcanic ejection, and its impact, and (3.3.2) earthquake, and its power—to be addressed in what follows, respectively.

Volcanic Ejection, and Its Impact

An impressive phenomenon about dynamics from the perspective of the nature has to do with volcanic ejection and its impact.

To understand volcanic ejection, it is important to know something about magma.

The Nature of Magma

By defintion, the word "magma" is "from Greek μάγμα" to mean "paste" and refers to "molten rock that is found beneath the surface of the Earth, and may also exist on other terrestrial planets. Besides molten rock, magma may also contain suspended crystals and gas bubbles. Magma often collects in a magma chamber inside a volcano." (WK 2010ff)

There are three ways that magma can affect its surroundings, namely, (a) "intrusion into adjacent rocks," (b) "extrusion onto the surface as lava," and (c) "explosive ejection as tephra to form pyroclastic rock." (WK 2010ff)

In general, magma is quite hot. For instance, because of its "complex high-temperature fluid substance," the "temperatures of most magmas are in the range 700 °C to 1300 °C (or 1300 °F to 2400 °F), but very rare carbonatite melts may be as cool as 600 °C, and komatiite melts may have been as hot as 1600 °C." (WK 2010ff) In this sense, it is no wonder that "all igneous rock is formed from magma." (WK 2010ff)

The locations where magmas are found are also diverse. For instance, "environments of magma formation and compositions" where magmas "are commonly"found include "subduction zones, continental rift zones, mid-oceanic ridges, and hotspots, some of which are interpreted as mantle plumes." (WK 2010ff)

However, "despite being found in such widespread locales, the bulk of the Earth's crust and mantle is not molten. Rather, most of the Earth takes the form of a rheid, a form of solid that can move or deform under pressure. Magma, as liquid, preferentially forms in high temperature, low pressure environments within several kilometers of the Earth's surface." (WK 2010ff)

An Example of Explosive Ejection

A good way to illustrative the amazing impact of magma on the formation of rocks over time is by way of understanding the "explosive ejection" of magma, which is one of three ways that magma can affect its surroundings—besides "intrusion into adjacent rocks" and "extrusion onto the surface as lava." (WK 2010ff)

Explosive ejection can best be illustrated by "volcanic ash," which refers to "small tephra, which are bits of pulverized rock and glass created by volcanic eruptions, less than 2 millimetres (0.1 in) in diameter." (WK 2010gg)

Volcanic eruptions can be understood in relation to its (a) formation, (b) composition, (c) spread, and (d) effects, as summarized below, in that order.

(a) Formation

Volcanic eruptions can occur due to "three mechanisms of volcanic ash formation," as shown below. (WK 2010gg; G. Heiken 1985)

- "Gas release under decompression causing magmatic eruptions"
- "Thermal contraction from chilling on contact with water causing phreatomagmatic eruptions"
- "Ejection of entrained particles during steam eruptions causing phreatic eruptions. The violent nature of volcanic eruptions involving steam results in the magma and solid rock surrounding the vent being torn into particles of clay to sand size."

In addition, volanic ash can be understood in terms of its composition.

(b) Composition

The composition of volcanic ash is precise enough. For instance, "any material explosively thrown out from a vent is tephra [unconsolidated] or pyroclastic debris [consolodiated]. Ash terminology is restricted to very fine rock and mineral particles less than 2 millimetres (0.079 in) in diameter which are ejected from a volcanic vent," as shown below, in comparing with other types of debris. (WK 2010gg; G. Heiken 1985)

"*Clast Size*	*Pyroclast*	*Tehpra*	*Pyroclastic Rock*"
"> 64 mm	Bomb, Block	Agglomerate	Pyroclastic breccia"
"< 64 mm	Lapillus	Lapilli Tephra	Lapilli Tuff'
"< 2 mm	Coarse Ash	Coarse Ash	Coarse (ash) Tuff'
"< 0.063 mm	Fine Ash	Fine Ash	Fine (ash) Tuff'

In this way, ash can be differentiated from magma, in that "ash is created when solid rock shatters and magma separates into minute particles during explosive volcanic activity. The usually violent nature of an eruption involving steam (phreatic eruption or phreatomagmatic eruption) results in the magma and solid rock surrounding the vent being torn into particles of clay to sand size." (WK 2010gg)

(c) Spread

The spread of volcanic ash is not trivial and can be quite expansive.

For instance, "plume that is often seen above an erupting volcano is composed primarily of ash and steam. The very fine particles may be carried for many miles, settling out as a dust-like layer across the landscape. This is known as an ashfall. If liquid magma is ejected as a spray, the particles will solidify in the air as small fragments of volcanic glass. Unlike the ash that forms from burning wood or other combustible materials, volcanic ash is hard and abrasive. It does not dissolve in water, and it conducts electricity, especially when it is wet." (WK 2010gg; MWD 2010c)

Ashfall so understood can be both desirable and undesirable. On the one hand, ashfall can be destructive, because, for example, "[i]n 1783, the Laki eruption killed about one-fifth of Iceland's population, and sent a huge toxic cloud of ash and sulphurous gases across Western Europe. In Britain alone, it has been estimated that 23,000 died from the poisoning." (WK 2010gg; BBC 2010; MSNBC 2010; D. Walker 2007)

On the other hand, ashfall can be constructive to life, and the reason is that, because "[a]shfall can become cemented together by heat to form a solid rock called tuff,...[it] breaks down over time, forming highly fertile soil, which has made many volcanic regions densely cultivated and inhabited despite the inherent dangers." (WK 2010gg; MWD 2010d; Skwirk 2010)

(d) Effects

The effects of volcanic ash can be many and powerful. Consider two examples below, for illustration.

The first example has to do with "atmospheric effects"; for instance, "[w]hen ash begins to fall during daylight hours, the sky turns hazy and a pale yellow color. The ashfall may become so dense that daylight turns the sky gray to pitch black, with the ash severely restricting visibility and deadening sound. A darkened ash sky lowers temperatures during daylight hours from what would otherwise be expected. Loud thunder, lightning, as well as the strong smell of sulfur accompany an ashfall." (WK 2010gg; USGS 2009a)

And the second example has to do with "hazards"; for instance, "[t]he most devastating effect of volcanic ash comes from pyroclastic flows. These occur when a volcanic eruption creates an 'avalanche' of hot ash, gases, and rocks that flow at high speed down the flanks of the volcano. These flows can be impossible to outrun. They can also be difficult to predict. In many cases prediction is based on the topography of a region, but a valley may fill and overflow. In 1902, the city of St. Pierre in Martinique was destroyed by a pyroclastic flow which killed over 29,000 people." (WK 2010gg; Zananas. 2008)

Some Critical Questions

However, the geological analysis of volcanic ejection is not without some critical issues for reflection. Consider a few criticisms below, for illustration.

Firstly, one major criticism of the geological analysis of volcanic ejection concerns the favorite number "three." In other words, there is a tendency of geologists to use the number "three" for their classification of geological materials.

For instance, in the current context, magma is classified into three ways of affecting its surroundings (i.e., intrusion, extrusion, and ejection), just as volcanic ash is classified into three major mechanisms of formation (i.e., gas release, thermal contraction, and particle ejection). And rocks are classified into three types (e.g., igneous, sedimentary, and metamorhic), as already summarized in *Sec. 1. 3*.

But why does the number "three" occur so often in geological classification—not four, or two, or five, or ten, or something else? This is true, whether they study statics (as in the classification of the types of rocks), or dynamics (as in the classification of the ways that magma affects its surroundings).

Secondly, another major criticsm of the geological analysis of volcanic ejection is that it is presented in such a seemingly "objective" way without any consideration of the role of the "mind" (as already discussed in the previous section, that is, *Sec. 3.2.2* on geology and the rediscovery of the mind).

And thirdly, still another major criticims of the geological analysis of volcanic ejection is that we still have a long way to understand fully its nature and impact, precisely because we have failed, for instance, to predict exactly its occurrence in advance to save lives and properties.

A Valuable Reflection

Lest misunderstanding occurs, these criticisms of the geological analysis of volcanic ejection should not be misconstrued as a suggestion of its uselessness.

Rather, they are introduced here to reveal the two opposing sides of the debate, such that the possibility and desirability of dynamics (from the perspective of nature with volcanic ejection as a case study here) are not to the extent that its respective defenders would like us to believe.

Furthermore, this analysis of volcanic ejection and its power can reflect something valuable about the ontological principles in existential dialectics, and good examples include the absoluteness-relativeness principle, the partiality-totality principle, the predictability-unpredictability principle, the explicability-inexplicability principle, the preciseness-vagueness principle, the simpleness-complicatedness principle, the openness-hiddenness principle, the denseness-emptiness principle, the change-constancy principle, the slowness-quickness principle, the expansion-contraction principle, the convention-novelty principle, the evolution-transformation principle, the symmetry-asymmetry principle, the softness-hardness principle, the regression-progression principle, and the same-difference principle.

For instance, in relation to the absoluteness-relativeness principle, if there is absoluteness (e.g., the absolute view of volcanic ejection as destructive, by those who have family members as victims of volcanic ejection), there is relativeness (e.g., what is bad about volcanic ejection for those who have family members as victims of volcanic ejection is not necessarily so for those who benefit from volcanic ejection, because "ashfall breaks down over time, forming highly fertile soil, which has made many volcanic regions densely cultivated and inhabited despite the inherent dangers"). And the reverse direction also holds true.

In relation to the predictability-unpredictability principle, if there is predictability (e.g,. the predictable tendency of volcanic ejection to result in ashfall, with both destructive and construcive consequences), there is unpredictability (e.g., the more difficult task to predict exactly when volcanic ejection occurs, in which particular way on a given occasion, because "[t]he…flows can be impossible to outrun," "[t]hey can also be difficult to predict," and "[i]n many cases prediction is based on the topography of a region, but a valley may fill and overflow"). And the reverse direction also holds true.

In relation to the explicability-inexplicability principle, if there is explicability (e.g., the explanation, by mainstream geologists, of the nature of magma and volcanic ejection on the basis of geological factors only, like their composition, formation, spread, and effects), there is inexplicability (e.g., the lack of sufficient explanation, by mainstream geologists, of why the role of "the mind" is necessarily ignored in undersanding and measuring natural events like volcanic ejection). And the reverse direction also holds true.

In relation to the preciseness-vagueness principle, if there is preciseness (e.g., the precise identification of different classifications about rocks, magmas, and volcanic mechanisms in terms of the number three), there is vagueness (e.g., the vaguness in the number three in regard to these classications, since it is not clear why they cannot be two, or four, or five, or another number). And the reverse direction also holds true.

In relation to the simpleness-complicatedness principle, if there is simpleness (e.g., the relatively simple analysis of volcanic ejection by geologists), there is complicatedness (e.g., the relatively more complicated analysis of volcanic ejection by raising some critical questions by the critics). And the reverse direction also holds true.

In relation to the openness-hiddenness principle, if there is openness (e.g., the open exploration of the nature of volcanic ejection in relation to its formation, composition, spread, and effects), there is hiddenness (e.g., the hidden bias in the mainstream geological analysis of volcanic ejection, because of its dismissal of the role of "the mind" in the analysis). And the reverse direction also holds true.

In relation to the denseness-emptiness principle, if there is denseness (e.g., "[t]he ashfall may become so dense that daylight turns the sky gray to pitch black"), there is emptiness (e.g., the relatively less dense, or more empty, concentration of ashfall in the sky, before its occurrence). And the reverse direction also holds true.

In relation to the change-constancy principle, if there is change (e.g., the dynamic nature of volcanic ejection, in relation to its different phases of formation, composition, spread, and effects), there is constancy (e.g., the ever constancy of the rock cycle, with volcanic ejection as one of its expressions). And the reverse direction also holds true.

In relation to the slowness-quickness principle, if there is slowness (e.g., the relatively slower speed of a rock running downhill in a normal situation), there is quickness (e.g., the relatively quicker speed of rocks, especially when reduced to molten forms into tiny pieces, as in "an 'avalanche' of hot ash, gases, and rocks" during "a volcanic eruption"). And the reverse direction also holds true.

In relation to the expansion-contraction principle, if there is expansion (e.g., the expansion of volcanic ash during the ejection), there is contraction (e.g., the contraction of the ash after the ejection, when it becomes cooler and solidified and can no longer expand, until the next rock cycle). And the reverse direction also holds true.

In relation to the convention-novelty principle, if there is convention (e.g., the conventional wisdom about the classifications about rocks, magmas, and volcanic mechanisms in terms of the number three), there is novelty (e.g., the alternative novel challenge to this conventional wisdom about the classifications, by questioning why the number three must be the case at all). And the reverse direction also holds true.

In relation to the evolution-transformation principle, if there is evolution (e.g,. the natural evolution of humans to encounter and study volcanic ejection in the state of nature), there is transformation (e.g., the technical transformation of humans to understand volcanic ejection by the invention of geosciences like geology). And the reverse direction also holds true.

In relation to the symmetry-asymmetry principle, if there is symmetry (e.g., the co-existence of different forms of ash, be they from burnng wood or from volcanic ejection), there is asymmetry (e.g., ash from burning wood is less abrasive than ash from volcanic ejection). And the reverse direction also holds true.

In relation to the softness-hardness principle, if there is softeness (e.g, the relatively softer nature of "ash that forms from burning wood or other combustible materials," as contrasted with ash from volcanic ejection), there is hardness (e.g., the relatively harder...nature of ash from volcanic ejection, which "is hard and abrasive" and "does not dissolve in water"). And the reverse direction also holds true.

In relation to the regression-progression principle, if there is regression (e.g, the regression made in volcanic ejection, as shown by the undesirable consequences like deaths and property damages), there is progression (e.g., the progression made in volcanic ejection, as shown by the formation of "highly fertile soil, which has made many volcanic regions densely cultivated and inhabited despite the inherent dangers"). And the reverse direction also holds true.

And in relation to the same-difference principle, if there is similarity in outcome (e.g., the occurrence of magma over time, regardless of whether it is due to intrusion, exrusion, or ejection), there is difference in outcome (e.g., "[a]shfall [from magma] can become cemented together by heat to form a solid rock called tuff,...[and] breaks down over time, forming highly fertile soil"—whereas lava from magma becomes cooler and solidifies into igneous rocks over time, not soil). And the reverse direction also holds true.

Earthquake, and Its Power

Another impressive phenomenon about dynamics from the perspective of the nature has to do with earthquake and its power—to be analyzed below (and summarized in *Table 3.2*).

About the Term "Earthquake"

By definition, the word "earthquake" has different names like "a quake, tremor, temblor or seismic activity" and refers to "the result of a sudden release of energy in the Earth's crust that creates seismic waves." (WK 2010hh)

In general, "earthquakes manifest themselves by shaking and sometimes displacing the ground. When a large earthquake epicenter is located offshore, the seabed sometimes suffers sufficient displacement to cause a tsunami. The shaking in earthquakes can also trigger landslides and occasionally volcanic activity." (WK 2010hh)

In the broadest sense, the word "earthquake" can also be used "to describe any seismic event—whether a natural phenomenon or an event caused by humans—that generates seismic waves. Earthquakes are caused mostly by rupture of geological faults, but also by volcanic activity, landslides, mine blasts, and nuclear experiments. An earthquake's point of initial rupture is called its focus or hypocenter. The term epicenter refers to the point at ground level directly above the hypocenter." (WK 2010hh)

And in terms of measurememt, "[e]arthquakes are measured with a seismometer; a device which also records is known as a seismograph. The moment magnitude (or the related and mostly obsolete Richter magnitude) of an earthquake is conventionally reported, with magnitude 3 or lower earthquakes being mostly imperceptible and magnitude 7 causing serious damage over large areas. Intensity of shaking is measured on the modified Mercalli scale." (WK 2010hh)

Naturally Occurring Earthquakes

A good illustration of earthquakes in general concern "naturally occurring earthquakes," not caused by humans.

For instance, "earthquakes will occur anywhere within the earth where there is sufficient stored elastic strain energy to drive fracture propagation along a fault plane. In the case of transform or convergent type plate boundaries, which form the largest fault surfaces on earth, they will move past each other smoothly and aseismically only if there are no irregularities or asperities along the boundary that increase the frictional resistance." (WK 2010hh)

However, "[m]ost boundaries do have such asperities and this leads to a form of stick-slip behaviour. Once the boundary has locked, continued relative motion between the plates leads to increasing stress and therefore, stored strain energy in the volume around the fault surface. This continues until the stress has risen sufficiently to break through the asperity, suddenly allowing sliding over the locked portion of the fault, releasing the stored energy." (WK 2010hh)

This stored energy is "released as a combination of radiated elastic strain seismic waves, frictional heating of the fault surface, and cracking of the rock, thus causing an earthquake. This process of gradual build-up of strain and stress punctuated by occasional sudden earthquake failure is referred to as the Elastic-rebound theory. It is estimated that only 10 percent or less of an earthquake's total energy is radiated as seismic energy. Most of the earthquake's energy is used to power the earthquake fracture growth or is converted into heat generated by friction." (WK 2010hh)

Earthquakes can be analyzed in terms of seven main features, as shown below.

(a) Earthquake Fault Types

The first main feature of earthquakes is that there are "three main types of fault that may cause an earthquake," namely, (i) "normal," (ii) "reverse (thrust)," and (iii) "strike-slip." (WK 2010hh)

On the one hand, "normal and reverse faulting are examples of dip-slip, where the displacement along the fault is in the direction of dip and movement on them involves a vertical component. Normal faults occur mainly in areas where the crust is being extended such as a divergent boundary. Reverse faults occur in areas where the crust is being shortened such as at a convergent boundary." (WK 2010hh)

On the other hand, "strike-slip faults are steep structures where the two sides of the fault slip horizontally past each other; transform boundaries are a particular type of strike-slip fault. Many earthquakes are caused by movement on faults that have components of both dip-slip and strike-slip; this is known as oblique slip." (WK 2010hh)

(b) Earthquakes away from Plate Boundaries

The second main feature of earthquakes is that they can occur away from tectonic plate boundaries.

For example, "[w]here plate boundaries occur within continental lithosphere, deformation is spread out over a much larger area than the plate boundary itself. In the case of the San Andreas fault continental transform, many earthquakes occur away from the plate boundary and are related to strains developed within the broader zone of deformation caused by major irregularities in the fault trace (e.g. the 'Big bend' region)." (WK 2010hh)

(c) Shallow-Focus and deep-Focus earthquakes

The third main feature of earthquakes is that they can be shallow or deep.

For example, "[t]he majority of tectonic earthquakes originate at the ring of fire in depths not exceeding tens of kilometers. Earthquakes occurring at a depth of less than 70 km are classified as 'shallow-focus' earthquakes, while those with a focal-depth between 70 and 300 km are commonly termed 'mid-focus' or 'intermediate-depth' earthquakes." (WK 2010hh)

In places like "subduction zones, where older and colder oceanic crust descends beneath another tectonic plate, deep-focus earthquakes may occur at much greater depths (ranging from 300 up to 700 kilometers)." (WK 201hh; USGS 2005)

(d) Earthquakes and Volcanic Activity

The fourth main feature of earthquakes is that earthquakes are related to volcanic eruptions in some ways.

For instance, "[e]arthquakes often occur in volcanic regions and are caused there, both by tectonic faults and the movement of magma in volcanoes. Such earthquakes can serve as an early warning of volcanic eruptions, as during the Mount St. Helens eruption of 1980." (WK 2010hh)

In addition, "earthquake swarms can serve as markers for the location of the flowing magma throughout the volcanoes. These swarms can be recorded by seismometers and tiltmeters (a device which measures the ground slope) and used as sensors to predict imminent or upcoming eruptions." (WK 2010hh; J. Watson 1998)

(e) Rupture dynamics

The fifth main feature of earthquakes is that it has a "rupture dynamics," which, unfortunately, is still not fully understood.

For instance, "a tectonic earthquake begins by an initial rupture at a point on the fault surface, a process known as nucleation. The scale of the nucleation zone is uncertain, with some evidence, such as the rupture dimensions of the smallest earthquakes, suggesting that it is smaller than 100 m while other evidence, such as a slow component revealed by low-frequency spectra of some earthquakes, suggest that it is larger." (WK 2010hh)

Also, "the possibility that the nucleation involves some sort of preparation process is supported by the observation that about 40% of earthquakes are preceded by foreshocks. Once the rupture has initiated it begins to propagate along the fault surface. The mechanics of this process are poorly understood, partly because it is difficult to recreate the high sliding velocities in a laboratory. Also the effects of strong ground motion make it very difficult to record information close to a nucleation zone." (WK 2010hh; NRS 2003)

Even then, with this lack of full understanding of rupture dynamics in mind—rupture propagations vary from case to case, not always at the same speed. For instance, "a small subset of earthquake ruptures appear to have propagated at speeds greater than the S-wave velocity. These supershear earthquakes have all been observed during large strike-slip events." (WK 2010hh)

On the other hand, there are slow earthquakes, as shown by the "particularly dangerous form of slow earthquake" like "the tsunami earthquake,…where the relatively low felt intensities, caused by the slow propagation speed of some great earthquakes, fail to alert the population of the neighbouring coast, as in the 1896 Meiji-Sanriku earthquake." (WK 2010hh; NRS 2003)

(f) Frequency, Magnitude, and Concentration of Occurrence

The sixth main feature of earthquakes is that they come in all shapes and sizes.

For instance, in terms of frequency, "[t]here are around 500,000 earthquakes each year. 100,000 of these can actually be felt." (WK 2010hh; USGS 2009b)

In terms of size (or magnitude), it is also noticed that "larger earthquakes occur less frequently….In the (low seismicity) United Kingdom, for example, it has been calculated that the average recurrences are: an earthquake of 3.7–4.6 every year, an earthquake of 4.7–5.5 every 10 years, and an earthquake of 5.6 or larger every 100 years. This is an example of the Gutenberg-Richter law." (WK 2010hh; R. Musson 2003)

And in terms of concentration, "[m]ost of the world's earthquakes (90%, and 81% of the largest) take place in the 40,000-km-long, horseshoe-shaped zone called the circum-Pacific seismic belt, known as the Pacific Ring of Fire, which for the most part bounds the Pacific Plate." (WK 2010hh) This is true, even though "earthquakes can occur almost anywhere." (WK 2010hh; USGS 2009c & 2009d)

(g) Different Causes

And the seventh main feature of earthquakes is that, although many of them are naturally caused, humans have also a role to play.

For instance, "[w]hile most earthquakes are caused by movement of the Earth's tectonic plates, human activity can also produce earthquakes. Four main activities contribute to this phenomenon: constructing large dams and buildings, drilling and injecting liquid into wells, and by coal mining and oil drilling." (WK 2010hh; A. Madrigal 2008)

As an illustration, "the greatest earthquake in Australia's history was...induced by humanity, through coal mining. The city of Newcastle was built over a large sector of coal mining areas. The earthquake was spawned from a fault which reactivated due to the millions of tonnes of rock removed in the mining process." (WK 2010hh;. B. Trembath 2007)

Some Problems with Studying Earthquakes

However, there are some problems associated with the sudy of earthquakes. Consider a few criticisms below, for illustration.

Firstly, one major criticism of the study of earthquakes is that it still is not fully understood, because it is very difficult to study them in a laboratory.

For instance, "[o]nce the rupture has initiated it begins to propagate along the fault surface. The mechanics of this process are poorly understood, partly because it is difficult to recreate the high sliding velocities in a laboratory. Also the effects of strong ground motion make it very difficult to record information close to a nucleation zone." (WK 2010hh; NRS 2003)

Secondly, another major criticism of the study of earthquakes is that not all earthquakes are easily detected.

For instance, a good example is "the tsunami earthquake,...where the relatively low felt intensities, caused by the slow propagation speed of some great earthquakes, fail to alert the population of the neighbouring coast, as in the 1896 Meiji-Sanriku earthquake." (WK 2010hh; NRS 2003)

And thirdly, still another major criticism of the study of earthquakes is that it does not give sufficient attention to the role of "the mind" in measuring earthquakes, in the context of the rediscovery of the mind as already discussed in *Sec. 3.2.2*.

A More Detached View

Yet, these criticisms of the study of earthquakes should not be misinterpreted as a confirmation of its invalidity.

Instead, they are introduced here to show us the two opposing sides of the debate, such that the possibility and desirability of dynamics (from the perspective of nature with earthquakes as a case study here) are not to the extent that its respective defenders would like us to believe.

In addition, this analysis of earthquake and its power can teach us something more detached about the ontological principles in existential dialectics, and good examples include the absoluteness-relativeness principle, the predictability-unpredictability principle, the explicability-inexplicability principle, the preciseness-vagueness principle, the simpleness-complicatedness principle, the openness-hiddenness principle, the denseness-emptiness principle, the order-chaos principle, the slowness-quickness principle, the expansion-contraction principle, the theory-praxis principle, the convention-novelty principle, the evolution-transformation principle, the symmetry-asymmetry principle, the softness-hardness principle, the regression-progression principle, and the same-difference principle.

For instance, in relation to the absoluteness-relativeness principle, if there is absoluteness (e.g., the absolute view of the measurement of an earthquake if placed at the "nucleation zone"), there is relativeness (e.g., what is right from the vantage point of a measurement if placed at the "nucleation zone" does not necessarily match the vantage point of a measurement if placed outside the "nucleation zone," because of the different effects of "strong ground motion" at different points in relation to the "nucleation zone"). And the reverse direction also holds true.

In relation to the predictability-unpredictability principle, if there is predictability (e.g., the predictable tendency of an earthquake "to propagate along the fault surface," once "the rupture has initiated"), there is unpredictability (e.g., the more difficult task to predict exactly "the scale of the nucleation zone" in relation to the propagation, which "is uncertain" in relation to a particular earthquake). And the reverse direction also holds true.

In relation to the explicability-inexplicability principle, if there is explanation (e.g., the explanation that "once the rupture [of an earthquake] has initiated it begins to propagate along the fault surface"), there is inexplicability (e.g., the lack of sufficient explanation about "[t]he mechanics of this process, which are poorly understood, partly because it is difficult to recreate the high sliding velocities in a laboratory" and also because "the effects of strong ground motion make it very difficult to record information close to a nucleation zone"). And the reverse direction also holds true.

In relation to the preciseness-vagueness principle, if there is preciseness (e.g., the precise identification of "around 500,000 earthquakes each year"), there is vagueness (e.g.., the vagueness in this number of 500,000, since it is not clear if there are other occurrences of earthquakes, which, however, are not successfully detected by seismologists, as the measurement depends on the quality and pervasiveness of the devices being used, among other criteria). And the reverse direction also holds true.

In relation to the simpleness-complicatedness principle, if there is simpleness (e.g., the relatively simple analysis of earthquakes by mainstream sesimologists), there is ocmplicatedness (e.g., the relatively more complicated analysis of earthquakes by also considering some critical problems pointed out by the critics). And the reverse direction also holds true.

In relation to the openness-hiddenness principle, if there is openness (e.g., the open exploration of the nature of rupture dynamics, like an earthquake "to propagate along the fault surface"), there is hiddenness (e.g., the hidden nature of the "rupture dynamics" to be revealed, since "the mechanics of this process are poorly understood"). And the reverse direction also holds true.

In relation to the denseness-emptiness principle, if there is denseness (e.g., the relatively denser concentration of earthquakes, about "90%, and 81% of the largest," along "the 40,000-km-long, horseshoe-shaped zone called the circum-Pacific seismic belt, known as the Pacific Ring of Fire, which for the most part bounds the Pacific Plate"), there is emptiness (e.g., the relatively less dense, or more empty, concentraton of earthquakes outside "the Pacific Ring of Fire"). And the reverse direction also holds true.

In relation to the order-chaos principle, if there is order (e.g., the relatively orderly nature of the movement of the ground, when there is no rupture of geological faults), there is chaos (e.g., the relatively chaotic nature of the movement of the ground, in case of an earthquake, when the propagation is associated with "high sliding velocities," in conjunction with "the effects of strong ground motion…close to a nucleation zone"). And the reverse direction also holds true.

In relation to the slowness-quickness principle, if there is slowness (e.g., the occurrence of "slow earthquakes" like "the tsunami earthquake,… caused by the slow propagation speed"), there is quickness (e.g., the "quick earthquakes" like "a small subset of earthquake ruptures…at speeds greater than the S-wave velocity"). And the reverse direction also holds true.

In relation to the expansion-contraction principle, if there is expansion (e.g., in "normal faults,…the crust is being extended such as at a divergent boundary"), there is contraction (e.g., in "reverse faults,…the crust is being shortened such as at a convergent boundary"). And the reverse direction also holds true.

In relation to the theory-praxis principle, if there is theory (e.g., the theoretical construction of seismology), there is praxis (e.g., the practical application of seismology for the purpose "to protect and prepare possible sites of earthquakes from severe damage," as in "earthquake engineering, earthquake preparedness, household seismic safety, seismic retrofit," etc.). (WK 2010hh) And the reverse direction also holds true.

In relation to the convention-novelty principle, if there is convention (e.g., the conventional wisdom in "the lifetime of the Greek philosopher Anaxagoras in the 5th century BCE to the 14th century CE," in that "earthquakes were usually attributed to 'air (vapors) in the cavities of the Earth'"), there is novelty (e.g., the alternative novel challenge to this ancient conventional wisdom with the contemporary view that earthqake is "the result of a sudden release of energy in the Earth's crust that creates seismic waves"). (WK 2010hh; EWEH 2003) And the reverse direction also holds true.

In relation to the evolution-transformation principle, if there is evolution (e.g., the natural evolution of humans in the state of nature to sense earthquakes by their senses), there is transformation (e.g., the technical transformation of human aiblity to detect earthquakes by the use of seismometers and other hi-tech devices). And the reverse direction also holds true.

In relation to the symmetry-asymmetry principle, if there is symmetry (e.g., the occurrence of about 500,000 earthquakes per year), there is asymmetry (e.g., some earthquakes are greater in magnitude than others). And the reverse direction also holds true.

In relation to the softness-hardness principle, if there is softness (e.g., the soft earthquakes with a 1.0 magnitude or less, being relatively harmess), there is hardness (e.g., the hard earthquakes with 8.0 magnitude or greater, causing relatively serious damages). And the reverse direction also holds true.

In relation to the regression-progression principle, if there is regression (e.g., the regression made in the study of earthquakes, as shown in the problems pointed out by the critics), there is progression (e.g,. the progress made in the study of earthquakes, to help us better understand the occurrences of earthquakes and their prevention). And the reverse direction also holds true.

And in relation to the same-difference principle, if there is similarity in outcome (e.g., the contribution to the molding and control of beliefs and values, regardless of whether this is achieved by way of seismology or by way of ancient mythology about earthquakes), there is difference in outcome (e.g., the contribution to the molding and control of beliefs and values by way of seismology for an instrumental/dominating way of dealing with nature—but the contribution to the molding and control of beliefs and values by way of ancient mythology about earthquakes for a substantive/reverent way of dealing with nature). And the reverse direction also holds true.

Dynamics and Society

Dynamics, when examined from the perspective of society, can illuminate a different light on its possibility and desirability, and this can be shown by way of two case studies, namely, (3.4.1) technology, and the debate on relative vs. absolute dating, and (3.4.2) engineering geology, and the protection of life and property—to be addressed in what follows, respectively.

Technology, and the Debate on Relative vs. Absolute Dating

In the context of understanding dynamics from the perspective of society, an interesting issue concerns the role of technology to determine the nature of geological time in relation to the debate on relative dating versus absolute dating.

To understand this debate, let's consider more closely (a) the nature of relative dating and then (b) the nature of relative dating—as shown below (and summarized in *Table 3.3*).

Relative Dating

In relation to "relative dating"—the development was closely associated with the beginnng history of geology "as a formal science," especially in the context of seven principles (to be summarized below).

Nowadays, "[g]eologists still use the [seven] principles…as a means to provide information about geologic history and the timing of geologic events." (WK 2010)

(a) The Principle of Intrusive Relationships

This first principle focuses on "crosscutting intrusions" in geology. (WK 2010)

For instance, "when an igneous intrusion cuts across a formation of sedimentary rock, it can be determined that the igneous intrusion is younger than the sedimentary rock. There are a number of different types of intrusions, including stocks, laccoliths, batholiths, sills and dikes." (WK 2010)

(b) The Principle of Cross-Cutting Relationships

This second principle, instead of focusing on "crosscutting intrusions" (as already covered by the principle of intrusive relationships), concerns "crosscutting relationships," which refer to "the formation of faults and the age of the sequences through which they cut." (WK 2010)

For instance "faults are younger than the rocks they cut; accordingly, if a fault is found that penetrates some formations but not those on top of it, then the formations that were cut are older than the fault, and the ones that are not cut must be younger than the fault. Finding the key bed in these situations may help determine whether the fault is a normal fault or a thrust fault." (WK 2010; P. Olsen 2001)

(c) The Principle of Inclusions and Components

This third principle refers to another relationship, in that, "with sedimentary rocks, if inclusions (or clasts) are found in a formation, then the inclusions must be older than the formation that contains them." (WK 2010)

For example, "in sedimentary rocks, it is common for gravel from an older formation to be ripped up and included in a newer layer. A similar situation with igneous rocks occurs when xenoliths are found. These foreign bodies are picked up as magma or lava flows, and are incorporated, later to cool in the matrix. As a result, xenoliths are older than the rock which contains them." (WK 2010)

(d) The Principle of Uniformitarianism

This fourth principle states that "the geologic processes observed in operation that modify the Earth's crust at present have worked in much the same way over geologic time." (WK 2010; R. Hooykaas 1963)

This "fundamental principle of geology" was "advanced by the 18th century Scottish physician and geologist James Hutton" and can be summarized in a few words: "The present is the key to the past." (WK 2010) As Hutton said, "the past history of our globe must be explained by what can be seen to be happening now." (WK 2010)

(e) The Principle of Original Horizontality

This fifth principle suggests that "the deposition of sediments occurs as essentially horizontal beds." (WK 2010) First proposed "by the Danish geological pioneer Nicholas Steno (1638-1686),...[t]his principle states that layers of sediment are originally deposited horizontally." (WK 2010ii)

For instance, "observation of modern marine and non-marine sediments in a wide variety of environments supports this generalization (although cross-bedding is inclined, the overall orientation of cross-bedded units is horizontal)." (WK 2010; P. Olsen 2001)

(f) The principle of Superposition

This sixth principle proposes that "a sedimentary rock layer in a tectonically undisturbed sequence is younger than the one beneath it and older than the one above it." (WK 2010)

This logic sounds sensible, because "a younger layer cannot slip beneath a layer previously deposited. This principle allows sedimentary layers to be viewed as a form of vertical time line, a partial or complete record of the time elapsed from deposition of the lowest layer to deposition of the highest bed." (WK 2010; P. Olsen 2001)

(g) The Principle of Faunal Succession

And this seventh principle refers to "the appearance of fossils in sedimentary rocks." (WK 2010)

For instance, "as organisms exist at the same time period throughout the world, their presence or (sometimes) absence may be used to provide a relative age of the formations in which they are found. Based on principles laid out by William Smith almost a hundred years before the publication of Charles Darwin's theory of evolution, the principles of succession were developed independently of evolutionary thought." (WK 2010)

Absolute Dating

By contrast to "relative dating" (above)—the technological advance since modern times has contributed to the rise of "absolute dating," with its dependence on hi-tech devices or instruments.

In general, the main advantage of "absolute dating" is that, with its hi-tech dependence, "geologists can also give precise absolute dates to geologic events. These dates are useful on their own, and can also be used in conjunction with relative dating methods or to calibrate relative dating methods." (WK 2010; R. Tucker 1998)

The development of "absolute dating" was regarded as "a large advance in geology in the advent of the 20th century," because of "the ability to give precise absolute dates to geologic events through radioactive isotopes and other methods. The advent of radiometric dating changed the understanding of geologic time. Before, geologists could only use fossils to date sections of rock relative to one another. With isotopic dates, absolute dating became possible, and these absolute dates could be applied to fossil sequences in which there was datable material, converting the old relative ages into new absolute ages." (WK 2010)

For instance, "isotope ratios are measured in minerals that give the amount of time that has passed since a rock passed through its particular closure temperature, the point at which different radiometric isotopes stop diffusing into and out of the crystal lattice." (WK 2010; H. Rollinson 1996; G. Faure 1998)

Absolute dating can come in different vesions, like "uranium-lead dating," "potassium-argon dating," "argon-argon dating," "uranium-thorium dating," and the like. (WK 2010)

Over the years, absolute dating has been used "for a variety of applications. Dating of lavas and ash layers can help to date stratigraphy and calibrate relative dating techniques. These methods can also be used to determine ages of pluton emplacement. Thermochemical techniques can be used to determine temperature profiles within the crust, the uplift of mountain ranges, and paleotopography." (WK 2010)

Some Problems with the Two Methods of Dating

However, there are some problems associated with the two methods of dating. Consider a few criticisms below, for illustration.

Firstly, one major criticism is that, in the case of relative dating, the principles are often not quite true.

For instance, in relation to "the principle of original horizontality," we now know that "not all sedimentary layers are deposited purely horizontally." (WK 2010ii)

Secondly, another major criticism is that, in the case of relative dating, reality can be more complicated than what the principles suggest.

For instance, in relation to "the principle of faunal succession," we now know that reality "becomes quite complex,...given the uncertainties of fossilization, the localization of fossil types due to lateral changes in habitat (facies change in sedimentary strata), and that not all fossils may be found globally at the same time." (WK 2010; S. Winchester)

Thirdly, still another major criticism is that, in the case of relative dating, there are disagreements about the assumptions used in the principles.

For instance, in relation to the principle of uniformitarianism, "Stephen Jay Gould's first scientific paper, *Is uniformitarianism necessary?* (1965), reduced the...interpretations [of uniformitarianism] to two, methodological and substantive uniformitarianism. He dismissed the first principle [viz., methodological uniformitarianism], which asserted spatial and temporal invariance of natural laws, as no longer an issue of debate. He rejected the second [viz., substantive uniformitarianism] as an unjustified limitation on scientific inquiry, as it constrains past geologic rates and conditions to those of the present." (WK 2010f)

In fact, "uniformitarianism...was originally proposed in contrast to catastrophism, which states that the distant past 'consisted of epochs of paroxysmal and catastrophic action interposed between periods of comparative tranquility.'" (WK 2010f; W. Whewell 1832)

Fourthly, one additional major criticism here is that, in the case of absolute dating, there are limitations in regard to its accuracy.

Consider two following illustrations to understand the limitations.

The first illustration has to do with "radiometric dating" (a version of absolute dating), and the business here is to know "the constant rate of decay of radioactive isotopes. Given an initial and a present quantity of such an isotope and its half-life, the time elapsed may be calculated." (WK 2010jj)

However, there are two general problems in "radiometric testing."

On the one hand, "[i]f a very short period of time has passed, as measured in number of half-lives, a particular technique will be less accurate and more susceptible to statistical fluctuations in the inherently random decay events." (WK 2010jj) For instance, in carbon-14 dating, "the half-life of carbon-14 is 5730 years," so this means that "radiocarbon is less useful to date some recent sites." (WK 2010jj)

On the other hand, "if many half lives of the isotope of interest have passed, too much of the sample may have decayed to provide an accurate reading." (WK 2010jj)

The second illustration has to do with "thermoluminesence testing" (another version of absolute dating), which is to "date...items to the last time they were heated. This technique is based on the principle that all objects absorb radiation from the environment. This process frees electrons within minerals that remain caught within the item. Heating an item to 350 degrees Celsius or higher releases the trapped electrons, producing light. This light can be measured to determine the last time the item was heated." (WK 2010jj)

But the problem with the reliance on "radiation" is that "radiation levels do not remain constant over time. Fluctuating levels can skew results—for example, if an item went through several high radiation eras, thermoluminesence will return an older date for the item. Many factors can spoil the sample before testing as well, exposing the sample to heat or direct light may cause some of the electrons to dissipate, causing the item to date younger. Because of these and other factors, Thermoluminesence is at the most about 15% accurate. It cannot be used to accurately date a site on its own." (WK 2010jj)

And fifthly, the last major criticism here is that, in the case of absolute dating, "absolute dates do not necessarily tell us when a particular cultural event happened," even though "they are invaluable in constructing a more specific sequence of events." (WK 2010jj)

A More Balanced Observation

However, these criticisms of the two methods of dating in geology should not be misunderstood as a rejection of them all.

Instead, they are introduced here to reveal the two opposing sides of the debate, such that the possibility and desirability of dynamics (from the perspective of society with the debate on relative vs. absolute dating as a case study here) are not to the extent that its respective defenders would like us to believe.

Furthermore, this analysis of technology and the debate on relative vs. absolute dating can shed further light on the ontological principles in existential dialectics, and good examples include the absoluteness-relativeness principle, the predictability-unpredictability principle, the explicability-inexplicability principle, the preciseness-vagueness principle, the simpleness-complicatedness principle, the openness-hiddenness principle, the denseness-emptiness principle, the change-constancy principle, the slowness-quickness principle, the expansion-contraction principle, the theory-praxis principle, the convention-novelty principle, the evolution-transformation principle, the symmetry-asymmetry principle, the softness-hardness principle, the regression-progression principle, and the same-difference principle.

For instance, in relation to the absoluteness-relativeness principle, if there is absoluteness (e.g., the "precise and absolute dates" in absolute dating), there is relativeness (e.g,. what is important for absolute dating in regard to its precise accuracy is not necessarily so for relative dating in regard to its focus on "a sequence of age"). (WK 2010jj) And the reverse direction also holds true.

In relation to the predictability-unpredictability principle, if there is predictability (e.g, the predictable tendency, from the perspective of "the principle of original horizontality," that "the deposition of sediments occurs as essentially horizontal beds"), there is unpredictability (e.g., the more difficult task to predict exactly if a particular deposition of sediments occurs as essentially horizontal beds, because we now know that "not all sedimentary layers are deposited purely horizontally"). And the reverse direction also holds true.

In relation to the explicability-inexplicability principle, if there is explicability (e.g., the explanation, from the perspective of "the principle of uniformitarianism," that "the geologic processes observed in operation that modify the Earth's crust at present have worked in much the same way over geologic time," such that "the present is the key to the past"), there is inexplicability (e.g., the lack of sufficient explanation, from the perspective of "the principle of uniformitarianism," of why "the distant past 'consisted of epochs of paroxysmal and catastrophic action interposed between periods of comparative tranquility'"). And the reverse direction also holds true.

In relation to the preciseness-vagueness principle, if there is preciseness (e.g., the precise identification of dates in absolute dating), there is vagueness (e.g., the vagueness in the precise dates in absolute dating, because "absolute dates do not necessarily tell us when a particular cultural event happened"). And the reverse direction also holds true.

In relation to the simpleness-complicatedness principle, if there is simpleness (e.g., the relatively simple views of the principles in relative dating), there is complicatedness (e.g., the relatively more complicated analyis of the principles in relative dating, as shown in the criticisms pointed out by the critics). And the reverse direction also holds true.

In relation to the openness-hiddenness principle, if there is openness (e.g, the open exploration of different geological relations as revealed by the principles in relative dating), there is hiddneness (e.g, the hidden biases in the principles, as shown by the problems and assumptions used in them, as pointed out by the critics). And the reverse direction also holds true.

In relation to the denseness-emptiness principle, if there is denseness (e.g., the relatively denser concentration of layers which are deposited on top of each other in a sedimentary rock…in a tectonically undisturbed sequence"), there is emptiness (e.g., the relatively less dense, or more empty, concentration of layers in an "ash," which "is created when solid rock shatters and magma separates into minute particles during explosive volcanic activity," before becoming cooler and solidified later). (WK 2010gg) And the reverse direction also holds true.

In relation to the change-constancy principle, if there is change (e.g., the ever new ways of measuring the age of earth materials in geology, like relative dating and relative dating), there is constancy (e.g, the ever constant existence of limitations in all methods, regardless of whether they are relative or absolute). And the reverse direction also holds true.

In relation to the slowness-quickness principle, if there is slowness (e.g., the relatively slower speed of an isotope to decay if it has a much longer half-life), there is quickness (e.g., the relatively quicker speed of an isotope to decay if it has a very short half-life). And the reverse direction also holds true.

In relation to the expansion-contraction principle, if there is expansion (e.g., the relative expansion of the influence of absolute dating "in geology in the advent of the 20th century"), there is contraction (e.g,. the relative contraction of the influence of relative dating in geology in the later part of the 20th century, with the rise of absolute dating). And the reverse direction also holds true.

In relation to the theory-praxis principle, if there is theory (e.g., the theoretical construction in chemistry and physics about "isotopes"), there is praxis (e.g., the practical application of the knowledge about "isotopes" for a variety of geological uses like "the thermochemical techniques…to determine temperature profiles within the crust, the uplift of mountain ranges," and so on). And the reverse direction also holds true.

In relation to the convention-novelty principle, if there is convention (e.g., the conventional wisdom about relative dating in the older days of geology), there is novelty (e.g,. the alternative novel challenge to the conventional wisdom about relative dating, with the new way known as absolute dating in the 20th century). And the reverse direction also holds true.

In relation to the evolution-transformation principle, if there is evolution (e.g,. the natural evolution of humans in the state of nature to understand earth materials with their senses), there is transformation (e.g, the technical transformation of human ability to understand earth materials by the hi-tech invention of "uranium-lead dating," "potassium-argon dating," "argon-argon dating," "uranium-thorium dating," and the like). And the reverse direction also holds true.

In relation to the symmetry-asymmetry principle, if there is symmetry (e.g., the existence of different versions of dating in geology, like relative dating and absolute dating), there is asymmetry (e.g., the usefulness of relative dating for a sequence of age but not for precise and absolute dates—but the usefulness of absolute dating for precise and absolute dates but not in cases where its limitations occur, like the limits when measured in number of half-lives or when the "fluctuating levels [of radiation] can skew results"). (WK 2010jj) And the reverse direction also holds true.

In relation to the softness-hardness principle, if there is softness (e.g., the co-existence of different "layers of sediment" on top of each other as "horizontal beds"), there is hardness (e.g, "an igneous intrusion cuts across a formation of sedimentary rock"). And the reverse direction also holds true.

In relation to the regression-progression principle, if there is regression (e.g, the regression made in absolute dating, as shown in the problems pointed out by the critics), there is progression (e.g,. the progress made in absolute dating for more precise dating in geology). And the reverse direction also holds true.

And in relation to the same-difference principle, if there is similarity in outcome (e.g., the contribution to the molding and control of beliefs and values, regardless of whether this is achieved by way of relative dating or by way of absolute dating), there is difference in outcome (e.g., the contribution to the molding and control of beliefs and values by way of relative dating for a relatively less certain view of the world—but the contribution to the molding and control of beliefs and values by way of absolute dating for a relatively more certain view of the world). And the reverse direction also holds true.

Engineering Geology, and the Protection of Life and Property

Another way to evaluate the possibility and desirability of dynamics from the perspective of society has to do with engineering geology and the protection of life and property in society.

The term "engineering geology" refers to "the application of the geologic sciences to engineering practice for the purpose of assuring that the geologic factors affecting the location, design, construction, operation and maintenance of engineering works are recognized and adequately provided for." (WK 2010kk)

More importantly, "[t]he principal objective of the engineering geologist is the protection of life and property against damage caused by geologic conditions." (WK 2010kk)

The Practice of Engineering Geology

The practice of engineering geology is modern in origin, because "the science and practice of engineering geology didn't begin as a recognized discipline until the late 1800s and early 1900s," and this is true, even though "the science of geology has been around since the 1700s." (WK 2010kk)

However, the need for engineering geology became more serious after 1928, "with the failure of the St. Francis dam in California and the loss of 426 lives. More engineering failures which occurred the following years also prompted the requirement for engineering geologists to work on large engineering projects." (WK 2010kk)

Because of its concern with human development to protect "life and property against damage caused by geologic conditions"—engineering geology focuses on "landforms and earth processes to identify potential geologic and related man-made hazards that may impact civil structures and human development." (WK 2010kk)

Thus, it is no wonder that "[n]early all engineering geologists are initially trained and educated in geology, primarily during their undergraduate education. This background in geology provides the engineering geologist with an understanding of how the earth works, which is crucial in mitigating for earth related hazards. Most engineering geologists also have graduate degrees where they have gained specialized education and training in soil mechanics, rock mechanics, geotechnics, groundwater, hydrology, and civil design. These two aspects of the engineering geologists' education provides them with a unique ability to understand and mitigate for hazards associated with earth-structure interactions." (WK 2010kk)

And engineering geology is also multidisciplinary in nature, as "engineering geologic practice is also closely related to the practice of geological engineering, geotechnical engineering, soils engineering, environmental geology and economic geology," for instance.

The Scope of Studies

Engineering geology studies are quite complicated; for instance, "[e]ngineering geologic studies may be performed during the planning, environmental impact analysis, civil or structural engineering design, value engineering and construction phases of public and private works projects, and during post-construction and forensic phases of projects. Works completed by engineering geologists include; geologic hazards, geotechnical, material properties, landslide and slope stability, erosion, flooding, dewatering, and seismic investigations, etc." (WK 2010kk)

More methodically, the scope of engineering geological studies can be illustrated below. (WK 2010kk)

- Ex: "for residential, commercial and industrial developments"
- Ex: "for governmental and military installations"
- Ex: "for public works such as a power plant, wind turbine, transmission line, sewage treatment plant, water treatment plant, pipeline (aqueduct, sewer, outfall), tunnel, trenchless construction, canal, dam, reservoir, building, railroad, transit, highway, bridge, seismic retrofit, airport and park"

- Ex: "for mine and quarry excavations, mine tailing dam, mine reclamation and mine tunneling"
- Ex: "for wetland and habitat restoration programs"
- Ex: "for coastal engineering, sand replenishment, bluff or sea cliff stability, harbor, pier and waterfront development"
- Ex: "for offshore outfall, drilling platform and sub-sea pipeline, sub-sea cable"

Of course, there can be other studies, but these example suffice for illustration here.

The Concern with Geohazards and Adverse Geo-Conditions

This broad scope of engineering geological studies is also related with its concern with "geohazards" and "adverse geo-conditions." (WK 2010kk)

For illustration, hereafter are some examples of the concern of engineering geological studies with "geologic hazards or other adverse conditions." (WK 2010kk)

- Ex: "fault rupture on seismically active faults"
- Ex: "seismic and earthquake hazards (ground shaking, liquefaction, lurching, lateral spreading, tsunami and seiche events)"
- Ex: "landslide, mudflow, rockfall, debris flow, and avalanche hazards"
- Ex: "unstable slopes and slope stability"
- Ex: "erosion"
- Ex: "slaking and heave of geologic formations"
- Ex: "ground subsidence (such as due to ground water withdrawal, sinkhole collapse, cave collapse, decomposition of organic soils, and tectonic movement)"
- Ex: "volcanic hazards (volcanic eruptions, hot springs, pyroclastic flows, debris flow, debris avalanche, gas emissions, volcanic earthquakes)"
- Ex: "non-rippable or marginally rippable rock requiring heavy ripping or blasting"
- Ex: "weak and collapsible soils, foundation bearing failures"
- Ex: "shallow ground water/seepage"

Of course, there can be other examples, but these suffice for illustration here.

The Diversity of Methods

Because engineering geology is broad scope and focuses on geohazards and adverse geo-conditions, it has to rely on a variety of methods to get things done.

Hereafter are some methods used engineering geology, for illustration. (WK 2010kk)

- Ex: "geologic field mapping of geologic structures, geologic formations, soil units and hazards"
- Ex: "the review of geologic literature, geologic maps, geotechnical reports, engineering plans, environmental reports, stereoscopic aerial photographs, remote sensing data, Global Positioning System (GPS) data, topographic maps and satellite imagery"
- Ex: "the excavation, sampling and logging of earth/rock materials in drilled borings, backhoe test pits and trenches, fault trenching, and bulldozer pits"
- Ex: "geophysical surveys (such as seismic refraction traverses, resistivity surveys, ground penetrating radar (GPR) surveys, magnetometer surveys, electromagnetic surveys, high-resolution sub-bottom profiling, and other geophysical methods)"
- Ex: "deformation monitoring as the systematic measurement and tracking of the alteration in the shape or dimensions of an object as a result of the application of stress to it manually or with an automatic deformation monitoring system"

Again, there can be other methods, but these example suffice for illustration here.

Some Problems in Engineering Geological Studies

In spite of its broad scope, its variety of methods, and its concern with geohazards and adverse geo-conditions, engineering geology does not often protect "life and property against damage caused by geologic conditions" as claimed. (WK 2010kk)

For instance, in 1960, there was the "tsunami damage in Hilo, Hawaii," and "the tsunami was generated by a magnitude 8.6 earthquake near Chile. Property damage in Hawaii was estimated at 24 million dollars." (NAG 2009)

In 1964, "the Government Hill School in Anchorage, AK" was greatly damaged "after the magnitude 8.5 earthquake of March 28." (NAG 2009) In fact, "a Federal Emergency Management Agency study considered just capital (damages to buildings and their contents) and income-related costs, and provided an estimate of $4.4 billion as the minimum average annualized loss due to earthquakes in the United States." (NAG 2009)

In 1983, "[t]he most costly landslide in U.S. history was a relatively slow-moving event in Thistle, Utah, in the spring….The landslide, caused by the wet El Nino winter of 1982-83, dammed the Spanish Fork River and buried U.S. Highway 6 and the main line of the Denver and Rio Grande Western Railroad. The town of Thistle was inundated under the floodwaters rising behind the landslide dam. Total losses were estimated at more than $400 million in 1983 dollars." (NAG 2009) In addition, "landslides constitute a major geologic hazard because they are widespread, occurring in all 50 States, and because they cause more than $2 billion in damages and more than 25 fatalities on average each year" in the U.S. (NAG 2009)

In 1989, "the Wahaula Visitor Center in Hawaii Volcanoes National Park was overrun by lava flows in June." (NAG 2009) And in 1994, "freeway interchange…collapsed in the…M6.9 Northridge, California earthquake." (NAG 2009)

So a good question to ask here is, Why did engineering geologists not succeed in protecting "life and property against damage caused by geologic conditions" in cases like this?

Consider three criticisms below, for illustration.

Firstly, one major criticism against engineering geological studies is that the geohazards and adverse geo-conditions are more complex than what engineering geologists, in spite of its broad scope and its variety of methods, can fully understand.

Secondly, another major criticism against engineering geological studies is that engineering geologists, because of their techncial education mostly in geology and related disciplines, often do not take into account the broader political, legal, socioeconomic, cultural, and other factors which can complicate the outcomes of engineering geological works.

And thirdly, one additional major criticism against engineering geological studies is that they have yet to "rediscover the mind" in geology so as to avoid the impoverishing tradition of "reductionism" so much criticized by H. Morowitz (1980)—as already discussed in *Sec. 3.2.2*.

The Larger Picture

Nevertheless, these criticisms of engineering geological studies do not mean that the field of engineering geology has made no contribution to the propection of "life and property against damage caused by geologic conditions."

Rather, they are introduced here to show us the two opposing sides of the debate, such that the possibility and desirability of dynamics (from the perspective of society with engineering geology as a case study here) are not to the extent that its respective defenders would like us to believe.

In addition, this analysis of engineering geology and the protection of life and property can show us the larger picture in regard to the ontological principles in existential dialectics, and good examples include the absoluteness-relativeness principle, the predictability-unpredictability principle, the partiality-totality principle, the explicability-inexplicability principle, the preciseness-vagueness principle, the simpleness-complicatedness principle, the openness-hiddenness principle, the denseness-emptiness principle, the slowness-quickness principle, the expansion-contraction principle, the theory-praxis principle, the convention-novelty principle, the evolution-transformation principle, the symmetry-asymmetry principle, the regression-progression principle, and the same-difference principle.

For instance, in relation to the absoluteness-relativeness principle, if there is absoluteness (e.g., the absolute orientation of engineering geologists, because of their narrow background, to not give sufficient consideration of societal, cultural, and mental factors when doing engineering geology), there is relativeness (e.g., what is appropriate for eningeering geologists in regard to their narrow outlook is not necsssarily so for the critics like H. Morowitz in regard to the impoverishing tradition of reductionism in geology and other hard sciences). And the reverse direction also holds true.

In relation to the predictability-unpredictability principle, if there is predictability (e.g, the predictable tendency of engineering geologists to pay attention to geohazards and adverse geo-conditions for the protection of life and property), there is unpredictability (e.g,. the more difficult task to predict exactly when engineering geologists will succeed on a particular occasion in protecting life and property, because of the history of failures as pointed out by the critics). And the reverse direction also holds true.

In relation to the partiality-totality principle, if there is partiality (e.g, the partial view of those engineering geologists who do not consider societal, cultural, and mental factors when studying geohazards), there is totality (e.g,. the holistic view of geohazards by combining not only geology and other related disciplines, but also societal, cultural, and mental factors, such that the outcomes are more uncertain than otherwise, in that the whole is not the sum of its parts). And the reverse direction also holds true.

In relation to the explicability-inexplicability principle, if there is explicability (e.g., the explanation, from the perspective of engineering geology, of its practice, on the basis that "[n]early all engineering geologists are initially trained and educated in geology"), there is inexplicability (e.g., the lack of sufficient explanation, from the perspective of engineering geology, of why engineering geologists should not give sufficient attention to cultural, societal, and mental factors in studying geohazards, as criticized by Morowitz in regard to reductionism). And the reverse direction also holds true.

In relation to the preciseness-vagueness principle, if there is preciseness (e.g., the precise identification of the studies and methods in engineering geology), there is vagueness (e.g., the vagueness in the identification of the studies and methods in engineerng geology, because it is not clear which ones of them are more important, useful, accurate, or else under which particular conditions). And the reverse direction also holds true.

In relation to the simpleness-complicatedness principle, if there is simpleness (e.g., the relatively simple anlaysis of the scope, concern, and methodology in engineering geology), there is complicatedness (e.g., the relatively more complicated analysis of the scope, concern, and methodology in engineering geology, by also taking into account the cultural, societal, and mental factors as pointed out by the critics). And the reverse direction also holds true.

In relation to the openness-hiddenness principle, if there is openness (e.g., the open exploration of the use of geology for studying geohazards and adverse geo-conditions in engineering geology), there is hiddenness (e.g., the hidden bias of reductionism in engineering geology, because of the lack of sufficient attention to cultural, societal, and mental factors as pointed out by the critics). And the reverse direction also holds true.

In relation to the denseness-emptiness principle, if there is denseness (e.g., the relatively denser concentration of some geohazards like "earthflows" to "occur on moderate slopes," as they "consist of saturated soil or fine-grained rock deposits that flow downhill"), there is emptiness (e.g., the relatively less dense, or more empty, concentration of some geohazards like "earthflows" to occur uphill). (NAG 2009) And the reverse direction also holds true.

In relation to the slowness-quickness principle, if there is slowness (e.g., some geohazards like "some volcanic eruptions are mild and slow"), there is quickness (e.g., some geohazards like "other [volcanic eruptions]...are powerful and dramatic" or fast). (NAG 2009) And the reverse direction also holds true.

In relation to the expansion-contraction principle, if there is expansion (e.g,. the relatively more developed ability of engineering geologists to study geohazards based on geological factors), there is contraction (e.g., the relatively less developed ability of engineering geologists to study geohazards based on cultural, societal, and mental factors). And the reverse direction also holds true.

In relation to the theory-praxis principle, if there is theory (e.g., the theoretical construction in geology about earthquakes, volcanic eruptions, etc.), there is praxis (e.g., the practical application of geology to engineering geology for the protection of life and property). And the reverse direction also holds true.

In relation to the convention-novelty principle, if there is convention (e.g., the conventional wisdom about studying geology without much attention to "human development" in the older days prior to the 20th century), there is novelty (e.g., the alternative novel challenge to the conventional wisdom about studying geology by the new way of focusing on geohazards with human development in mind, as in engineering geology since the 20th century). And the reverse direction also holds true.

In relation to the evolution-transformation principle, if there is evolution (e.g.., the natural evolution of human ability to study geohazards by their senses in the state of nature), there is transformation (e.g., the technical transformation of human ability to study geohazards by the invention of engineering geology with its broadness of scope and its variety of methods). And the reverse direction also holds true.

In relation to the symmetry-asymmetry principle, if there is symmetry (e.g., the chronic occurrences of different landslides in the U.S. over time), there is asymmetry (e.g., "[t]he most costly landslide in U.S. history" happened in Thistle, Utah in the spring of 1983, whereas other landslides, by contrast, were less costly). And the reverse direction also holds true.

In relation to the regression-progression principle, if there is regression (e.g., the regression made in engineering geology, as shown in the history of failures to protect "life and property against damage caused by geologic conditions"), there is progression (e.g., the progress made in engineering geology, to help us understand geohazards and, under the right conditions, for the protection of life and property). And the reverse direction also holds true.

And in relation to the same-difference principle, if there is similarity in outcome (e.g., the contribution to the molding and control of beliefs and values, regardless of whether this is achieved by way of reductionistic engineering geological studies or by way of non-reductionistic engineering geological studies), there is difference in outcome (e.g., the contribution to the molding and control of beliefs and values by way of reductionistic engineering geological studies for a relativley more simplistic and certain view of reality—but the contribution to the molding and control of beliefs and values by way of non-reductionistic engineering geological studies for a more complicated and uncertain view of reality). And the reverse direction also holds true.

Dynamics and Culture

Dynamics, when explored from the perspective of culture, reveals in a distinctive way its possibility and desirability, and this can be shown by way of two case studies, namely, (3.5.1) historical geology, and the nature of cultural contingency, and (3.5.2) geological time, and the use/misuse of metaphors—to be addressed below, in that order.

Historical Geology, and the Nature of Cultural Contingency

A refreshing way to evaluate the possibility and desirability of dynamics from the perspective of culture has to do with the history of geological thoughts and its cultural contingency, which belongs to the field of "historical geology." (WK 2010ll)

The Definition of a Term

The term "historical geology" refers to "the use of the principles of geology to reconstruct and understand the history of the Earth. It focuses on geologic processes that change the Earth's surface and subsurface; and the use of stratigraphy, structural geology and paleontology to tell the sequence of these events. It also focuses on the evolution of plants and animals during different time periods in the geological timescale. The discovery of radioactivity and the development of a variety of radiometric dating techniques in the first half of the 20th century provided a means of deriving absolute versus relative ages of geologic history." (WK 2010ll; H. Levin 2003)

Historical geology is important for other fields. For instance, "economic geology, the search for and extraction of energy and raw materials, is heavily dependent on an understanding of the geological history of an area. Environmental geology, including most importantly the geologic hazards of earthquakes and volcanism, must also include a detailed knowledge of geologic history." (WK 2010ll)

The Cultural Contingency of Geological Thoughts

More fundamentally, the history of geological thoughts in historical geology also reveals to us their cultural contingency, with each era having its own peculiar concern with particular ideas and holding particular values.

For illustration, consider four different examples, namely, the four examples representing the four eras of the 17th, 18th, 19th, and 20th centuries, respectively in that order—as shown below (and summarized in *Table 3.4*).

(a) Fossil Records

The first example seems innocent enough, since its appearance in the 17th century challenged the ancient mystical stories about geological objects at the time, which then led to the beginning of geology as a modern science. (WK 2010nn)

This challenge came from the work of Nicolaus Steno, "also known as Niels Stensen," who is nowadays often regarded as "the father of geology and stratigraphy" and "was the first to observe and propose some of the basic concepts of historical geology. One of these concept was that fossils originally came from living organisms." (WK 2010ll & 2010mm; T. Woods 2005)

Steno's discovery came in 1666, when "two fishermen caught a huge female shark near the town of Livorno, and Ferdinando II de' Medici, Grand Duke of Tuscany, ordered its head to be sent to Steno. Steno dissected the head and published his findings in 1667. He noted that the shark's teeth bore a striking resemblance to certain stony objects, found embedded within rock formations, that his learned contemporaries were calling glossopetrae or 'tongue stones.'" (WK 2010mm)

Before Steno, people held some unscientific (mystical) views about stones like this. For example, "ancient authorities, such as the Roman author Pliny the Elder, in his *Naturalis Historiae*, had suggested that these stones fell from the sky or from the Moon. Others were of the opinion, also following ancient authors, that fossils naturally grew in the rocks. Steno's contemporary Athanasius Kircher, for example, attributed fossils to a 'lapidifying virtue diffused through the whole body of the geocosm,' consided an inherent characteristic of the earth—an Aristotelian approach." (WK 2010mm)

Steno held a different view and followed in the footsteps of Fabio Colonna, who "had already shown in a convincing way that glossopetrae are shark teeth, in his treaty *De Glossopetris Dissertatio* published in 1616. Steno added to Colonna's theory a discussion on the differences in composition between glossopetrae and living sharks' teeth, arguing that the chemical composition of fossils could be altered without changing their form, using the contemporary corpuscular theory of matter." (WK 2010mm)

Then, "Steno's work on shark teeth led him to the question of how any solid object could come to be found inside another solid object, such as a rock or a layer of rock. The 'solid bodies within solids' that attracted Steno's interest included not only fossils, as we would define them today, but minerals, crystals, encrustations, veins, and even entire rock layers or strata. He published his geologic studies in *De solido intra solidum naturaliter contento dissertationis prodromus,* or *Preliminary discourse to a dissertation on a solid body naturally contained within a solid* in 1669. Steno was not the first to identify fossils as being from living organisms; his contemporaries Robert Hooke and John Ray also argued that fossils were the remains of once-living organisms." (WK 2010mm)

(b) Uniformitarianism

The second example seems quite bold, since its appearance in the 18th century challenged the religious account of earth's history at the time.

This challenge came from the works of James Hutton and Charles Lyell, when they "contributed to early understanding of the Earth's history with their observations at Edinburgh in Scotland concerning angular unconformity in a rock face and it was in fact Lyell that influenced Charles Darwin greatly in his theory of evolution by speculating that the present is the key to the past." (WK 2010ll)

In fact, it was Hutton who "first proposed the theory of uniformitarianism, which is now a basic principle in all branches of geology. Hutton also supported the idea that the Earth was very old as opposed to prevailing concept of the time which said the Earth had only been around a few millennia. Uniformitarianism describes an Earth created by the same natural phenomenon that are at work today." (WK 2010ll)

Hutton's idea was developed in 1785, when he proposed "a self-maintaining infinite cycle," in that "[t]he solid parts of the present land appear in general, to have been composed of the productions of the sea, and of other materials similar to those now found upon the shores." (WK 2010f)

Hutton then concluded with the following three claims. (WK 2010f)

- "That the land on which we rest is not simple and original, but that it is a composition, and had been formed by the operation of second causes."
- "That before the present land was made, there had subsisted a world composed of sea and land, in which were tides and currents, with such operations at the bottom of the sea as now take place."
- "That while the present land was forming at the bottom of the ocean, the former land maintained plants and animals; at least the sea was then inhabited by animals, in a similar manner as it is at present."

Then, Hutton "sought evidence to support his idea that there must have been repeated cycles, each involving deposition on the seabed, uplift with tilting and erosion then undersea again for further layers to be deposited. At Glen Tilt in the Cairngorm mountains he found granite penetrating metamorphic schists, in a way which indicated to him that the presumed primordial rock had been molten after the strata had formed." (WK 2010mm; R. Macfarlane 2003; SG 2010)

In addition, he "had read about angular unconformities as interpreted by Neptunists, and found Hutton's Unconformity at Jedburgh where layers of greywacke in the lower layers of the cliff face have been tilted almost vertically before being eroded to form a level plane, under horizontal layers of Old Red Sandstone. In the Spring of 1788 he took a boat trip along the Berwickshire coast with John Playfair and the geologist Sir James Hall, and found a dramatic unconformity showing the same sequence at Siccar Point." (WK 2010mm; JO 2010)

With Hutton's work in mind, Charles Lyell published his multi-volume *Principles of Geology* in the early 1830's and "developed Hutton's idea that the earth was shaped entirely by slow-moving forces still in operation today, acting over a very long period of time." (WK 2010mm)

(c) Catastrophism

The third example concerns an opposing view against uniformitarianism, which resulted in a heated debate during the 19th century, in that "[t]he prevailing concept of the 18th century was that of a very short Earth history dominated by catastrophic events. This view was strongly supported by religious thinkers based on a largely literal interpretation of biblical passages." (WK 2010ll)

In other words, "[t]he concept of uniformitarianism met with considerable resistance and the catastrophism vs. gradualism debate of the 19th century resulted." (WK 2010ll)

After all, one should remember that, "[u]ntil the 19th century the dominant scientific beliefs in Europe were founded on the biblical narratives of Creation and the universal deluge. Other ancient deluge myths have been discovered since then, explaining why the flood story was 'stated in scientific methods with surprising frequency among the Greeks,' an example being Plutarch's account of the Ogygian flood." (WK 2010e)

In this way, "Earth's history was viewed as the result of an accumulation of catastrophic events over a relatively short time period, before the depth of geological time was appreciated. In this way they were able to explain the observations of early geologists within the framework of a short Earth history." (WK 2010e)

For example, in England "during the early nineteenth century, a group of geologists that included William Buckland and Robert Jameson" linked "the biblical flood" with geological events." (WK 2010e)

Jameson, for one, wrote about Georges Cuvier's work on "fossil quadrapeds that discussed his ideas on castastrophic extinction into English and published it under the title 'Theory of the Earth'. He added extensive editorial notes to the translation that explicitly linked the latest of Cuvier's revolutions with the biblical flood." (WK 2010e; M. Rudwick 1972)

And Buckland, for another, "spent much of his early career trying to demonstrate the reality of the biblical flood with geological evidence. He frequently cited Cuvier's work even though Cuvier had proposed an inundation of limited geographic extent and extended duration, and Buckland, to be consistent with the biblical account, was advocating a universal flood of short duration." (WK 2010e; M. Rudwick 1972)

(d) A Compromise

And the fourth example is more contemporary in origin (especially since the later 20th century), since it sides neither with uniformitarianism nor with catastrophism and opts for a middle road, in that "a variety of discoveries in the 20th century provided ample evidence that Earth history is a product of both gradual incremental processes and sudden cataclysmic events." (WK 2010ll)

For instance, "[v]iolent events such as meteorite impacts and large volcanic explosions do shape the Earth's surface along with gradual processes such as weathering, erosion and deposition much as they have throughout Earth history. The present is the key to the past—includes catastrophic as well as gradual processes." (WK 2010ll)

More recently, with the advance in hi-tech equipments for studying geology, there is shift of paradigm for "a more integrative approach, considering the Earth in a broader context encompassing the atmosphere, biosphere and hydrosphere." (WK 2010nn; BGS 2006)

Good examples of the advance in hi-tech equipments for this shift are "satellites located in space that take wide scope photographs of the Earth" which "provide such a perspective. In 1972, The Landsat Program, a series of satellite missions jointly managed by NASA and the U.S. Geological Survey, began supplying satellite images that can be geologically analyzed. These images can be used to map major geological units, recognize and correlate rock types for vast regions and track the movements of Plate Tectonics. A few applications of this data include the ability to produce geologically detailed maps, locate sources of natural energy and predict possible natural disasters caused by plate shifts." (WK 2010nn; L. Rocchio 2010)

Looking Beyond

This brief history of geological thoughts, with the four examples for illustration here, does not mean to favor any one of them over the others.

Instead, they are introduced here to show us how much geological thoughts in historical geology have their own cultural contingency, such that each era has its own peculiar concern with particular ideas and holding particular values.

In so doing, the examples also show us the opposing sides of the debate, such that the possibility and desirability of dynamics (from the perspective of culture with historical geology as a case study here) are not to the extent that its respective defenders would like us to believe, because ideas and values change over time, with the arrival of a new era.

Furthermore, this analysis of historical geology and the nature of cultural contingency can teach us how to look beyond the history of an era, in regard to the ontological principles in existential dialectics, and good examples include the absoluteness-relativeness principle, the predictability-unpredictability principle, the partiality-totality principle, the explicability-inexplicability principle, the preciseness-vagueness principle, the simpleness-complicatedness principle, the openness-hiddenness principle, the denseness-emptiness principle, the slowness-quickness principle, the expansion-contraction principle, the theory-praxis principle, the convention-novelty principle, the evolution-transformation principle, the symmetry-asymmetry principle, the regression-progression principle, and the same-difference principle.

For instance, in relation to the absoluteness-relativeness principle, if there is absoluteness (e.g., the absolute view of creationist catastrophists in regard to a short Earth history), there is relativeness (e.g., what is true for creationist catastrophists in regard to a short Earth history is not so for uniformitarianists in regard to a long Earth history). And the reverse direction also holds true.

In relation to the predictability-unpredictability principle, if there is predictability (e.g., the predictable tendency of the history of geological thoughts to change over time in relation to its own peculiar concern with particular ideas and holding particular values in a given era), there is unpredictability (e.g., the more difficult task to predict exactly what the peculiar concerns with particular ideas and holding particular values in the future eras will be). And the reverse direction also holds true.

In relation to the partiality-totality principle, if there is partiality (e.g, the partial view of each era with its own peculiar concern with particular ideas and holding particular values), there is totality (e.g, the totalistic view of all eras with their own peculiar concern with particular ideas and holding particular values, such that it is not certain what the whole will be like in the future even if one adds up the parts, so that the whole is not the sum of its parts). And the reverse direction also holds true.

In relation to the explicability-inexplicability principle, if there is explicability (e.g, the explanation, within uniformitarianism, of why the present is key to the past, on the basis of "an Earth created by the same natural phenomenon that are at work today"), there is inexplicability (e.g, the lack of sufficient explanation, within uniformitarianism, of the existence of abrupt changes in Earth history as shown by the occurences of catastrophes). And the reverse direction also holds true.

In relation to the preciseness-vagueness principle, if there is preciseness (e.g., the precise identification of "the present" as "key to the past" in uniformitarianism), there is vagueness (e.g., the vagueness in the "present" as "key to the past," since it is not clear to what extent this then means that the present and the past share "the same kind" of causes, "the same energy," "the same effects," and the same "geological circumstances"). (WK 2010f) And the reverse direction also holds true.

In relation to the simpleness-complicatedness principle, if there is simpleness (e.g., the relatively simple view of a short Earth history by creationist catastrophists), there is complicatedness (e.g., the relatively more complicated view of Earth history in both gradual and abrupt ways by many contemporary geologists). And the reverse direction also holds true.

In relation to the openness-hiddenness principle, if there is openness (e.g., the relatively open exploration of abrupt changes in Earth history by creationist catastrophists), there is hiddenness (e.g., the hidden bias in creationist catastrophism because of its religious ideology). And the reverse direction also holds true.

In relation to the denseness-emptiness principle, if there is denseness (e.g, the relatively denser concentration of scholars with mystical beliefs about geological objects in antiquity), there is emptiness (e.g., the relatively less dense, or more empty, concentrations of scholars with mystical beliefs about geological objects nowadays in the scientific community). And the reverse direction also holds true.

In relation to the slowness-quickness principle, if there is slowness (e.g., "gradual processes such as weathering, erosion and deposition" in Earth history), there is quickness (e.g., "[v]iolent events such as meteorite impacts and large volcanic explosions" in Earth history). And the reverse direction also holds true.

In relation to the expansion-contraction principle, if there is expansion (e.g., the relative expansion of the influence of uniformitarianism since modern times), there is contraction (e.g., the relative contraction of the influence of mytical beliefs about geological objects since modern times). And the reverse direction also holds true.

In relation to the theory-praxis principle, if there is theory (e.g, the religious construction about "the biblical flood"), there is praxis (e.g., the practical application of "the biblical flood" to creationist catastrophism in geology). And the reverse direction also holds true.

In relation to the convention-novelty principle, if there is convention (e.g, the conventional wisdom of understanding geological objects with mystical beliefs in antiquity), there is novelty (e.g, the alternative novel challenge to the conventional wisdom of understanding geological objects with mystical beliefs in antiquity, by the new way of studying geology by fossil records, as proposed by Steno in the 17th century). And the reverse direction also holds true.

In relation to the evolution-transformation principle, if there is evolution (e.g., the natural evolution of humans to understand geological objects with mystical beliefs in relatively primitive historical eras), there is transformation (e.g,. the technical transformation of human understanding of geological objects by the invention of a scientific system of ideas known as modern geology, starting in the 17th century onwards). And the reverse direction also holds true.

In relation to the symmetry-asymmetry principle, if there is symmetry (e.g., the existence of different geological beliefs in different historical eras of geology), there is asymmetry (e.g., the popularity of mystical beliefs about geological objects in antiquity—but the acceptance of an integrated approach towards uniformitarianism and catastrophism nowadays). And the reverse direction also holds true.

In relation to the regression-progression principle, if there is regression (e.g., the regression made in uniformitarianism, as shown in the problems pointed out by Stephen Jay Gould, as already summarized in Chapter One), there is progression (e.g., the progress made in uniformitarianism, to help us understand better the role of gradual changes in Earth history). And the reverse direction also holds true.

And in relation to the same-difference principle, if there is similarity in outcome (e.g., the contribution to the molding and control of beliefs and values, regardless of whether this is achieved by way of mystical ideas about geological objects or by way of scientific ideas about them), there is difference in outcome (e.g., the contribution to the molding and control of beliefs and values by way of mystical ideas about geological objects for a relatively more reverent relation with nature—but the contribution to the molding and control of beliefs and values by way of of scientific ideas about geolotical objects for a relatively more dominating relation with nature). And the reverse direction also holds true.

Geological Time, and the Use/Misuse of Metaphors

The discussion of creationist catastrophism (in the previous section) reveals the role of religion in geology, which is another intriguing way to evaluate the possibility and desirability of dynamics in geology from the perspective of culture.

The role of religion in geology is related in a different issue, this time, in the context of the study of geological time with the use and misuse of metaphors.

The Role of Metaphors

Stephen Jay Gould (1987) in *Time's Arrow, Time's Cycle* offered his own account of "the discovery of geological time" in terms of the metaphors of "time's arrow" (for uniqueness and unpredictable change) and "time's cycle" (for uniformity and predictability). (F. Sulloway 1987)

The term "geological time" in geology is impressive enough, since "geological time is so immense compared with the human experience of time that we can only hope to grasp it dimly through analogies." (F. Sulloway 1987)

Precisely because of this immense nature of geological time, it was, for instance, for a while in its history embedded with "biblical standards," where religion made its mark on geology. (F. Sulloway 1987)

This historical account of geological time by Gould "should be seen as part of the generation of historians who have been affected by T.S.

Kuhn's *Structure of Scientific Revolutions* (1962). Kuhn argued, in part, that science is a social activity and that theories are intellectual constructions imposed on data, not demanded by them. The views of Kuhn and other philosophers and sociologists of science have helped historians of science to recognize, as Gould emphatically does, that mental constructs (metaphors, analogies, personal philosophies, imaginative leaps)—not empirical discoveries—are what bring about scientific advance. 'Facts' are so embedded in theory that they simply do not have the kind of independent probative power they were once supposed to possess." (F. Sulloway 1987)

In the footsteps of this historicist (Kuhnian) tradition, "Gould pinpoints a powerful pair of metaphors—time's arrow and time's cycle—by which humankind has always tried to grasp the concept of time. Time's arrow captures the uniqueness and distinctive character of sequential events, whereas time's cycle provides these events with another kind of meaning by evoking lawfulness and predictability. Gould notes that this metaphorical pair is common not only in the thinking of ancient and preliterate peoples but also in the Judeo-Christian tradition, in which time's arrow nevertheless began to predominate." (F. Sulloway 1987)

Gould focused on three major thinkers in the history of geology who, in one way or another, were influenced by metaphors in their works on geological time, namely, (a) Thomas Burent, (b) James Hutton, and (c) Charles Lyell—to be summarized below, in that order (and also in *Table 3.5*).

Thomas Burnet as the Anti-Hero

For illustration, Gould focused first on the work of Thomas Burnet (1965), who becomes the anti-hero in the current context of deconstructing metaphorical construction of geological time.

For Thomas Burnet, as shown in his main work *Sacred Theory of the Earth* (published in 1684 in English), "Christ, at the top, has his left foot on the earth as it was in the beginning of the creation ('without form and void'). Earth history moves clockwise, recording the perfect (featureless) earth of Eden, the Flood (with Noah's ark floating just above the center), the present state of the earth, the coming conflagration that shall consume and purify the earth once more, and finally the earth transformed into a star after the righteous have ascended to heaven. Above Christ is the inscription from the Book of Revelation, 'I am the alpha and omega,' that is, the beginning and the end." (F. Sulloway 1987)

The example of Burnet's work is important for Gould's account, because "Burnet's theory illustrates the metaphors of time's arrow and time's cycle in unmistakable form. His is a one-cycle theory in which biblical narrative (time's arrow) runs its course within a wider conception of 'the great year' and 'great circle of time and fate' that bring about the return of Paradise. It is precisely this literal belief in Scripture that has made Burnet a pariah in the history of geology." (F. Sulloway 1987)

However, "Burnet, Gould demonstrates, was hardly the religious fanatic he is supposed to have been when he is placed within the context of contemporary scientific thought. Compared with the textbook legend, Burnet was, ironically, adamant about explaining the history of the earth as recorded in Scriptures entirely within the frame of natural science, devoid of all appeals to miracles or divine intervention. Whereas his contemporaries had to call upon God to create new and vast sources of water for the Flood, for example, Burnet tried to avoid such external interventions by positing an underground source of water released onto the earth's surface through a fault in the crust. Similarly, Burnet believed that Vesuvius and Etna would provide the sources of fire that would ultimately consume and purify the world prior to the second coming of Christ." (F. Sulloway 1987)

In fact, "in a revealing exchange, Burnet in 1681 argued with Isaac Newton over the length of the original 'days' of creation. Newton saw a way out of the difficulty of assuming God had made the world in a week. He believed that the 'days' of Genesis might have been much longer than present ones, and that God, when the job was finally done, intervened in order to speed up the earth's rotation. Burnet regarded such a theory as totally unacceptable precisely because it required divine intervention. Thus the 'bad guy' of geological textbook history was actually more devoted to rational, miracle free science than the greatest scientist of his age." (F. Sulloway 1987)

James Hutton as the Champion of Endless Cycles

Gould then examined the work of James Hutton as the next figure who combined science with metaphors in his study of geological time.

For Gould, "Hutton's genius was to introduce the concept of repair into geology and, with it, the notion of deep time. The textbooks, of course, see this as a triumph of science and empiricism over religion, but it was nothing of the kind. Ironically, Hutton's entire theory of the earth was an a priori conception inspired jointly by religious considerations and 'the most rigid and uncompromising version of time's cycle ever developed by a geologist.'" (F. Sulloway 1987)

The work of Hutton "grew out of a problem—what may be called 'the paradox of the soil.' A gentleman farmer, Hutton was well aware that good soil, the product of the 'denudation,' or eroding, of rock strata, eventually loses its richness to the plant life it sustains. Were there to be no geological source for continual new soil, Hutton believed, then the world would bear the intolerable stamp of an imperfectly designed abode for man's existence. Hutton's homocentric and teleological concept of the world therefore demanded that the soil, new soil, should never run out. This requirement in turn demanded the uplift of new strata to become the sources for soil replenishment." (F. Sulloway 1987)

With this motivation in mind, "Hutton, belatedly in his career, set out to find evidence for uplift (which he naturally did, since he was already looking for it). In fact, he found evidence for repeated uplifts of the earth's crust. This realization led him inexorably to the discovery of deep time. In Hutton's final theory the earth became an entirely self-regulating machine cycling its way, over and over again, through three geological stages: (1) denudation and decay; (2) the deposition of new marine strata; and (3) the melting, expansion, and uplift of the lowest strata as 'igneous' rock ready to be broken down again for future plants. Hutton's recognition that certain rocks had solidified from molten magma was a particularly powerful new insight." (F. Sulloway 1987)

In the process, the metaphorical zealousness of Hutton took a hold of him, such that he became "[s]o rigid [with the] vision of an endlessly cycling earth having 'no vestige of a beginning' and 'no prospect of an end' that he lost all interest in the historical nature of geological change. The divine benevolence entailed in these cycles was everything to Hutton, who, in a Newtonian rather than a historical image, compared them to the planets revolving ceaselessly about the sun. 'Through the thousand pages of Hutton's treatise [the 2 volumes *The Theory of the Earth* in 1795],' Gould poignantly remarks, 'we find not a single sentence that treats the different ages and properties of strata as interesting in themselves—as markers of distinction for particular times.'" (F. Sulloway 1987)

Charles Lyell as the Ideologue of Uniformity

Gould then moved on to the work of Charles Lyell as the next figure who combined science with metaphors in his study of geological time.

For Gould, "[i]t is important to bear in mind that Charles Lyell was trained as a lawyer. His rhetorical skills were considerable indeed, and, as Gould makes clear, they are crucial to understanding his impact upon the history of geology. When pleading for his favorite client, which became known as the 'uniformitarian' theory of geology, he portrayed the previous history of his discipline as a gradual overcoming of primitive superstitions, wild speculations, and biblical allegiances." (F. Sulloway 1987)

But this is misleading, because, for Gould, Lyell also "created his own legend, much as Sigmund Freud did, as an archempiricist free of all bias and preconception. But Lyell was not selling just evidence and fieldwork over previous dogma and speculative theory. Rather he foisted upon his contemporaries a 'fascinating and particular theory rooted in...time's cycle.'" (F. Sulloway 1987)

To see the hidden bias in Lyell's work with his obsession with the metaphoric views of time's arrow and time's cycle, for Gould, one needs to examine closely Lyell's theory, "which set itself up against the prevailing geological 'catastrophism'" and in the process "confused a number of distinct meanings under the banner of 'uniformitarianism.' First, Lyell argued for the uniformity of nature's laws (that is, the notion that laws do not change with time or place). Second, he argued for the uniformity of process, which simply means always explaining past changes by currently known causes as long as these will suffice." (F. Sulloway 1987)

In fact, Lyell's obession with his metaphorical beliefs about time's arrow and time's cycles was so intense that he even went so far as to making the uncritical claims that the "rates of geological change are always uniform (that is, gradual) and that the general state of the world also remains uniform (that is, there is no progression or directionality in the long run). The last of these claims was the most peculiar of all within Lyell's vision of earth history. It led him to deny all evidence of progression in the fossil record and hence to reject not only Lamarck's theory of evolution but also contemporary...notions, in which 'higher' organisms were thought to replace 'lower' ones after mass extinctions." (F. Sulloway 1987)

In other words, "[s]o wedded was Lyell to his conception of nondirectional change occurring within great geological cycles that he even believed the more temperate climate of the Carboniferous period might one day return and, with it, the great dinosaurs of that age. 'The huge iguanodon,' Lyell argued, 'might reappear in the woods, and the ichthyosaur in the sea, while the pterodactyle might flit again through umbrageous groves of tree-ferns.'" (F. Sulloway 1987)

The Problems with Gould's Work

For Gould, his work *Time's Arrow, Time's Cycle* is to show us that "arrows and cycles are 'eternal metaphors' in the understanding of time. In a thoughtful complement to his discussion of the history of geology, he shows how these two metaphors have figured in the art and sculpture associated with major biblical themes. Both metaphors, he concludes, are needed 'for any comprehensive view of history.'" (F. Sulloway 1987)

But Gould's work is not without its critics. For illustration, consider a few criticisms as shown below (and summarized in *Table 3.5*).

Firstly, one major criticism of Gould's work is that the obsession with metphors that he attributed to those opponents (like Burnet, Hutton, and Lyell) can be turned against his work too.

For instance, the critics are quick to point out that "Gould pushes his temporal metaphors too far. For example, he occasionally tends, for the sake of his central argument, to play down the religious nature of the controversies surrounding the discovery of deep time." (F. Sulloway 1987)

Secondly, another major criticism of Gould's work is that the same historicist critique can also be directed against Gould himself.

In other words, when Gould told us, according to Sulloway (1987), that "scientists are actually influenced by history all the time,"—this same critique can also be directed against Gould himself, because in this postmodernist age of ours, there is the obsession with revisionist (deconstructivist) critcisms, to the extent that there is also a need to go beyond the "postmodernist" discourse. This is something that I already originally proposed as "after-postmodernity" in *The Future of Human Civilization* (2000), in that the postmodernist project is to be transcended too.

Thirdly, still another major criticism of Gould's work is that the truth may lie somewhere between the opposing views.

For instance, for many geologists nowadays, "Earth's history is a slow, gradual process punctuated by occasional natural catastrophic events that have affected Earth and its inhabitants"—such that the works of Burnet, Hutton, and Lyell can have substantial scientific merits, in spite of the criticisms by Gould about their metaphorical constructions. (WK 2010f; CE 2007)

A Sober Reflection

However, these criticisms of Gould's work do not mean to endorse the works of his three opponents.

On the contrary, the criticisms are introduced here to show us the two opposing sides of the debate, such that the possibility and desirability of dynamics (from the perspective of culture with geological time as a case study here) are not to the extent that its respective defenders would like us to believe.

In addition, this analysis of geological time and the use/misuse of metaphors can lead us to a sober reflection about the subject, in regard to the ontological principles in existential dialectics, and good examples include the absoluteness-relativeness principle, the predictability-unpredictability principle, the partiality-totality principle, the explicability-inexplicability principle, the preciseness-vagueness principle, the simpleness-complicatedness principle, the openness-hiddenness principle, the denseness-emptiness principle, the change-constancy principle, the slowness-quickness principle, the expansion-contraction principle, the theory-praxis principle, the convention-novelty principle, the evolution-transformation principle, the symmetry-asymmetry principle, the softness-hardness principle, the regression-progression principle, and the same-difference principle.

For instance, in relation to the absoluteness-relativeness principle, if there is absoluteness (e.g., the absolute view of Burnet in regard to his "one-cycle theory in which biblical narrative [time's arrow] runs its course within a wider conception of 'the great year' and 'great circle of time and fate' that bring about the return of Paradise"), there is relativeness (e.g., what is true for Burnet is questionable for Gould in his critique of Burnet's obsession with the metaphorical construction of geological time). And the reverse direction also holds true.

In relation to the predictability-unpredictability principle, if there is predictability (e.g,. the predictable tendency of geological time in terms of "an entirely self-regulating machine cycling its way, over and over again," in different stages of "soil replenishment," according to Hutton), there is unpredictability (e.g,. the more difficult task to predict, if looked only within Hutton's "entirely self-regulating machine," the "particular" and "distinct" features and "properties of strata" as "interesting in themselves—as markers of distinction for particular times," because Hutton was so obsessed with the "endless cycles" that "he lost all interest in the historical nature of geological change"). And the reverse direction also holds true.

In relation to the partiality-totality principle, if there is partiality (e.g., the parital view of each perspective from the work of Burnet, Hutton, or Lyell), there is totality (e.g., the totalistic view of all perspectives from the works of Burnet, Hutton, Lyell, and all other geologists, such that it is not certain what the future of geological time will be like, so that the whole is not the sum of its parts). And the reverse direction also holds true.

In relation to the explicability-inexplicability principle, if there is explicability (e.g,. the explanation of "soil replenishment" by Hutton in terms of three stages, namely, "denudation and decay," "the deposition of new marine strata," and "the melting, expansion, and uplift of the lowest strata as 'igneous' rock ready to be broken down again for future plants"), there is inexplicability (e.g,. the lack of sufficient explanation, if looked only within Hutton's "entirely self-regulating machine," of why there exist abrupt changes or catastrophes in Earth history like volcanic eruptions which can affect the soils on Earth). And the reverse direction also holds true.

In relation to the preciseness-vagueness principle, if there is preciseness (e.g., the precise identification of three stages of soil replenishment by Hutton), there is vagueness (e.g., the vagueness in the identification of three stages by Hutton, because he became "[s]o rigid [with the] vision of an endlessly cycling earth having 'no vestige of a beginning' and 'no prospect of an end' that he lost all interest in the historical nature of geological change" and did not treat "the different ages and properties of strata as interesting in themselves—as markers of distinction for particular times"). And the reverse direction also holds true.

In relation to the simpleness-complicatedness principle, if there is simpleness (e.g., the relatively simple analysis of "geological time" by Burnet, Hutton, and Lyell, on the basis of "time's arrow" and "time's cycle"), there is complicatedness (e.g., the relatively more complicated analysis of "geological time" by also considering the different problems in the works of the three thinkers in regard to their metaphorical constructions, as pointed out by Gould). And the reverse direction also holds true.

In relation to the openness-hiddenness principle, if there is openness (e.g., the open exploration of a short Earth history on the basis of The Book of Revelation by Burnet), there is hiddenness (e.g., the hidden bias in the work by Burnet, because of biblical contamination). And the reverse direction also holds true.

In relation to the denseness-emptiness principle, if there is denseness (e.g., the relatively denser concentration of geologists with "biblical standards" in the older days), there is emptiness (e.g., the relatively less dense, or more empty, concentration of geologists with "biblical standards" nowadays). And the reverse direction also holds true.

In relation to the change-constancy principle, if there is change (e.g., the ever new ways of understanding geological time, as shown in the different views by Burnet, Hutton, Lyell, and Gould in this history of geology), there is constancy (e.g,. the ever constant existence of different views about geological time in the history of geology). And the reverse direction also holds true.

In relation to the slowness-quickness principle, if there is slowness (e.g,. the relatively slower changes in Earth history, like the different stages of soil replenishment as pointed out by Hutton), there is quickness (e.g., the relativey quicker changes in Earth history, like abrupt volcanic eruptions and dramatic earthquakes). And the reverse direction also holds true.

In relation to the expansion-contraction principle, if there is expansion (e.g, the relative expansion of the influence of uniformitarianism in the 19th century and beyond), there is contraction (e.g, the relative decline of the view of creationist catastrophism in the 19th century and beyond). And the reverse direction also holds true.

In relation to the theory-praxis principle, if there is theory (e.g., the theoretical construction of creationism), there is praxis (e.g., the practical application of creationism to geology by Burnet). And the reverse direction also holds true.

In relation to the convention-novelty principle, if there is convention (e.g., the conventional wisdom in the older days about the use of creationism in the study of geological time, like the one by Burnet), there is novelty (e.g., the alternative novel challenge to the conventional wisdom about the use of creationism in the study of geological time, as shown in the new view by Lyell in regard to his uniformitarianism). And the reverse direction also holds true.

In relation to the evolution-transformation principle, if there is evolution (e.g., the natural evolution of humans to understand soil replenishment, even if in a primitive way, in the state of nature), there is transformation (e.g., the technical transformation of human understanding of soil replenishment by the invention of the metaphorical construction of geological time by Hutton in terms of "endless cycles" in geology). And the reverse direction also holds true.

In relation to the symmetry-asymmetry principle, if there is symmetry (e.g., the existence of different views about geological time in geology), there is asymmetry (e.g., the focus on "biblical standards" in the view by Burnet—but the focus on "uniformity" in the view by Lyell). And the reverse direction also holds true.

In relation to the softness-hardness principle, if there is softness (e.g,. the inclusion of those views of others like Burnet, Hutton, and Lyell for analysis by Gould, insofar as they support what Gould wanted to say), there is hardness (e.g., the exclusion of those views of others like Burnet, Hutton, and Lyell for analysis by Gould, insofar as they did not support what Gould wanted to say, because "he occasionally tends, for the sake of his central argument, to play down the religious nature of the controversies surrounding the discovery of deep time"). And the reverse direction also holds true.

In relation to the regression-progression principle, if there is regression (e.g., the regression made in Gould's analysis, as shown by the criticisms pointed out by the criticis), there is progression (e.g., the progression made in Gould's analysis, to help us better understand the nature of metaphorical construction in geology). And the reverse direction also holds true.

And in relation to the same-difference principle, if there is similarity in outcome (e.g., the contribution to the molding and control of beliefs and values, regardless of whether this is achieved by way of Burnet's work on geological time or by way of Gould's work on geological time), there is difference in outcome (e.g., the contribution to the molding and control of beliefs and values by way of Burnet's work on geological time for a stewardship of nature in relation to a supernatural being—but the contribution to the molding and control of beliefs and values by way of Gould's work on geological time for a relatively less certain view of nature). And the reverse direction also holds true.

The Weakness of Dynamics

This comprehensive analysis of the possibility and desirability of dynamics, from the perspectives of the mind, nature, society, and culture, is important to show that dynamics is both possible and desirable, but not to the extent that the spokespersons from each side would like us to believe. And this can apply too to statics (as already addressed in ChapterTwo).

Dynamics thus has its weakness, just as it has its strength too, as another side of the same mirror—just as, if statics has its own changeability, it also has its own durability.

These dual analyses of dynamics and statics are important, because, as already pointed out in *Sec. 1.3*, the structure (or statics) and process (or dynamics) of the planet's physical matter are intricately related.

With this in mind, now that this intellectual journey has come to an end, what then can be concluded?

This is the topic to which we now turn, in the next chapter for the future of geology.

Table 3.1. Dynamics and Its Restriction

- **Dynamics and the Mind**
  —Ex: geology, and the controversy about feminine intuition
  —Ex: geology, and the rediscovery of the mind

- **Dynamics and Nature**
  —Ex: volcanic ejection, and its impact
  —Ex: earthquake, and its power

- **Dynamics and Society**
  —Ex: technology, and the debate on relative vs. absolute dating
  —Ex: engineering geology, and the protection of life and property

- **Dynamics and Culture**
  —Ex: historical geology, and the nature of cultural contingency
  —Ex: geological time, and the use/misuse of metaphors

Notes: The examples in the categories are solely illustrative (not exhaustive), and the comparison is relative (not absolute), nor are they necessarily mutually exclusive. And some can be easily re-classified elsewhere. As generalities, they allow exceptions.
Source: A summary of Ch.3 of *FPHGEOL*

Table 3.2. The Controversy about Earthquakes

- **Main Features**
  - —Ex: earthquake fault types
  - —Ex: earthquakes away from Plate Boundaries
  - —Ex: shallow-focus and deep-focus earthquakes
  - —Ex: earthquakes and volcanic activity
  - —Ex: rupture dynamics
  - —Ex: frequency, magnitude, and concnetration of occurrence
  - —Ex: different causes

- **Problems with Studying Earthquakes**
  - —Firstly, one major criticism of the study of earthquakes is that it still is not fully understood, because it is very difficult to study them in a laboratory.
  - —Secondly, another major criticism of the study of earthquakes is that not all earthquakes are easily detected.
  - —And thirdly, still another major criticism of the study of earthquakes is that it does not give sufficient attention to the role of "the mind" in measuring earthquakes, in the context of the rediscovery of the mind as already discussed in *Sec. 3.2.2*.

Notes: The examples in the categories are solely illustrative (not exhaustive), and the comparison is relative (not absolute), nor are they necessarily mutually exclusive. And some can be easily re-classified elsewhere. As generalities, they allow exceptions.
Source: From *Sec. 3.3.2* of *FPHGEOL*. See book for citations, like WK (2010hh), and others.

Table 3.3. The Debate on Relative vs. Absolute Dating (Part I)

- **Relative Dating**
  —Ex: the principle of intrusive relationships
  —Ex: the principle of cross-cutting relationships
  —Ex: the principle of inclusions and components
  —Ex: the principle of uniformitarianism
  —Ex: the principle of original horizontality
  —Ex: the principle of superposition
  —Ex: the principle of faunal succession

- **Absolute Dating**
  —The development of "absolute dating" was regarded as "a large advance in geology in the advent of the 20th century," because of "the ability to give precise absolute dates to geologic events through radioactive isotopes and other methods. The advent of radiometric dating changed the understanding of geologic time. Before, geologists could only use fossils to date sections of rock relative to one another. With isotopic dates, absolute dating became possible, and these absolute dates could be applied to fossil sequences in which there was datable material, converting the old relative ages into new absolute ages." (WK 2010)
  —For instance, "isotope ratios are measured in minerals that give the amount of time that has passed since a rock passed through its particular closure temperature, the point at which different radiometric isotopes stop diffusing into and out of the crystal lattice." (WK 2010; H. Rollinson 1996; G. Faure 1998)

(continued on next page)

Table 3.3. The Debate on Relative vs. Absolute Dating (Part II)

- **Problems with the Two Methods of Dating**
  —Firstly, one major criticism is that, in the case of relative dating, the principles are often not quite true.
  —Secondly, another major criticism is that, in the case of relative dating, reality can be more complicated than what the principles suggest.
  —Thirdly, still another major criticism is that, in the case of relative dating, there are disagreements about the assumptions used in the principles.
  —Fourthly, one additional major criticism here is that, in the case of absolute dating, there are limitations in regard to its accuracy.
  —And fifthly, the last major criticism here is that, in the case of absolute dating, "absolute dates do not necessarily tell us when a particular cultural event happened," even though "they are invaluable in constructing a more specific sequence of events." (WK 2010jj)

Notes: The examples in the categories are solely illustrative (not exhaustive), and the comparison is relative (not absolute), nor are they necessarily mutually exclusive. And some can be easily re-classified elsewhere. As generalities, they allow exceptions.
Source: From *Sec. 3.4.1* of *FPHGEOL*. See book for citations, like WK (2010), and others.

Table 3.4. The Cultural Contingency of Geological Thoughts

- **Fossil Records**
  —The first example seems innocent enough, since its appearance in the 17th century challenged the ancient mystical stories about geological objects at the time, which then led to the beginning of geology as a modern science. (WK 2010nn)

- **Uniformitarianism**
  —The second example seems quite bold, since its appearance in the 18th century challenged the religious account of earth's history at the time.

- **Catastrophism**
  —The third example concerns an opposing view against uniformitarianism, which resulted in a heated debate during the 19th century, in that "[t]he prevailing concept of the 18th century was that of a very short Earth history dominated by catastrophic events. This view was strongly supported by religious thinkers based on a largely literal interpretation of biblical passages." (WK 2010ll) In other words, "[t]he concept of uniformitarianism met with considerable resistance and the catastrophism vs. gradualism debate of the 19th century resulted." (WK 2010ll)

- **A Compromise**
  —And the fourth example is more contemporary in origin (especially since the later 20th century), since it sides neither with uniformitarianism nor with catastrophism and opts for a middle road, in that "a variety of discoveries in the 20th century provided ample evidence that Earth history is a product of both gradual incremental processes and sudden cataclysmic events." (WK 2010ll)

Notes: The examples in the categories are solely illustrative (not exhaustive), and the comparison is relative (not absolute), nor are they necessarily mutually exclusive. And some can be easily re-classified elsewhere. As generalities, they allow exceptions.
Source: From *Sec. 3.5.1* of *FPHGEOL*. See book for citations, like WK (2010ll, 2010mm & 2010nn), and others.

Table 3.5. The Quarrel on Geological Time (Part I)

• **Thomas Burnet—The Sacred View of the Earth**

—"Burnet's theory illustrates the metaphors of time's arrow and time's cycle in unmistakable form. His is a one-cycle theory in which biblical narrative (time's arrow) runs its course within a wider conception of 'the great year' and 'great circle of time and fate' that bring about the return of Paradise. It is precisely this literal belief in Scripture that has made Burnet a pariah in the history of geology." (F. Sulloway 1987)

• **James Hutton—The View of Endless Cycles**

—"In Hutton's final theory the earth became an entirely self-regulating machine cycling its way, over and over again, through three geological stages: (1) denudation and decay; (2) the deposition of new marine strata; and (3) the melting, expansion, and uplift of the lowest strata as 'igneous' rock ready to be broken down again for future plants....[He became] so rigid [with the] vision of an endlessly cycling earth having 'no vestige of a beginning' and 'no prospect of an end' that he lost all interest in the historical nature of geological change." (F. Sulloway 1987)

• **Charles Lyell—The View of Uniformity**

—Lyell's obession with his metaphorical beliefs about time's arrow and time's cycles was so intense that he claimed that the "rates of geological change are always uniform (that is, gradual) and that the general state of the world also remains uniform (that is, there is no progression or directionality in the long run). The last of these claims was the most peculiar of all within Lyell's vision of earth history. It led him to deny all evidence of progression in the fossil record and hence to reject not only Lamarck's theory of evolution but also contemporary...notions, in which 'higher' organisms were thought to replace 'lower' ones after mass extinctions....So wedded was Lyell to his conception of nondirectional change occurring within great geological cycles that he even believed the more temperate climate of the Carboniferous period might one day return and, with it, the great dinosaurs of that age." (F. Sulloway 1987)

(continued on next page)

Table 3.5. The Quarrel on Geological Time (Part I)

- **Stephen Gould—The View of Metahpors**
  —His work *Time's Arrow, Time's Cycle* is to show us that "arrows and cycles are 'eternal metaphors' in the understanding of time. In a thoughtful complement to his discussion of the history of geology, he shows how these two metaphors have figured in the art and sculpture associated with major biblical themes. Both metaphors, he concludes, are needed 'for any comprehensive view of history.'" (F. Sulloway 1987)
  —Gould's work is not without its critics. Firstly, one major criticism of Gould's work is that the obsession with metphors that he attributed to those opponents (like Burnet, Hutton, and Lyell) can be turned against his work too. Secondly, another major criticism of Gould's work is that the same historicist critique can also be directed against Gould himself. And thirdly, still another major criticism of Gould's work is that the truth may lie somewhere between the opposing views.

Notes: The examples in the categories are solely illustrative (not exhaustive), and the comparison is relative (not absolute), nor are they necessarily mutually exclusive. And some can be easily re-classified elsewhere. As generalities, they allow exceptions.
Source: From *Sec. 3.5.2* of *FPHGEOL*. See book for citations, like F. Sulloway (1987), and others.

• PART FOUR •

Conclusion

CHAPTER 4
CONCLUSION—THE FUTURE OF GEOLOGY

> The study of life at all levels, from social to molecular behavior, has in modern times relied on reductionism as the chief explanatory concept. This approach to knowledge tries to comprehend one level of scientific phenomena in terms of concepts at a lower and presumably more fundamental level....And in geology, the formations and properties of minerals are described using the features of the constituent crystals. The essence of these cases is seeking explanation in underlying structures and activities.
> —Harold Morowitz (1981)

Beyond Uniformity, Catastrophe, and Reivison

The analyses of statics (in Chapter Two) and dynamics (in Chapter Three) play the vital role for the understanding of the extent of their possibility and desirability in geology, in a small sense, and in life, in a broad sense.

These analyses are timely enough, in light of the two views in contention (as already described in *Sec. 1.1*). In other words, the revisionist account by the critics like Stephen Jay Gould can be contrasted by an opposing view in the older days, when Charles Lyell, for example, "portrayed the previous history of his discipline as a gradual overcoming of primitive superstitions, wild speculations, and biblical allegiances. In doing so he created his own..., as an archempiricist free of all bias and preconception." (F. Sulloway 1987)

Contrary to these two opposing views (and others as already discussed in this book), the history of geology is neither so metaphorical nor so empirical to the extent that the respective defenders would like us to believe.

But this rejection of the opposing views in contention does not logically suggest that geology is therefore not worth studying, or that it is useless. Neither to say, neither of these two extreme views is reasonable either.

As already indicated in *Sec. 1.1*, this book offers an alternative (better) way to understand the future of geology, especially in the context of statics and dynamics—while learning from different approaches in the literature but without favoring any one of them (nor integrating them, since they are not necessarily compatible with each other).

Thus, this book offers a new theory (that is, *the resettlement theory of geology*) to go beyond the existing approaches in the literature on geology in a new way not conceived before—and has four distinctive features to remember.

Firstly, my theory makes good use of all theoretical approaches in the literature on geology, be they about the catastrophe argument, the uniformity argument, and the revision argument—especially from the most comprehensive combined perspectives of the mind, nature, society, and culture (as already discussed in Chapter Two and Chapter Three).

Secondly, just like many other theories of mine in my previous books, my theory here does not heavily favor any specific theory over others in the literature, nor trying to integrate them (as they are not necessarily compatible with each other).

Thirdly, my theory treats the issue of geology in the distinctive dialectic context of statics and dynamics in geology—especially in the larger dialectic context of no statics without dynamics, and vice versa, together with the subsequent transcendence of both.

And fourthly, it contains sixteen major theses, namely, (a) the first thesis: the absoluteness-relativeness principle, (b) the second thesis: the predictability-unpredictability principle, (c) the third thesis: the explicability-inexplicability principle, (d) the fourth thesis: the preciseness-vagueness principle, (e) the fifth thesis: the simpleness-complicatedness principle, (f) the sixth thesis: the openness-hiddenness principle, (g) the seventh thesis: the denseness-emptiness principle, (h) the eighth thesis: the slowness-quickness principle, (i) the nineth thesis: the expansion-contraction principle, (j) the tenth thesis: the theory-praxis principle, (k) the eleventh thesis: the convention-novelty principle, (l) the twelveth thesis: the evolution-transformation principle, (m) the thirteenth thesis: the symmetry-asymmetry principle, (n) the fourteenth thesis: the regression-progression principle, (o) the fifteenth thesis: the sameness-difference principle, and (p) the sixteenth thesis: the post-human resettlement, in relation to the three great future transformations of geology—to be elaborated in the rest of the book, with a summary in the concluding chapter.

Of course, as this is something that I often emphasized in my previous books, other principles (besides the 15 as cited above) are also relevant, but these 15 are the most relevant in the current case study (in terms of the number of citations of each principle in the book).

Even then, in some cases, the difference between any two given principles, for instance, in terms of the number of citations in a book, is rather small, so the reason in those cases is more aesthetic (than otherwise), because it looks nicer to list only 15 theses for 15 principles (than 21 theses for all of the 21 principles) in the Table of Contents.

This is true, even if different studies of the same kind can yield different views about the degree of relevance for each principle, depending on the specific nature of a research in question, needless to say. So, if a different author analyzes the same subject matter in a different way, the relevance of the principles will be different.

With this clarification in mind—the seminal project here, if successful, will fundamentally change the way that we think about geology, from the combined perspectives of the mind, nature, society, and culture, with enormous implications for the human future and what I originally called its "post-human" fate.

In the Category of Method

Firstly, in regard to the formalness-informalness principle (on the formal requirements of logical systems), if there is formalness (e.g., the formal logical requirement of completeness and soundness in a system of ideas, in *Sec. 2.3.2*; the formal logical requirement of a system of ideas to be sound, in *Sec. 2.4.2*; the formal logical requirement of a system of ideas to be complete and sound, in *Sec. 2.5.1*; the formal requirement of completeness in a system of ideas, in *Sec. 2.5.2*; the formal logical requirement of completeness and soundness in a system of ideas, in *Sec. 3.2.1*; and the formal logical requirment of completedness in a system of ideas, in *Sec. 3.2.2*), there is informalenss too (e.g., the nonformal existence of incompleteness in the system of rock classification, as there remain "many obscure points about their history" to be understood, in *Sec.*

2.3.2; the nonformal existence of some sort of inaccuracy in hydrogeological works by numerical methods, because of their assumption "under more general conditions to an approximation," especially in computer simulation, in *Sec. 2.4.2*; the nonformal existence of some degree of incompleteness and unsoundness in the methods used in geology, like the inherent assumptions in regard to the unreality of control in lab methods and the contamination and unfairness of randomization in field methods, in *Sec. 2.5.1*; the nonformal existence of incompleteness in the creationist view, since Duane Gish himself admitted that "[w]e do not know how the Creator created....This is why we refer to creation as special creation. We cannot discover by scientific investigation anything about the creative processes used by the Creator," in *Sec. 2.5.2*; the nonformal existence of incompleteness in the claim by Andrews about intuition, since intuition cannot be proved, in *Sec. 3.2.1*; and the nonformal existence of the contamination, due to the role of the observer, in measuring a natural event in quantum mechanics, with negative implications for the unrealistic search for an objective Archmedian point, in *Sec. 3.2.2*). And the reverse direction also holds true.

And secondly, in regard to the partiality-totality principle (on the relationships between whole and parts), if there is partial analysis (e.g., the partial view of the simple parts of the HCS by the students, such that they did not recognize the HCS, in *Sec. 2.2.1*; the partial view of a metamorphic rock in an old cemetery in America, in *Sec. 2.2.2*; the partial view of each of the possible factors in hydrogeology, be it chemical, physical, biological, legal, political, etc., in *Sec. 2.4.2*; the partial view of a geologist with a masculine focusedness to decode the rocks by reading each separately first, in *Sec. 3.2.1*; the partial view of those in the hard sciences in the older days, when the role of the observer was not taken into account when measuring natural events, in *Sec. 3.2.2*; the partial view of those engineering geologists who do not consider societal, cultural, and mental factors when studying geohazards, in *Sec. 3.4.2*; the partial view of each era with its own peculiar concern with particular ideas and holding particular values, in *Sec. 3.5.1*; and the parital view of each perspective from the work of Burnet, Hutton, or Lyell, in *Sec. 3.5.2*), there is also holistic analysis (e.g., the holistic view of the HCS as a whole by the students later, thanks to the use of the Prägnanz law in gestalt psychology, such that the whole is not the sum of its parts, in *Sec. 2.2.1*; the holistic view of a metamorphic rock coupled with some creepy images, in an old American cemetery, such that the whole is not the sum of its parts, in *Sec. 2.2.2*; the holistic view of all the possible factors in hydrogeology as "the different facets of a multi-component system," such that the interactions of them all are highly complex, with uncertain outcomes, so that the whole is not the sum of its parts, in *Sec. 2.4.2*; the holistic view of a geologist with a feminine diffusiveness to decode the rocks by "first build[ing] an overview rather than immediately try[ing] to decode each separate lamination," in *Sec. 3.2.1*; the totalistic view of someone like Morowitz who wants to include all factors into the equation, like the mind-event integration, in *Sec. 3.2.2*; the holistic view of geohazards by combining not only geology and other related disciplines, but also societal, cultural, and mental factors, such that the outcomes are more uncertain than otherwise, in that the whole is not the sum of its parts, in *Sec. 3.4.2*; the totalistic view of all eras with their own peculiar concern with particular ideas and holding particular values, such that it is not certain what the whole will be like in the future even if one adds up the parts, so that the whole is not the sum of its parts, in *Sec. 3.5.1*; and the totalistic view of all perspectives from the works of Burnet, Hutton, Lyell, and all other geologists, such that it is not certain what the future of geological time will be like, so that the whole is not the sum of its parts, in *Sec. 3.5.2*). And the reverse direction also holds true.

In the Category of Structure

Firstly, in regard to the finiteness-transfiniteness principle (on the nature of numbers), if there is finiteness (e.g., the finite number of the types of rocks in Earth, that is, the three types, in *Sec. 2.3.2*; the finite number of the types of natural resources in Earth, that is, the two types known as biotic and abiotic resources, in *Sec. 2.4.1*; and the finite number of ways that "contaminants move with groundwater," like the three ways via "hydrodynamic dispersion," "molecular diffusion," and "retardation by absorption," in *Sec. 2.4.2*), there is also transfiniteness (e.g., the transfinite number of all the indiviudal rocks in Earth, in *Sec. 2.3.2*; the transfinite number of all the indiviudal units of biotic and abiotic resources in Earth, be they measured in terms of gallons, kilograms, etc., in *Sec. 2.4.1*; and the transfinite number of contaminants like "molecules" and "small particles in gases and liquids" that move within groundwater, in *Sec. 2.4.2*). And the reverse direction also holds true.

In the Category of Process

Firstly, in regard to the change-constancy principle (on the alteration of things), if there is change (e.g., the ever new ways of building cemeteries, like the use of slates, granites, etc., in *Sec. 2.2.2*; the ever new ways to do geology in its history, be they about field methods, lab methods, analytical methods, etc., in *Sec. 2.5.1*; the ever new ways of classification of ancient organisms in the history of paleontology, like "phenetics," "gradistics," "cladistics," "Linnean taxonomy," "creationism," "molecular systematics," etc., in *Sec. 2.5.2*; the dynamic nature of volcanic ejection, in relation to its different phases of formation, composition, spread, and effects, in *Sec. 3.3.1*; the ever new ways of measuring the age of earth materials in geology, like relative dating and relative dating, in *Sec. 3.4.1*; and the ever new ways of understanding geological time, as shown in the different views by Burnet, Hutton, Lyell, and Gould in this history of geology, in *Sec. 3.5.2*), there is constancy too (e.g., the constancy of an uneasy feeling among visitors to cemeteries, regardless of which kiind of rocks will be used to build them, in *Sec. 2.2.2*; the ever constant existence of some methods governing the field of geology at a given point in time, in *Sec. 2.5.1*; the ever constant existence of some classifications of ancient organisms in the history of paleontology, in *Sec. 2.5.2*; the ever constancy of the rock cycle, with volcanic ejection as one of its expressions, in *Sec. 3.3.1*; the ever constant existence of limitations in all methods, regardless of whether they are relative or absolute, in *Sec. 3.4.1*; and the ever constant existence of different views about geological time in the history of geology, in *Sec. 3.5.2*). And the reverse direction also holds true.

And secondly, in relation to the order-chaos principle (on the pattern of things), if there is order (e.g., the relativley orderly appearance of cemeteries with the use of slates, when they are newly built, in *Sec. 2.2.2*; the relatively orderly arrangement of organisms in a classificatory scheme like the Linnean taxonomy, in *Sec. 2.5.2*; the relatively orderly, neat world in Newtonian mechanics, in *Sec. 3.2.2*; and the relatively orderly nature of the movement of the ground, when there is no rupture of geological faults, in *Sec. 3.3.2*), there is chaos too (e.g., the relatively chaotic appearance of cemeteries with the use of slates, after some centuries when the slates are "cracked, faded, pitted and streaked," in *Sec. 2.2.2*; the relatively chaotic view of organisms before the Linnean taxonomy, when the field of "botany" suffered "from the chaos of new forms produced by horticulturalists," in *Sec. 2.5.2*; the relatively chaotic, uncertain world in quantum mechanics, in *Sec. 3.2.2*; and the relatively chaotic nature of the movement of the ground, in case of an earthquake, when the propagation is associated with "high sliding velocities," in conjunction with "the effects of strong ground motion…close to a nucleation zone," in *Sec. 3.3.2*). And the reverse direction also holds true.

In the Category of Agency

Firstly, in regard to the softness-hardness principle (on the force of change), on the one hand, there is softness (e.g., slate is "relatively soft and easy to carve," in *Sec. 2.2.2*; the sliding of two plates along each other without collision in "transform boundaries, in *Sec. 2.3.1*; "the gradual accumulation of sediment," such that different layers of materials lie on top of each other peacefully, as "more and more material is deposited on top," in *Sec. 2.3.2*; the call for harmonious co-existence with nature, in *Sec. 2.4.1*; diffusion and dispersion to help the spread of contaminants within goundwater, in *Sec. 2.4.2*; the inclusion of the role of the mind in studying nature by anti-reductionist scientists, in *Sec. 3.2.2*; the relatively softer nature of "ash that forms from burning wood or other combustible materials," as contrasted with ash from volcanic ejection, in *Sec. 3.3.1*; the soft earthquakes with a 1.0 magnitude or less, being relatively harmess, in *Sec. 3.3.2*; the co-existence of different "layers of sediment" on top of each other as "horizontal beds," in *Sec. 3.4.1*; and the inclusion of those views of others like Burnet, Hutton, and Lyell for analysis by Gould, insofar as they support what Gould wanted to say, in *Sec. 3.5.2*).

On the other hand, there is also its hardness (e.g., "granite is so much harder than slate," in *Sec. 2.2.2*; the collision between two continental plates in "convergent boundaries," in *Sec. 2.3.1*; the use of "sufficiently high" pressure and heat to burn and change "the composition and fabric of the rock," in *Sec. 2.3.2*; the call for exploitative domination over nature, in *Sec. 2.4.1*; the retardation by absorption to oppose, or at least to slown down, the spread of contaminants within goundwater, in *Sec. 2.4.2*; the exclusion of the role of the mind in studying nature by reductionist scientists, in *Sec. 3.2.2*; the relatively harder…nature of ash from volcanic ejection, which "is hard and abrasive" and "does not dissolve in water," in *Sec. 3.3.1*; the hard earthquakes with 8.0 magnitude or greater, causing relatively serious damages, in *Sec. 3.3.2*; "an igneous intrusion cuts across a formation of sedimentary rock," in *Sec. 3.4.1*; and the exclusion of those views of others like Burnet, Hutton, and Lyell for analysis by Gould, insofar as they did not support what Gould wanted to say, because "he occasionally tends, for the sake of his central argument, to play down the religious nature of the controversies surrounding the discovery of deep time," in *Sec. 3.5.2*).

And the reverse direction also holds true.

1st Thesis: The Absoluteness-Relativeness Principle

The first thesis refers to the absoluteness-relativeness principle (on the multiplicity of things) in the category of "method" in existential dialectics, which was first proposed in *FPHK*, in that there is the multiplicity of things in reality, be they about entities, qualities (or properties), and relationships. If there is something absolute, there is likewise something relative. And there is no absoluteness without relativeness—and vice versa.

Both absoluteness and relativeness here are also relevant to different modalities often cited in the literature on ontology, such as possibility (e.g., something "can" happen) and its opposite (e.g., impossibility), probability (e.g., something "will" happen) and its opposite (e.g., improbability), and necessity (e.g., something "should" happen) and its opposite (e.g., contingency).

For instance, on the one hand, there are absolute viewpoints (e.g., the absolute viewpoint by Favera and Medeiros that "the former researchers—mainly the late Rodi Ávila Medeiros, who was a keen sedimentologist—failed in not describing HCS, because they did not have a past experience to see it as a unit defined by the whole," in *Sec. 2.2.1*; the absolute view by those who believe in the "older is creepier" notion, in *Sec. 2.2.2*; the absolute view by some geologists that the "excess density of the oceanic lithosphere sinking in subduction zones is the most powerful source of plate motion," in *Sec. 2.3.1*; the absolute view by petrologists that rocks can be classified in terms of three types, in *Sec. 2.3.2*; the absolute view of each type of mineral resource classification as the right one, in *Sec. 2.4.1*; the absolute view of analytical methods, because of their mathematical operation "under a simplified set of conditions exactly," in *Sec. 2.4.2*; the absolute view of each type of methods in geology as the right one, in *Sec. 2.5.1*; the absolute view of creationism in regard to "special creation," in *Sec. 2.5.2*; the absolute claim by Andrews about feminine intuition, without seriously questioning its validity, in *Sec. 3.2.1*; the absolute anti-reductinist view of Morowitz in regard to his uncritical integration of the mind factor into the study of the hard sciences, in *Sec. 3.2.2*; the absolute view of volcanic ejection as destructive, by those who have family members as victims of volcanic ejection, in *Sec. 3.3.1*; the absolute view of the measurement of an earthquake if placed at the "nucleation zone," in *Sec. 3.3.2*; the "precise and absolute dates" in absolute dating, in *Sec. 3.4.1*; the absolute orientation of engineering geologists, because of their narrow background, to not give sufficient consideration of societal, cultural, and mental factors when doing engineering geology, in *Sec. 3.4.2*; the absolute view of creationist catastrophists in regard to a short Earth history, in *Sec. 3.5.1*; and the absolute view of Burnet in regard to his "one-cycle theory in which biblical narrative [time's arrow] runs its course within a wider conception of 'the great year' and 'great circle of time and fate' that bring about the return of Paradise," in *Sec. 3.5.2*).

On the other hand, there are likewise relative counterparts (e.g., what is true for Favera and Medeiros in regard to the failure of former researchers to recognize HCS because of the lack of past experience is not necessarily so for the critics, because this explanation is too simplistic, in *Sec. 2.2.1*; what is true for those who regard older cemeteries as more troubling is not necessarily so for those who regard newer or modern cemeteries as more troubling instead, in *Sec. 2.2.2*; what is true for some geologists in regard to the "excess density of the oceanic lithosphere sinking in subduction zones" as "the most powerful source of plate motion" is not necessarily so for others, because "it is still a matter of some debate;" as geologists still disagree about the "relative significance of each mechanism," in *Sec. 2.3.1*; what is true for petrologists in regard to the classification of rocks into three types is not necessarily so for the critics, who are more cautious, because of "many obscure points about their history upon which experiment may yet throw light," in *Sec. 2.3.2*; what is right for one mineral resource classification like "the Canadian CIM classification" is not necessarily so for another mineral resource classification like "the South African Code for the Reporting of Mineral Resources and Mineral Reserves," in *Sec. 2.4.1*; what is the right focus for analytical methods in regard to exactness is not necessarily so for numerical methods in regard to approximation, in *Sec. 2.4.2*; what is right for fieldwork is not necesssarily so for lab work, since each has its own merits and problems, in *Sec. 2.5.1*; what is true for the creationists like Gish in relation to "special creation" is not so for Darwin in regard to "evolution" which is "blind," without the creator, in *Sec. 2.5.2*; what is right for Andrews in regard to intuition is not necessarily so for the critics, who seriously question its validity, in *Sec. 3.2.1*; what is right for Morowitz in regard to his uncritical advocation for the integration of the mind factor into the study of the hard sciences is not necessarily so for those critics in the hard sciences who are less enthusiastic about the integration, since the hard sciences, for the critics, are much different from the humanities, although they have some commonalities, in *Sec. 3.2.2*; what is bad about volcanic ejection for those who have family members as victims of volcanic ejection is not necessarily so for those who benefit from volcanic ejection, because "ashfall breaks down over time, forming highly fertile soil, which has made many volcanic regions densely cultivated and inhabited despite the inherent dangers," in *Sec. 3.3.1*; what is right from the vantage point of a measurement if placed at the "nucleation zone" does not necessarily match the vantage point of a measurement if placed outside the "nucleation zone," because of the different effects of "strong ground motion" at different points in relation to the "nucleation zone," in *Sec. 3.3.2*; what is important for absolute dating in regard to its precise accuracy is not necessarily so for relative dating in regard to its focus on "a sequence of age," in *Sec. 3.4.1*; what is appropriate for eningeering geologists in regard to their narrow outlook is not necsssarily so for the critics like H. Morowitz in regard to the impoverishing tradition of reductionism in geology and other hard sciences, in *Sec. 3.4.2*; what is true for creationist catastrophists in regard to a short Earth history is not so for uniformitarianists in regard to a long Earth history, in *Sec. 3.5.1*; and what is true for Burnet is questionable for Gould in his critique of Burnet's obsession with the metaphorical construction of geological time, in *Sec. 3.5.2*).

And the reverse direction also holds true.

However, it should be stressed (as this is something that I regularly did in my previous books, *almost verbatim*) that there are different shades of gray (or different degrees of truth) in the two opposites, and the classification is not necessarily mutually exclusive either. In fact, even when some combinations of the two occur, they only end up sharing the same dialectic relationship, but in a different degree.

Also, there are some other relationships (as a kind of family resemblance) which have something in common with the principle, but they are not exactly the same but only more or less comparable, which varies from case to case.

Good examples of family resemblance in relation to the principle are uniformity-diversity, internalness-externalness, immanence-transcendence, and so forth.

2nd Thesis: The Predictability-Unpredictability Principle

The second thesis is called the predictability-unpredictability principle (on the occurrence of events) in the category of "method" in existential dialectics (which was first proposed in *FC* and, later, other books of mine), in that both predictability and unpredictability have a major role to play in the occurrence of things, so that neither determinism nor indeterminism wins the centuries-old fight. There is no predictability without unpredictability—and vice versa.

There are events which are predictable, just as there are those which are not. Or what is regarded as unpredictable at one point in time may turn out to be predictable later, and, conversely, what is deemed as predictable may turn out to not be so predictable. Even in predictability, outcomes are subject to uncertainty, the degree of which varies from case to case.

For instance, on the one hand, there is predictability (e.g., the predictable tendency of an individual, according to gestalt psychology, to "perceive, decode and process an image or a concept," "only from the perception of the totality," in *Sec. 2.2.1*; the predictable tendency of slates to be "susceptible to elemental weathering," like those in the historic Market Street Burying Ground, in *Sec. 2.2.2*; the predictable tendency of some major driving forces of plate motions, like "friction," "gravitation," and "tidal forces," in *Sec. 2.3.1*; the predictable tendency of petrologists to classify rocks into three types, in *Sec. 2.3.2*; the predictable tendency of mineral resources to fall into one of the three types in the mineral resource classification, in *Sec. 2.4.1*; the predictable tendency of a hydrogeologist to "predict…future behavior of an aquifer system, based on analysis of past and present observations," in *Sec. 2.4.2*; the predictable tendency of a geologist doing field work to go out to the field for "the collection of raw data," in *Sec. 2.5.1*; the predictable tendency of ranking in a Linnean classification, in *Sec. 2.5.2*; the predictable tendency of a geologist with a feminine diffusiness to study rocks by intuition than by analysis, in *Sec. 3.2.1*; the predictable tendency of an anti-reductionist like Morowitz to include the mind in studying nature, in *Sec. 3.2.2*; the predictable tendency of volcanic ejection to result in ashfall, with both destructive and construcive consequences, in *Sec. 3.3.1*; the predictable tendency of an earthquake "to propagate along the fault surface," once "the rupture has initiated," in *Sec. 3.3.2*; the predictable tendency, from the perspective of "the principle of original horizontality," that "the deposition of sediments occurs as essentially horizontal beds," in *Sec. 3.4.1*; the predictable tendency of engineering geologists to pay attention to geohazards and adverse geo-conditions for the protection of life and property, in *Sec. 3.4.2*; the predictable tendency of the history of geological thoughts to change over time in relation to its own peculiar concern with particular ideas and holding particular values in a given era, in *Sec. 3.5.1*; and the predictable tendency of geological time in terms of "an entirely self-regulating machine cycling its way, over and over again," in different stages of "soil replenishment," according to Hutton, in *Sec. 3.5.2*).

On the other hand, there is unpredictability (e.g., the more difficult task to predict which specific law of gestalt, or which specific combination of them, an individual will use or not use in regard to an image or a concept on a particular occasion, in *Sec. 2.2.1*; the more difficult task to predict whether or to what extent older cemeteries with the use of metamorphic rocks like slates are more troubling than modern cemeteries with the use of hard-to-crack granites, for some particular visitors on a given occasion, in *Sec. 2.2.2*; the more difficult task to predict "what degree each process contributes to the motion of each tectonic plate," in *Sec. 2.3.1*; the more difficult task to predict the nature of "allied rocks" in a given case, because "[t]here are…no hard and fast boundaries between allied rocks," in *Sec. 2.3.2*; the more difficult task to predict exactly what future mineral resource classification will be like, since classifications change over time and there are also different types of classification available in each historical era, in *Sec. 2.4.1*; the more difficult task for a hydrogeologist to predict exactly the outcomes of an aquifer system, because of the multiple factors involved, whose interactions are highly complex, in *Sec. 2.4.2*; the more difficult task to predict exactly the validity of a given fieldwork on a particular occasion, because of the inherent problems associated with field methods, in *Sec .2.5.1*; the more difficult task to predict the specifc ranks any taxa belong to, since "the number of generally recognized ranks is limited," which "means that many taxa must go unnamed because no ranks are available for them," in *Sec. 2.5.2*; the more difficult task to predict to what extent exactly a particular intuition on a given occasion will yield accurate result, in *Sec. 3.2.1*; the more difficult task to predict exactly the degree of relevance of this mind-event integration for a given study in the hard sciences, which are different from the humanities, even if the role of the mind is relevant to both, in *Sec. 3.2.2*; the more difficult task to predict exactly when volcanic ejection occurs, in which particular way on a given occasion, because "[t]he…flows can be impossible to outrun," "[t]hey can also be difficult to predict," and "[i]n many cases prediction is based on the topography of a region, but a valley may fill and overflow," in *Sec. 3.3.1*; the more difficult task to predict exactly "the scale of the nucleation zone" in relation to the propagation, which "is uncertain" in relation to a particular earthquake, in *Sec. 3.3.2*; the more difficult task to predict exactly if a particular deposition of sediments occurs as essentially horizontal beds, because we now know that "not all sedimentary layers are deposited purely horizontally," in *Sec. 3.4.1*; the more difficult task to predict exactly when engineering geologists will succeed on a particular occasion in protecting life and property, because of the history of failures as pointed out by the critics, in *Sec. 3.4.2*; the more difficult task to predict exactly what the peculiar concerns with particular ideas and holding particular values in the future eras will be, in *Sec. 3.5.1*; and the more difficult task to predict, if looked only within Hutton's "entirely self-regulating machine," the "particular" and "distinct" features and "properties of strata" as "interesting in themselves—as markers of distinction for particular times," because Hutton was so obsessed with the "endless cycles" that "he lost all interest in the historical nature of geological change," in *Sec. 3.5.2*).

And the reverse direction also holds true.

Again, it should be stressed, however, that there are different shades of gray (or different degrees of truth) in the two opposites, and the classification is not necessarily mutually exclusive either. In fact, even when some combinations of the two occur, they only end up sharing the same dialectic relationship, but in a different degree.

Also, there are some other relationships (as a kind of family resemblance) which have something in common with the principle, but they are not exactly the same but only more or less comparable, which varies from case to case.

Good examples of family resemblance in relation to the principle are sureness-arbitrariness, and so forth.

3rd Thesis: The Explicability-Inexplicability Principle

The third thesis is called the explicability-inexplicability principle (on the underlying mechanisms of things) in the category of "method" in existential dialectics (which was proposed in *FPHU* and other books of mine), in that both explicability and inexplicability are part of the understanding of things. There is no explicability without inexplicability—and vice versa.

This principle tells us the dual nature of the research dilemma, in that, if reality can be explained in some ways, it also has its other ways which are not quite explainable, at a given point in time.

For instance, on the one hand, there is explicability (e.g., the explanation by Favera of his ability to successfully recognize the HCS in Brazil because of the appeal to his past experience from his previous trip in Texas, in *Sec. 2.2.1*; the explanation of the creepy look of older cemeteries on the basis of the use of slates, which have been "cracked, faded, pitted and streaked," in *Sec. 2.2.2*; the explanation, within the theory of plate tectonics, of plate motions on the basis of the three main driving forces, in *Sec. 2.3.1*; the explanation, by petrologists, of rock classification on the basis of "mineral and chemical composition,...the texture of the constituent particles and...the processes that formed them," in *Sec. 2.3.2*; the explanation, by economic geologists, of the importance of earth materials because of their usefulness for economic and/industrial purposes, in *Sec. 2.4.1*; the explanation, if viewed within analytical methods, of the need for "a simplified set of conditions," because of the mathematical requirement of "exactness," in *Sec. 2.4.2*; the explanation of the preference for fieldwork, by those specializing in this method as shown by the examples earlier in Chapter One, on the basis that "[g]eology has a long history of field experiments, since the time of Avicenna," as a kind of methodological tradition, in *Sec. 2.5.1*; the explanation, within molecular systematics, of its classification as a better alternative to the mainstream evolutionary "phylogenetic taxonomy," on the basis of "genetic details" in relation to their molecular structure," in *Sec. 2.5.2*; the explanation by Andrews that women bring unique talents to geology on the basis of her claim about feminine intuition, in *Sec. 3.2.1*; the explanation, by Morowitz, of the importance of the mind in studying nature on the basis of the contamination of the role of the observer in measuring natural events as shown in quantum mechanics, in *Sec. 3.2.2*; the explanation, by mainstream geologists, of the nature of magma and volcanic ejection on the basis of geological factors only, like their composition, formation, spread, and effects, in *Sec. 3.3.1*; the explanation that "once the rupture [of an earthquake] has initiated it begins to propagate along the fault surface," in *Sec. 3.3.2*; the explanation, from the perspective of "the principle of uniformitarianism," that "the geologic processes observed in operation that modify the Earth's crust at present have worked in much the same way over geologic time," such that "the present is the key to the past," in *Sec. 3.4.1*; the explanation, from the perspective of engineering geology, of its practice, on the basis that "[n]early all engineering geologists are initially trained and educated in geology," in *Sec. 3.4.2*; the explanation, within uniformitarianism, of why the present is key to the past, on the basis of "an Earth created by the same natural phenomenon that are at work today," in *Sec. 3.5.1*; and the explanation of "soil replenishment" by Hutton in terms of three stages, namely, "denudation and decay," "the deposition of new marine strata," and "the melting, expansion, and uplift of the lowest strata as 'igneous' rock ready to be broken down again for future plants," in *Sec. 3.5.2*).

On the other hand, there is inexplicability (e.g., the lack of sufficient explanation by Favera about this appeal to his past experience, since it did not explain how Favera came to learn about the HCS in his previous contact with the HCS during a SEPM field trip in Texas in 1978 in the first place, in *Sec. 2.2.1*; the lack of sufficient explanation, if understood only within the "older is creepier" notion, of why, for some other individuals, what is more troubling is not older cemeteries but the modern ones, in *Sec. 2.2.2*; the lack of convincing explanation, within the theory of plate tectonics, of why there cannot be any other forces which should be accounted for—not just friction, gravitation, and tidal forces, in *Sec. 2.3.1*; the lack of sufficient explanation of why rock classification should not include other criteria except the ones of "mineral and chemical composition,...the texture of the constituent particles and...the processes that formed them," given the fact that there remain "many obscure points about their history" to be understood, in *Sec. 2.3.2*; the lack of sufficient explanation by economic geologists of why this exploitation of earth materials for economic and/or industrial purposes is necessarily desirable, since, for the critics like the environmentalists and the anti-colonialists, the exploitation has its dark sides, which must be redressed, in *Sec. 2.4.1*; the lack of convincing explanation, if viewed within analytical methods, of why this assumption for "a simplified set of conditions" is necessarily needed, since there can be other ways of doing hydrogeology, like the numerical methods "under more general conditions to an approximation," in *Sec. 2.4.2*; the lack of sufficient explanation, if looked within this methodological tradition, of why fieldwork should be the preferred method, in light of historical hindsight nowadays that there are other methods available in our times, in *Sec .2.5.1*; the lack of sufficient explanation, within molecular systematics, of why it is a better alternative, since it does not always work, as shown in the case about animals, "because in mammals mitochondria are inherited only from the mother," which means that "inheritance in the paternal line might not be detected," in *Sec. 2.5.2*; the lack of sufficient explanation by Andrews of why this so-called unique talent of women is not really an embarassing handicap at all, since, for the critics, her work can instead explain the opposite conclusion, namely, that women are not as good at science as men, because the "capacity to process...input into a product is masculine," which means that women are not naturally good as men in analysis, in *Sec. 3.2.1*; the lack of sufficient explanation, by Morowitz, of why the role of the observer is equally or more or less relevant to both the humanities and the hard sciences, while the two are quite different from each other, even though they have something in common, in *Sec. 3.2.2*; the lack of sufficient explanation, by mainstream geologists, of why the role of "the mind" is necessarily ignored in undersanding and measuring natural events like volcanic ejection, in *Sec. 3.3.1*; the lack of sufficient explanation about "[t]he mechanics of this process, which are poorly understood, partly because it is difficult to recreate the high sliding velocities in a laboratory" and also because "the effects of strong ground motion make it very difficult to record information close to a nucleation zone," in *Sec. 3.3.2*; the lack of sufficient explanation, from the perspective of "the principle of uniformitarianism," of why "the distant past 'consisted of epochs of paroxysmal and catastrophic action interposed between periods of comparative tranquility,'" in *Sec. 3.4.1*; the lack of sufficient explanation, from the perspective of engineering geology, of why engineering geologists should not give sufficient attention to cultural, societal, and mental factors in studying geohazards, as criticized by Morowitz in regard to reductionism, in *Sec. 3.4.2*; the lack of sufficient explanation, within uniformitarianism, of the existence of abrupt changes in Earth history as shown by the occurences of catastrophes, in *Sec. 3.5.1*; and the lack of sufficient explanation, if looked only within Hutton's "entirely self-regulating machine," of why there exist abrupt changes or catastrophes in Earth history like volcanic eruptions which can affect the soils on Earth, in *Sec. 3.5.2*).

And the reverse direction also holds true.

Once more, it should be stressed, however, that there are different shades of gray (or different degrees of truth) in the two opposites, and the classification is not necessarily mutually exclusive either. In fact, even when some combinations of the two occur, they only end up sharing the same dialectic relationship, but in a different degree.

Also, there are some other relationships (as a kind of family resemblance) which have something in common with the principle, but they are not exactly the same but only more or less comparable, which varies from case to case.

Good examples of family resemblance in relation to the principle are underlyingness-regularness, causation-regularity, causation-correlation and so forth.

4th Thesis: The Preciseness-Vagueness Principle

The fourth thesis is called the preciseness-vagueness principle (on the refinement of things) in the category of "structure" in existential dialectics (which was first out worked out in *FIA*), in that both preciseness and vagueness are important, not that one is better than the other, but that both are used, in different degrees of preference, in accordance to the contextual application from the perspectives of nature, the mind, culture, and society. There is no preciseness without vagueness—and vice versa.

For instance, on the one hand, there is preciseness (e.g., the precise identification of the 6 laws of gestalt by Favera and Medeiros for their application to the discovery of the HCS in Brazil, in *Sec. 2.2.1*; the precise identification of the fear of death as the reason for some visitors to find old cemeteries creepy, in *Sec. 2.2.2*; the precise measurement of an "average oceanic lithosphere" to be "100 km thick," in *Sec. 2.3.1*; the precise identification of three types of rocks in rock classification, in *Sec. 2.3.2*; the precise identification of the three types of Earth materials in the mineral resource classification, in *Sec. 2.5.1*; the precise identification of the six physical properties of an aquifer system, in *Sec. 2.4.2*; the precise control of the conditions in a lab experiment, in *Sec. 2.5.1*; the precise identification by Carolus Linnaeus, of five and later seven ranks in his classification, in *Sec. 2.5.2*; the precise identification of intuition as the unique talent that women bring to geology, according to Andrews, in *Sec. 3.2.1*; the precise identification of the role of the mind by Morowitz for a better way to do geology and other sciences in general, in *Sec. 3.2.2*; the precise identification of different classifications about rocks, magmas, and volcanic mechanisms in terms of the number three, in *Sec. 3.3.1*; the precise identification of "around 500,000 earthquakes each year," in *Sec. 3.3.2*; the precise identification of dates in absolute dating, in *Sec. 3.4.1*; the precise identification of the studies and methods in engineering geology, in *Sec. 3.4.2*; the precise identification of "the present" as "key to the past" in uniformitarianism, in *Sec. 3.5.1*; and the precise identification of three stages of soil replenishment by Hutton, in *Sec. 3.5.2*).

On the other hand, there is vagueness (e.g., the vagueness in the identification, since it is not really clear how the individual concepts in the laws like "similarity," "proximity," etc. are supposed to mean, since they can mean different things to different individuals and are applied in different ways on different occassions, in *Sec. 2.2.1*; the vagueness in this identification of the fear of death, since it is not clear what this fear of death really means to different individuals, as it may refer to the inevitability of death, the awfulness of death, the relief in death, etc., in *Sec. 2.2.2*; the vagueness in the measurement of the average oceanic lithosphere, because it is not clear how thicker it becomes each year, how much the rate varies from one specific location to another, etc., in *Sec. 2.3.1*; the vagueness in the rock classification, since it is not clear which "selected points [more or less arbitrary]" a given rock can belong to, "in a continuously graduated series," in *Sec. 2.3.2*; the vagueness in the classification of the three types, since there is no consensus on the classification and there are different versions like "the Canadian CIM classification," "the Australasian Joint Ore Reserves Committee Code," and "the South African Code for the Reporting of Mineral Resources and Mineral Reserves," in *Sec. 2.5.1*; the vagueness in understanding the properties and the solutions in a given aquifer system, because, in numerical methods, for instance, the outcomes are "under more general conditions to an approximation," in *Sec. 2.4.2*; the vagueness about the validity of the outcomes of a lab experiment, if its conditions are so precisely controlled and the results become not realistic, such that their validity is not clear, in *Sec. 2.5.1*; the vagueness in the five or seven ranks by Linnaeus, since it is not clear of why there must be five or seven, and the number was later increased to 21 after his death, in *Sec. 3.5.2*; the vagueness in the identification about intuition, since what is intuitive for one person may not be so for another, so it can mean different things to different individuals, in *Sec. 3.2.1*; the vagueness in what is exactly meant by "the mind," since it can refer to different things to different individuals, in *Sec. 3.2.2*; the vaguness in the number three in regard to these classications, since it is not clear why they cannot be two, or four, or five, or another number, in *Sec. 3.3.1*; the vagueness in this number of 500,000, since it is not clear if there are other occurrences of earthquakes, which, however, are not successfully detected by seismologists, as the measurement depends on the quality and pervasiveness of the devices being used, among other criteria, in *Sec. 3.3.2*; the vagueness in the precise dates in absolute dating, because "absolute dates do not necessarily tell us when a particular cultural event happened," in *Sec. 3.4.1*; the vagueness in the identification of the studies and methods in engineerng geology, because it is not clear which ones of them are more important, useful, accurate, or else under which particular conditions, in *Sec. 3.4.2*; the vagueness in the "present" as "key to the past," since it is not clear to what extent this then means that the present and the past share "the same kind" of causes, "the same energy," "the same effects," and the same "geological circumstances," in *Sec. 3.5.1*; and the vagueness in the identification of three stages by Hutton, because he became "[s]o rigid [with the] vision of an endlessly cycling earth having 'no vestige of a beginning' and 'no prospect of an end' that he lost all interest in the historical nature of geological change" and did not treat "the different ages and properties of strata as interesting in themselves—as markers of distinction for particular times," in *Sec. 3.5.2*).

And the reverse direction also holds true.

Once more, it should be stressed, however, that there are different shades of gray (or different degrees of truth) in the two opposites, and the classification is not necessarily mutually exclusive either. In fact, even when some combinations of the two occur, they only end up sharing the same dialectic relationship, but in a different degree.

Also, there are some other relationships (as a kind of family resemblance) which have something in common with the principle, but they are not exactly the same but only more or less comparable, which varies from case to case.

Good examples of family resemblance in relation to the principle are clarity-ambiguity, directness-indirectness, quantitativeness-qualitativeness, specificity-obscurity, describability-nondescribability, specificity-generality, thickness-thinness, concretness-abstractness, and the like.

5th Thesis: The Simpleness-Complicatedness Principle

The fifth thesis refers to the simpleness-complicatedness principle (on the interconnection among things) in the category of "structure" in existential dialectics (which was first out worked out in *FIA*), in that both simpleness and complicatedness are vital, without favoring one over the other, and each is utilized, depending on the basis of the perspectives of nature, the mind, culture, and society. There is no simpleness without complicatedness—and vice versa.

For instance, on the one hand, there is simple analysis (e.g., the relatively simple analysis by Favera and Medeiros of the discovery of the HCS in Brazil on the basis of the laws of gestalt, in *Sec. 2.2.1*; the relatively simplistic account of old cemeteries on the basis of the "older is creepier" notion, in *Sec. 2.2.2*; the relatively simple analysis of plate tectonics on the basis of its features and the driving forces, in *Sec. 2.3.1*; the relatively simple analysis of rock classification into three types, in *Sec. 2.3.2*; the relatively simple analysis of mineral resource classifcation into three types, in *Sec. 2.4.1*; the relatively simplistic account for an aquifer system in analytical methods, because they operate "under a simplified set of conditions," in *Sec. 2.4.2*; the relatively simple treatment of a method by its own proponents, in *Sec. 2.5.1*; the relatively simple classification of ancient organisms on the basis of the evolutionary principle of common descent, in *Sec. 2.5.2*; the relatively simple account for intuition by Andrews, in *Sec. 3.2.1*; the relatively simple analysis of the problem of reductionism by Morowitz, in *Sec. 3.2.2*; the relatively simple analysis of volcanic ejection by geologists, in *Sec. 3.3.1*; the relatively simple analysis of earthquakes by mainstream sesimologists, in *Sec. 3.3.2*; the relatively simple views of the principles in relative dating, in *Sec. 3.4.1*; the relatively simple anlaysis of the scope, concern, and methodology in engineering geology, in *Sec. 3.4.2*; the relatively simple view of a short Earth history by creationist catastrophists, in *Sec. 3.5.1*; and the relatively simple analysis of "geological time" by Burnet, Hutton, and Lyell, on the basis of "time's arrow" and "time's cycle," in *Sec. 3.5.2*).

On the other hand, there is also complicated counterpart (e.g., the relatively more complicated analysis of the discovery of the HCS in Brazil by way of showing some problems in their claim, in *Sec. 2.2.1*; the relatively more complicated analysis of the "older is creepier" notion, by also considering alternative reactions by other visitors, who may find modern, not older, cemeteries more troubling instead, in *Sec. 2.2.2*; the relatively more complicated analysis of plate tectonics by considering some of the problems still unresolved, especially for the critics, in *Sec. 2.3.1*; the relatively more complicated analysis of rock classification by considering some difficult problems associated with the classification, in *Sec. 2.3.2*; the relatively more complicated analysis of mineral resource classification by also considering the problems associated with it, as pointed out by the critics, in *Sec. 2.4.1*; the relatively more complicated account for an aquifer system by also considering the different problems associated with analytic methods, in *Sec. 2.4.2*; the relatively complicated treatment of a method by its critics, with the relentless consideration of fundamental problems associated with the method, in *Sec. 2.5.1*; the relatively complicated analysis of the classification by also considering alternative classifications which question the mainstream evolutionary versions, in *Sec. 2.5.2*; the relatively more complicated account for intuition by also considering the critical problems pointed out by the critics, in *Sec. 3.2.1*; the relatively more complicated analysis of the problem of reductionism by also considering the problems associated with the view by Morowitz himself, in *Sec. 3.2.2*; the relatively more complicated analysis of volcanic ejection by raising some critical questions by the critics, in *Sec. 3.3.1*; the relatively more complicated analysis of earthquakes by also considering some critical problems pointed out by the critics, in *Sec. 3.3.2*; the relatively more complicated analyis of the principles in relative dating, as shown in the criticisms pointed out by the critics, in *Sec. 3.4.1*; the relatively more complicated analysis of the scope, concern, and methodology in engineering geology, by also taking into account the cultural, societal, and mental factors as pointed out by the critics, in *Sec. 3.4.2*; the relatively more complicated view of Earth history in both gradual and abrupt ways by many contemporary geologists, in *Sec. 3.5.1*; and the relatively more complicated analysis of "geological time" by also considering the different problems in the works of the three thinkers in regard to their metaphorical constructions, as pointed out by Gould, in *Sec. 3.5.2*).

And the reverse direction also holds true.

Again, it should be stressed, however, that there are different shades of gray (or different degrees of truth) in the two opposites, and the classification is not necessarily mutually exclusive either. In fact, even when some combinations of the two occur, they only end up sharing the same dialectic relationship, but in a different degree.

Also, there are some other relationships (as a kind of family resemblance) which have something in common with the principle, but they are not exactly the same but only more or less comparable, which varies from case to case.

Good examples of family resemblance in relation to the principle are inflexibility-flexibility, standardization-specialization, imperfectness-perfectness, superficiality-depth, shallowness-deepness, economicalness-elaboratedness, plainness-circumspection, and the like.

6th Thesis: The Openness-Hiddenness Principle

The sixth thesis refers to the openness-hiddenenss principle (on the detection of things) in the category of "structure" in existential dialectics (which was already worked out in my previous books, especially in *FPHU*), in that reality has its hidden face, just as it is open to outside view in some other ways. There is no openness without hiddenness—and vice versa.

For instance, on the one hand, there is openness (e.g., the open exploration of the relationship between gestalt psychology and the discovery of the HCS in Brazil by Favera and Medeiros, in *Sec. 2.2.1*; the open exploration of the relationship between geological rocks and their use for graves, in *Sec. 2.2.2*; the open exploration of three main types of plate boundaries, in *Sec. 2.3.1*; the open exploration of rocks in terms of their classification into three different types, in *Sec. 2.3.2*; the open exploration of mineral resource classification, in accordance to a particular version like "the Canadian CIM classification," "the Australasian Joint Ore Reserves Committee Code," or "the South African Code for the Reporting of Mineral Resources and Mineral Reserves," in *Sec. 2.4.1*; the open exploration of analytical methods for the solutions to an aquifer system, in *Sec. 2.4.2*; the open exploration of the usefulness of fieldwork in geology, for example, in regard to "geological mapping," "surveying of topographic features," etc., in *Sec. 2.5.1*; the open exploration of another classification of ancient organisms from the creationist viewpoint, in *Sec. 2.5.2*; the open exploration of the role of feminine intuition in geology, in *Sec. 3.2.1*; the open exploration of the role of the observer in debunking reductionism by Morowitz, in *Sec. 3.2.2*; the open exploration of the nature of volcanic ejection in relation to its formation, composition, spread, and effects, in *Sec. 3.3.1*; the open exploration of the nature of rupture dynamics, like an earthquake "to propagate along the fault surface," in *Sec. 3.3.2*; the open exploration of different geological relations as revealed by the principles in relative dating, in *Sec. 3.4.1*; the open exploration of the use of geology for studying geohazards and adverse geo-conditions in engineering geology, in *Sec. 3.4.2*; the relatively open exploration of abrupt changes in Earth history by creationist catastrophists, in *Sec. 3.5.1*; and the open exploration of a short Earth history on the basis of The Book of Revelation by Burnet, in *Sec. 3.5.2*).

On the other hand, there is also hiddenness (e.g., the hidden bias in the analysis by Favera and Medeiros because of their obsession with gestalt psychology in understanding the reasons behind the discovery of the HCS in Brazil, while ignoring other possible causes, in *Sec. 2.2.1*; the hidden bias in the "older is creepier" notion, because of its obsession with older cemeteries, without giving sufficient attention to new cemeteries, in *Sec. 2.2.2*; the hidden bias of the exploration of the three main types of plate boundaries, because it often does not give sufficient attention to the "plate boundary zones," which are "where the effects of the interactions are unclear and the broad belt boundaries are not well defined," iin *Sec. 2.3.1*; the hidden bias of rock classification, since its obsession with three types does not give sufficient attention to "many obscure points about their history upon which experiment may yet throw light," in *Sec. 2.3.2*; the hidden bias in each version of mineral resource classification, since there is no consensus on one classification as the best for all, in *Sec. 2.4.1*; the hidden bias in the analytical methods because of their simplistic assumption for "a simplified set of conditions exactly," in *Sec. 2.4.2*; the hidden bias in fieldwork, as shown by the disadvantages pointed out by the critics, in *Sec. 2.5.1*; the hidden bias in the creationist view, because it cannot be proved, as Duane Gish himself said that "[w]e do not know how the Creator created," and also because it is baised against its nemesis, the evolutionary theory of common descent, in *Sec. 2.5.2*; the hidden bias of this claim about femine intuition in geology, because of its feminist ideological motivation to "overthrow…the paradigm that women aren't as good at science as men," in *Sec. 3.2.1*; the hidden bias in the analysis of Morowitz, because of its one-dimensional enthusiasm against reductionism, without due consideration of the problems associated with reverse-reductionism, in *Sec. 3.2.2*; the hidden bias in the mainstream geological analysis of volcanic ejection, because of its dismissal of the role of "the mind" in the analysis, in *Sec. 3.3.1*; the hidden nature of the "rupture dynamics" to be revealed, since "the mechanics of this process are poorly understood," in *Sec. 3.3.2*; the hidden biases in the principles, as shown by the problems and assumptions used in them, as pointed out by the critics, in *Sec. 3.4.1*; the hidden bias of reductionism in engineering geology, because of the lack of sufficient attention to cultural, societal, and mental factors as pointed out by the critics, in *Sec. 3.4.2*; the hidden bias in creationist catastrophism because of its religious ideology, in *Sec. 3.5.1*; and the hidden bias in the work by Burnet, because of biblical contamination, in *Sec. 3.5.2*).

And the reverse direction also holds true.

Again, it should be stressed, however, that there are different shades of gray (or different degrees of truth) in the two opposites, and the classification is not necessarily mutually exclusive either. In fact, even when some combinations of the two occur, they only end up sharing the same dialectic relationship, but in a different degree.

Also, there are some other relationships (as a kind of family resemblance) which have something in common with the principle, but they are not exactly the same but only more or less comparable, which varies from case to case.

Good examples of family resemblance in relation to the principle are overtness-covertness, publicness-privateness, openness-closedness, transparency-secrecy, openness-biasedness, and so on.

7th Thesis: The Denseness-Emptiness Principle

The seventh thesis concerns the denseness-emptiness principle (on the distribution of entities in space) in the category of "structure" in existential dialectics, which is first proposed in *FPHUP*, in that both density and void are needed, in relation to the mind, nature, culture, and society, albeit in different ways. There is no denseness without emptiness—and vice versa.

Lest any misunderstanding occurs, the term "void" is used here only as an approximation of emptiness (depending on the degree of the lack of density), since, in physics, it is well known that "empty" space is not really empty all the way, because it can be full of energy (e.g., random quantum fluctuations at the sub-atomic level, and, for that matter, dark energy in the universe) and matter (e.g., different versions of sub-atomic particles, and, for that matter, dark matter in the universe). (F. Wilczek 2008)

For instance, on the one hand, there is denseness (e.g., the relativley denser concentration of the HCS in the Parnaíba Basin as discovered by Favera and his team, in *Sec. 2.2.1*; the relatively denser concentration of the use of slates in older cemeteries like the historic Market Street Burying Ground, in *Sec. 2.2.2*; the relatively denser nature of "cool oceanic lithosphere," which "is significantly denser than the hot mantle material from which it is derived," in *Sec. 2.3.1*; the relatiely denser nature of the sediment which is buried, as " more and more material is deposited on top," before "the sediment will become so dense that it is essentially rock," in *Sec. 2.3.2*; the relatively denser concentration of minerals with economic value for extraction in "mineral reserves," in *Sec. 2.4.1*; the relatively denser concentration of contaminants like "molecules and small particles in gases and liquids" within groundwater, when they are held back by the "retardation by absorption" to the soil, in *Sec. 2.4.2*; the relatively denser concentration of people doing fieldwork in geology in the older days, in *Sec. 2.5.1*; the relatively denser concentration of "mitochondria" in mammals along the "maternal" line to be detected by molecular systematics, because "in mammals mitochondria are inherited only from the mother," in *Sec. 2.5.2*; the relatively denser concentration of feminists who accept the claim by Andrews that women brings unique talents in geology by feminine intuition, in *Sec. 3.2.1*; the relatively denser concentration of reductionist scientists in the hard sciences to not include the role of the observer in studying nature, in *Sec. 3.2.2*; "[t]he ashfall may become so dense that daylight turns the sky gray to pitch black," in *Sec. 3.3.1*; the relatively denser concentration of earthquakes, about "90%, and 81% of the largest," along "the 40,000-km-long, horseshoe-shaped zone called the circum-Pacific seismic belt, known as the Pacific Ring of Fire, which for the most part bounds the Pacific Plate," in *Sec. 3.3.2*; the relatively denser concentration of layers which are deposited on top of each other in a sedimentary rock...in a tectonically undisturbed sequence," in *Sec. 3.4.1*; the relatively denser concentration of some geohazards like "earthflows" to "occur on moderate slopes," as they "consist of saturated soil or fine-grained rock deposits that flow downhill," in *Sec. 3.4.2*; the relatively denser concentration of scholars with mystical beliefs about geological objects in antiquity, in *Sec. 3.5.1*; and the relatively denser concentration of geologists with "biblical standards" in the older days, in *Sec. 3.5.2*).

On the other hand, there is emptiness (e.g., the relatively less dense, or more empty, concentration of the HCS outside the Parnaíba Basin that Favera and his team did not succeed in discovering much, in *Sec. 2.2.1*; the relatively less dense, or more empty, concentration of the use of slates in modern cemeteries in America, in *Sec. 2.2.2*; the relatively less dense nature of "hot mantle material" when compared with "cool oceanic lithosphere," in *Sec. 2.3.1*; the relatively less dense, or more empty, nature of the recent material which is deposited on top of the sediment buried far beneath, in *Sec. 2.3.2*; the relatively less dense, or more empty, concentration of minerals with economic value for extraction in "mineral occurrences," in *Sec. 2.4.1*; the relatively less dense, or more empty, concentration of contaminants like "molecules and small particles in gases and liquids" within groundwater, when they are not held back by the "retardation by absorption" to the soil, in *Sec. 2.4.2*; the relatively less dense, or more empty, concentration of people doing computer simulation in geology in the older days, in *Sec. 2.5.1*; the relatively less dense, or more empty, concentration of "mitochondria" in mammals along the "paternal" line to be detected by molecular systematics, because "in mammals mitochondria are inherited only from the mother," in *Sec. 3.5.2*; the relatively less dense, or more empty, concentration of the critics who accept the claim by Andrews that women brings unique talents in geology by feminine intuition, in *Sec. 3.2.1*; the relatively less dense, or more empty, concentration of anti-reductionist scientists like Morowitz to not include the role of the observer in studying nature, in *Sec. 3.2.2*; the relatively less dense, or more empty, concentration of ashfall in the sky, before its occurrence, in *Sec. 3.3.1*; the relatively less dense, or more empty, concentraton of earthquakes outside "the Pacific Ring of Fire," in *Sec. 3.3.2*; the relatively less dense, or more empty, concentration of layers in an "ash," which "is created when solid rock shatters and magma separates into minute particles during explosive volcanic activity," before becoming cooler and solidified later, in *Sec. 3.4.1*; the relatively less dense, or more empty, concentration of some geohazards like "earthflows" to occur uphill, in *Sec. 3.4.2*; the relatively less dense, or more empty, concentrations of scholars with mystical beliefs about geological objects nowadays in the scientific community, in *Sec. 3.5.1*; and the relatively less dense, or more empty, concentration of geologists with "biblical standards" nowadays, in *Sec. 3.5.2*).

And the reverse direction also holds true.

Again, it should be stressed, however, that there are different shades of gray (or different degrees of truth) in the two opposites, and the classification is not necessarily mutually exclusive either. In fact, even when some combinations of the two occur, they only end up sharing the same dialectic relationship, but in a different degree.

Also, there are some other relationships (as a kind of family resemblance) which have something in common with the principle, but they are not exactly the same but only more or less comparable, which varies from case to case.

Good examples of family resemblance in relation to the principle are fullness-voidness, nearness-farness, concentration-dispersion, and the like.

8th Thesis: The Slowness-Quickness Principle

The eighth thesis concerns the slowness-quickness principle (on the speed of change) in the category of "process") in existential dialectics (which was first worked out in *FIA*), in that both slowness and quickness co-exist, with their own internal tension, to the extent that each fights for its own relevance with the other, in accordance to the perspectives of nature, the mind, culture, and society, without one being the victor and the other being the vanquished in the long haul. There is no slowness without quickness—and vice versa.

For instance, on the one hand, there is slowness (e.g., the relatively slower readiness of the students to recognize the HCS in Brazil at the beginning, when they did not benefit from the help of the laws of gestalt like the Prägnanz law, in *Sec. 2.2.1*; the relatively slower speed of granites to be "cracked, faded, pitted and streaked" by weathering, in *Sec. 2.2.2*; the relatively slower speed of plate motions in the Mid-Atlantic Ridge, with "a typical 10–40 mm/a [per year]…about as fast as fingernails grow," in *Sec. 2.3.1*; the relatively slower speed of magma, in the case of intrusive rock formation, when it "cools and crystallizes slowly within the Earth's crust," in *Sec. 2.3.2*; the relatively slower speed of minerals to be transported to the right place for mining, if there is no well developed transport system, in *Sec. 2.4.1*; the relative slowing-down of the speed of a contaminant in the case of the "retardation by absorption" to the soil within groundwater, in *Sec. 2.4.2*; the relatively slower readiness of a geologist doing computer lab work to go out to the field to collect raw data, in *Sec. 2.5.1*; the relatively slower readiness of a creationist to classify organisms by common descent, in *Sec. 2.5.2*; the relatively slower speed of trying to understand something by someone who uses analytical proof, not by intuition, in *Sec. 3.2.1*; the relatively slower readiness of a hard-core scientist to take full consideration of the role of the mind in studying natural events, in *Sec. 3.2.2*; the relatively slower speed of a rock running downhill in a normal situation, in *Sec. 3.3.1*; the occurrence of "slow earthquakes" like "the tsunami earthquake,... caused by the slow propagation speed," in *Sec. 3.3.2*; the relatively slower speed of an isotope to decay if it has a much longer half-life, in *Sec. 3.4.1*; some geohazards like "some volcanic eruptions are mild and slow," in *Sec. 3.4.2*; "gradual processes such as weathering, erosion and deposition" in Earth history, in *Sec. 3.5.1*; and the relatively slower changes in Earth history, like the different stages of soil replenishment as pointed out by Hutton, in *Sec. 3.5.2*).

On the other hand, there is quickness (e.g., the relatively quicker readiness of Favera to recognize the HCS in Brazil because of his previous trip in Texas, thanks to the use of the law on the past experience, in *Sec. 2.2.1*; the relatively quicker speed of slates to be "cracked, faded, pitted and streaked" by weathering, in *Sec. 2.2.2*; the relatively faster speed of plate motions in the Nazca Plate, with "about 160 mm/a [per year]...about as fast as hair grows," in *Sec. 2.3.1*; the relatively faster speed of magma, in the case of extrusive rock formation, when it "reach[es]...the surface either as lava or fragmental ejecta," in *Sec. 2.3.2*; the relatively faster speed of minerals to be transported to the right place for mining, if there are well developed transport systems, in *Sec. 2.4.1*; the relative quicker speed of a contaminant in the cases of of "diffusion and dispersion, which simply help spreading the contaminant" within groundwater, in *Sec. 2.4.2*; the relatively quicker readiness of a geologist doing the excavation of fossil material to go out to the field to collect raw data, in *Sec. 2.5.1*; the relatively quicker readiness of a specialist in phylogenetic taxonomy to classify organisms by common descent, in *Sec. 2.5.2*; the relatively quicker speed of trying to understand something by someone who uses intuition, in *Sec. 3.2.1*; the relatively quicker readiness of a humanist like a Taoist to consider the role of the mind in studying nature, in *Sec. 3.2.2*; the relatively quicker speed of rocks, especially when reduced to molten forms into tiny pieces, as in "an 'avalanche' of hot ash, gases, and rocks" during "a volcanic eruption," in *Sec. 3.3.1*; the "quick earthquakes" like "a small subset of earthquake ruptures...at speeds greater than the S-wave velocity," in *Sec. 3.3.2*; the relatively quicker speed of an isotope to decay if it has a very short half-life, in *Sec. 3.4.1*; some geohazards like "other [volcanic eruptions]...are powerful and dramatic" or fast, in *Sec. 3.4.2*; "[v]iolent events such as meteorite impacts and large volcanic explosions" in Earth history, in *Sec. 3.5.1*; and the relativey quicker changes in Earth history, like abrupt volcanic eruptions and dramatic earthquakes, in *Sec. 3.5.2*).

And the reverse direction also holds true.

Again, it should be stressed, however, that there are different shades of gray (or different degrees of truth) in the two opposites, and the classification is not necessarily mutually exclusive either. In fact, even when some combinations of the two occur, they only end up sharing the same dialectic relationship, but in a different degree.

Also, there are some other relationships (as a kind of family resemblance) which have something in common with the principle, but they are not exactly the same but only more or less comparable, which varies from case to case.

Good examples of family resemblance in relation to the principle are inconvenience-convenience, passiveness-activenessness, gradualness-abruptness, deceleration-acceleration, and the like.

9th Thesis: The Expansion-Contraction Principle

The nineth thesis is called the expansion-contraction principle in the category of "process" in existential dialectics, in that entities in the world can both expand in some ways and contract in other ones, as part of their nature. There is no expansion without contraction—and vice versa.

This principle, although not so explicitly called, was already used in my previous works on different topics (e.g., the theory of floating consciousness in *FCD* and *FPHC*, the union of the unions in *BWT*, and the cyclical progression of system fragmentation and integration in *FCD*).

For instance, on the one hand, there is expansion (e.g., the relative expansion of the influence of gestalt psychology in relation to the work of Favera and his team, after they discovered the linkage between the laws of gestalt and the perception of geological structures, in *Sec. 2.2.1*; the relative expansion of the use of granites in modern American cemeteries, in *Sec. 2.2.2*; the relatively more expansive nature of plate motions in "divergent boundaries," when contrasted with "convergent boundaries," in *Sec. 2.3.1*; the relative expansion of magma, in the case of extrusive rock formation, when it "reach[es]...the surface either as lava or fragmental ejecta," in *Sec. 2.3.2*; the relatively more developed ability to exploit the resources in nature for economic and/or industrial purposes, in *Sec. 2.4.1*; the relatively more developed ability in those specializing in analytic methods to exploit mathematics for "simple, elegant" solutions, in *Sec. 2.4.2*; the relatively more developed ability of a geologist doing computer lab work to engage in computer simulation, in *Sec. 2.5.1*; the relatively more developed ability of a creationist to classify things on the basis of religious revelation, in *Sec. 2.5.2*; the relatively more developed ability of a woman who is endowed with the feminine intuition to see things "diffusively," in *Sec. 3.2.1*; the relatively more developed ability of a Taoist to be in harmony with nature by practising the mind-nature integration, in *Sec. 3.2.2*; the expansion of volcanic ash during the ejection, in *Sec. 3.3.1*; in "normal faults,...the crust is being extended such as at a divergent boundary," in *Sec. 3.3.2*; the relative expansion of the influence of absolute dating "in geology in the advent of the 20th century," in *Sec. 3.4.1*; the relatively more developed ability of engineering geologists to study geohazards based on geological factors, in *Sec. 3.4.2*; the relative expansion of the influence of uniformitarianism since modern times, in *Sec. 3.5.1*; and the relative expansion of the influence of uniformitarianism in the 19th century and beyond, in *Sec. 3.5.2*).

On the other hand, there is contraction (e.g., the relative decline of the influence of the "atomistic" practice in science in relation to the work of Favera and his team, after they discovered the linkage between the laws of gestalt and the perception of geological structures, in *Sec. 2.2.1*; the relative decline of the use of slates in modern Amerian cemeteries, in *Sec. 2.2.2*; the relatively more contractive nature of plate motions in "convergent boundaries," when contrasted with "divergent boundaries," in *Sec. 2.3.1*; the relative contraction of a sediment, when it is buried and becomes "compacted as more and more material is deposited on top," in *Sec. 2.3.2*; the relatively less developed ability to exist harmoniously with nature without dominating or exploiting it, as advocated by the critics against industrialization in the back-to-nature movement, in *Sec. 2.4.1*; the relatively less developed ability in those specializing in analytic methods to develop more realistic models "under more general conditions to an approximation," in *Sec. 2.4.2*; the relatively less developed ability of a geologist doing computer lab work to engage in well drilling and logging, in *Sec. 2.5.1*; the relatively less developed ability of a creationist to classify things on the basis of scientific investigation, in *Sec. 2.5.2*; the relatively less developed ability of a woman who is endowed with the feminine intuition to "analyze" things with focusedness, in *Sec. 3.2.1*; the relatively less developed ability of a Taoist to intrude into nature by dominating it as a hard-core scientists would like to do with no scruples, in *Sec. 3.2.2*; the contraction of the ash after the ejection, when it becomes cooler and solidified and can no longer expand, until the next rock cycle, in *Sec. 3.3.1*; in "reverse faults,...the crust is being shortened such as at a convergent boundary," in *Sec. 3.3.2*; the relative contraction of the influence of relative dating in geology in the later part of the 20th century, with the rise of absolute dating, in *Sec 3.4.1*; the relatively less developed ability of engineering geologists to study geohazards based on cultural, societal, and mental factors, in *Sec. 3.4.2*; the relative contraction of the influence of mytical beliefs about geological objects since modern times, in *Sec. 3.5.1*; and the relative decline of the view of creationist catastrophism in the 19th century and beyond, in *Sec. 3.5.2*).

And the reverse direction also holds true.

Again, it should be stressed, however, that there are different shades of gray (or different degrees of truth) in the two opposites, and the classification is not necessarily mutually exclusive either. In fact, even when some combinations of the two occur, they only end up sharing the same dialectic relationship, but in a different degree.

Also, there are some other relationships (as a kind of family resemblance) which have something in common with the principle, but they are not exactly the same but only more or less comparable, which varies from case to case.

Good examples of family resemblance in relation to the principle are conquest-autarky, rise-fall (or up-down), spread-shrink, extendingness-shorteningness, and so forth.

10th Thesis: The Theory-Praxis Principle

The tenth thesis is the theory-praxis principle (on the duality of knowledge) in the category of "agency" in existential dialectics (which was first worked out in *FPHE*), in that, if there is theoretical construction, there is likewise its practical application, both technical and normative. There is no theory without praxis—and vice versa.

For instance, on the one hand, there is theory (e.g., the theoretical construction of gestalt psychology like the laws of gestalt, in *Sec. 2.2.1*; the theoretical construction of the nature of rocks in geology, in *Sec. 2.2.2*; the theoretical construction of geology for the understanding of Earth materials, in *Sec. 2.4.1*; the theoretical works of computer science, in *Sec. 2.4.2;* the theoretical construction of electronics, chemistry, etc., in *Sec. 2.5.1*; the theoretical construction of molecular biology, in *Sec. 2.5.2*; the Jungian theory of psychology, in *Sec. 3.2.1*; the theory of quantum mechanics, in *Sec. 3.2.2*; the theoretical construction of seismology, in *Sec. 3.3.2*; the theoretical construction in chemistry and physics about "isotopes," in *Sec. 3.4.1*; the theoretical construction in geology about earthquakes, volcanic eruptions, etc., in *Sec. 3.4.2*; the religious construction about "the biblical flood," in *Sec. 3.5.1*; and the theoretical construction of creationism, in *Sec. 3.5.2*).

On the other hand, there is praxis (e.g., the practical application of the laws of gestalt for the discovery of the HCS in Brazil by Favera and his team, in *Sec. 2.2.1*; the practical application of the understanding of rocks in geology for the use of building cemeteries, in *Sec. 2.2.2*; the practical application of geology for economic and/or industrial purposes, as in economic geology, in *Sec. 2.4.1*; the use of computers for the practical application of hydrogeology, like the use of numerical methods for the solutions of an aquifer system, with the help of computer simulation and calculation, in *Sec. 2.4.2*; the practical application of electronics, chemistry, etc. for the field of petrology with the use of the "electron microprobe," for instance, in *Sec. 2.5.1*; the practical application of molecular biology to molecular systematics for the classification of organisms, in *Sec. 2.5.2*; the practical application of the Jungian psychology for the study of the feminine intuition in geology by Andrews and her followers, in *Sec. 3.2.1*; the practical application of quantum mechanics, by Morowizt and other antireductionists, for the mind-nature integration in the hard sciences, in *Sec. 3.2.2*; the practical application of seismology for the purpose "to protect and prepare possible sites of earthquakes from severe damage," as in "earthquake engineering, earthquake preparedness, household seismic safety, seismic retrofit," etc., in *Sec. 3.3.2*; the practical application of the knowledge about "isotopes" for a variety of geological uses like "the thermochemical techniques...to determine temperature profiles within the crust, the uplift of mountain ranges," and so on, in *Sec. 3.4.1*; the practical application of geology to engineering geology for the protection of life and property, in *Sec. 3.4.2*; the practical application of "the biblical flood" to creationist catastrophism in geology, in *Sec. 3.5.1*; and the practical application of creationism to geology by Burnet, in *Sec. 3.5.2*).

And the reverse direction also holds true.

Again, it should be stressed, however, that there are different shades of gray (or different degrees of truth) in the two opposites, and the classification is not necessarily mutually exclusive either. In fact, even when some combinations of the two occur, they only end up sharing the same dialectic relationship, but in a different degree.

Also, there are some other relationships (as a kind of family resemblance) which have something in common with the principle, but they are not exactly the same but only more or less comparable, which varies from case to case.

Good examples of family resemblance in relation to the principle are discovery-application, knowledge-action, invention-innovation, and so forth.

11th Thesis: The Convention-Novelty Principle

The eleventh thesis is the convention-novelty principle (on the nature of creative thinking) in the category of "agency" in existential dialectics (which was first worked out in *FPHCT*), in that, if there is conventional wisdom, there is likewise novel challenge, to the extent that both convergent and divergent thinking are part of life. There is no convention without novelty—and vice versa.

As srummarized from *FPHCT* (almost verbatim here), there are (a) "creative techniques" and (b) "creative traits," which, when satisfied—in relation to the larger context of the mind, nature, society, and culture—can be used to enhance creative works.

In addition, creative thinking has its own possibilities and limits (in relation to invention), just as it has its own promises and pitfalls (in relation to innovation)— as already analyzed in *FPHCT*).

In the end, creative thinking has its own desirability and dark sides (as also already analyzed in *FPHCT*).

With these clarifications in mind—there are good empirical examples for the convention-novelty princple.

For instance, on the one hand, there is convention (e.g., the conventional wisdom in regard to "the 'atomistic'...practice in science, which advocates the analysis of isolated parts of the problem in order to reduce its complexity," in *Sec. 2.2.1*; the conventional wisdom about using slates to build graves in colonial America, in *Sec. 2.2.2*; the conventional wisdom before the theory of plate tectonics, especially "in the late 19th and early 20th centuries," when "geologists assumed that the Earth's major features were fixed, and that most geologic features such as mountain ranges could be explained by vertical crustal movement, through geosynclinal theory," in *Sec. 2.3.1*; the conventional wisdom of studying rocks by fieldwork, in *Sec. 2.3.2*; the conventional wisdom in modern times about the need to use Earth materials for economic and/or industrial uses, in *Sec. 2.4.1*; the conventional wisdom about doing calculation by hand in the older days, like the example concerning Richardson in the 1920s, in *Sec. 2.4.2*; the conventional wisdom of doing fieldwork in much of the history of geology until recently, in *Sec. 2.5.1*; the conventional wisdom about classifying organisms by Carolus Linnaeus in the 18th century, on the basis of kingdoms and ranks, for insance, in *Sec. 2.5.2*; the conventional wisdom that "women aren't as good at science as men," in *Sec. 3.2.1*; the conventinal wisdom about reductionism in the older days in the hard sciences, in *Sec. 3.2.2*; the conventional wisdom about the classifications about rocks, magmas, and volcanic mechanisms in terms of the number three, in *Sec. 3.3.1*; the conventional wisdom in "the lifetime of the Greek philosopher Anaxagoras in the 5th century BCE to the 14th century CE," in that "earthquakes were usually attributed to 'air (vapors) in the cavities of the Earth,'" in *Sec. 3.3.2*; the conventional wisdom about relative dating in the older days of geology, in *Sec. 3.4.1*; the conventional wisdom about studying geology without much attention to "human development" in the older days prior to the 20th century, in *Sec. 3.4.2*; the conventional wisdom of understanding geological objects with mystical beliefs in antiquity, in *Sec. 3.5.1*; and the conventional wisdom in the older days about the use of creationism in the study of geological time, like the one by Burnet, in *Sec. 3.5.2*).

On the other hand, there is novelty (e.g., the alternative novel challenge to this conventional wisdom about the "atomistic" practice in science, by way of a new practice proposed by Favera and Medeiros along the line of gestalt psychology, in *Sec. 2.2.1*; the alternative novel challenge to this conventional wisdom, by way of the new method of using granites to build graves in modern America, in *Sec. 2.2.2*; the alternative novel challenge to the conventional wisdom of "geosynclinal theory," by the proposal of the new theory of plate tectonics, for plate motions, in *Sec. 2.3.1*; the alternative novel challenge to this conventional wisdom of doing geology by the new method of "experimental petrology" in our times which "employs high-pressure, high-temperature apparatus to investigate the geochemistry and phase relations of," say, "rocks of the lower crust and upper mantle that rarely survive the journey to the surface in pristine condition," in *Sec. 2.3.2*; the alternative novel challenge to this modern industrialist conventional wisdom about Earth materials, by the anti-industrial and back-to-nature critics for a more harmonious co-existence with nature, without the instrumental attempt to exploit or dominate it, in *Sec. 2.4.1*; the alternative novel challenge to this conventional wisdom of doing calculation by hand, with the use of computers instead in our times, in *Sec. 2.4.2*; the alternative novel challenge to the conventional wsidom of doing fieldwork in geology, by the new methods like computer lab work, analytical methods, etc., in *Sec. 2.5.1*; the novel challenge to this conventional wisdom by the new approach known as the evolutionary classification by common descent, in *Sec. 2.5.2*; the alternative challenge to this conventional wisdom by Andrews that "women bring unique talents to geology," in *Sec. 3.2.1*; the alternative novel challenge to this conventional wisdom about reductionism, by the contemporary anti-reductionist movement by Morowitz and others in the hard sciences, in *Sec. 3.2.2*; the alternative novel challenge to this conventional wisdom about the classifications, by questioning why the number three must be the case at all, in *Sec. 3.3.1*; the alternative novel challenge to this ancient conventional wisdom with the contemporary view that earthqake is "the result of a sudden release of energy in the Earth's crust that creates seismic waves," in *Sec. 3.3.2*; the alternative novel challenge to the conventional wisdom about relative dating, with the new way known as absolute dating in the 20th century, in *Sec. 3.4.1*; the alternative novel challenge to the conventional wisdom about studying geology by the new way of focusing on geohazards with human development in mind, as in engineering geology since the 20th century, in *Sec. 3.4.2*; the alternative novel challenge to the conventional wisdom of understanding geological objects with mystical beliefs in antiquity, by the new way of studying geology by fossil records, as proposed by Steno in the 17th century, in *Sec. 3.5.1*; and the alternative novel challenge to the conventional wisdom about the use of creationism in the study of geological time, as shown in the new view by Lyell in regard to his uniformitarianism, in *Sec. 3.5.2*).

And the reverse direction also holds true.

Again, it should be stressed, however, that there are different shades of gray (or different degrees of truth) in the two opposites, and the classification is not necessarily mutually exclusive either. In fact, even when some combinations of the two occur, they only end up sharing the same dialectic relationship, but in a different degree.

Also, there are some other relationships (as a kind of family resemblance) which have something in common with the principle, but they are not exactly the same but only more or less comparable, which varies from case to case.

Good examples of family resemblance in relation to the principle are convergence-divergence (or convergent thinking vs. divergent thinking), normalness-nonnormalness, conformity-nonconformity, and so on.

12th Thesis: The Evolution-Transformation Principle

The twelveth thesis refers to the evolution-transformation principle (on the multiple kinds of agency). in the category of "agency" in existential dialectics (which was first worked out in *FAE* and then in other books of mine).

This principle—and the symmetry-asymmetry principle, for instance—are both about agency. More precisely, the word "agency," in a formal definition, refers to "a person or thing through which power is exerted or an end is achieved." (MWD 2007) It therefore does not have to necessarily involve an intelligent lifeform.

Because of this dual meaning in agency, the evolution-transformation principle is more concerned with *the kind of agency*, that is, both about the *evolution* in the state of nature (e.g., an object of natural beauty) and the *transformation* in the world of intelligent lifeforms (e.g., a work of art). There is no evolution without transformation—and vice versa.

In classical Darwinian evolutionary theory (as more detailedly analyzed in *BNN*), evolution is "blind." But in the human world, change often takes place because of the conscious intervention of humans in transforming society and culture, just to cite two instances.

And the transformative part of the principle precisely refers to the other dimension in the dual meaning of agency, in giving technology (as invented by intelligent lifeforms like humans and, soon, post-humans) a major role to play in the change of the world, which is something that I extensively analyzed in *FHC* in the context of the technophilic lifeworld, espeically though not exclusively since modern times.

For instance, on the one hand, there is evolution (e.g., the natural evolution of humans to make observations about nature in a primitive way, in the state of nature, in *Sec. 2.2.1*; the natural evolution of humans to handle the death of individuals in a primitive way in the state of nature, in *Sec. 2.2.2*; the natural evolution of humans to make observations about nature in a non-scientific way in the state of nature, in *Sec. 2.3.1*; the natural evolution of humans to observe nature in a primitive way in the state of nature, in *Sec. 2.3.2*; the natural evolution of humans to use Earth resources for existence at the subsistence level in the state of nature, in *Sec. 2.4.1*; the natural evolution of humans to count in a simple way in the state of nature, in *Sec. 2.4.2*; the natural evolution of humans to observe rocks with their senses in the state of nature, in *Sec. 2.5.1*; the natural evolution of humans in the state of nature to observe animals and other things in nature, in a rather primitive way, in *Sec. 2.5.2*; the natural evolution of human males and females in the state of nature to relate with each other in a relaively primitive way, in *Sec. 3.2.1*; the natural evolution of humans to study nature in a primitive way in the state of nature, in *Sec. 3.2.2*; the natural evolution of humans to encounter and study volcanic ejection in the state of nature, in *Sec. 3.3.1*; the natural evolution of humans in the state of nature to sense earthquakes by their senses, in *Sec. 3.3.2*; the natural evolution of humans in the state of nature to understand earth materials with their senses, in *Sec. 3.4.1*; the natural evolution of human ability to study geohazards by their senses in the state of nature, in *Sec. 3.4.2*; the natural evolution of humans to understand geological objects with mystical beliefs in relatively primitive historical eras, in *Sec. 3.5.1*; and the natural evolution of humans to understand soil replenishment, even if in a primitive way, in the state of nature, in *Sec. 3.5.2*).

On the other hand, there is transformation (e.g., the technical transformation of human ability to observe nature by way of the invention of geology, with the discovery of HCS and others, in *Sec. 2.2.1*; the technical transformation of humans to handle the death of individuals by the invention of modern cemeteries, with the use of granites, in *Sec. 2.2.2*; the technical transformation of human ability to understand nature by the invention of geology like the theory of plate tectonics, in *Sec. 2.3.1*; the technical transformation of human ability to observe nature by the invention of "high-pressure, high-temperature apparatus" for the study of, say, "the geochemistry and phase relations" of rocks, in *Sec. 2.3.2*; the technical transformation of human ability to use Earth resources by the invention of machines and new systems of knowledge like economic geology for large-scale exploitation of nature, in *Sec. 2.4.1*; the technical transformation of human ability to count by way of the invention of supercomputers nowadays, in *Sec. 2.4.2*; the technical transformation of human ability to observe rocks by the invention of "optical microscopy" and "electron microprobe," for instance, in *Sec. 2.5.1*; the technical transformation of human abiltity to understand animals and things in nautre, by the invention of molecular systematics, in *Sec. 2.5.2*; the technical transformation of human relations between the two genders by way of the invention of different ideologies like feminism, sexism, etc., in *Sec. 3.2.1*; the technical transformation of human ability to study nature by the invention of quantum mechanics, information theory, etc. in the 20th century, in *Sec. 3.2.2*; the technical transformation of humans to understand volcanic ejection by the invention of geosciences like geology, in *Sec. 3.3.1*; the technical transformation of human aiblity to detect earthquakes by the use of seismometers and other hi-tech devices, in *Sec. 3.3.2*; the technical transformation of human ability to understand earth materials by the hi-tech invention of "uranium-lead dating," "potassium-argon dating," "argon-argon dating," "uranium-thorium dating," and the like, in *Sec. 3.4.1*; the technical transformation of human ability to study geohazards by the invention of engineering geology with its broadness of scope and its variety of methods, in *Sec. 3.4.2*; the technical transformation of human understanding of geological objects by the invention of a scientific system of ideas known as modern geology, starting in the 17th century onwards, in *Sec. 3.5.1*; and the technical transformation of human understanding of soil replenishment by the invention of the metaphorical construction of geological time by Hutton in terms of "endless cycles" in geology, in *Sec. 3.5.2*).

And the reverse direction also holds true.

Again, it should be stressed, however, that there are different shades of gray (or different degrees of truth) in the two opposites, and the classification is not necessarily mutually exclusive either. In fact, even when some combinations of the two occur, they only end up sharing the same dialectic relationship, but in a different degree.

Also, there are some other relationships (as a kind of family resemblance) which have something in common with the principle, but they are not exactly the same but only more or less comparable, which varies from case to case.

Good examples of family resemblance in relation to the principle are inorganicness-volition, nonwillingness-willingness, naturalness-technologicalness, naturalness-nonnaturalness, and so on.

13th Thesis: The Symmetry-Asymmetry Principle

The thirteenth thesis is labeled as the symmetry-asymmetry principle (on the relationships among existents) in the category of "agency" in existential dialectics (which was also already worked out in my previous books), in that there is no asymmetry without symmetry—and vice versa.

For instance, the Same can be symmetric and asymmetric towards the Others. But in case of asymmetry (as analyzed in *BDPD*), oppression and self-oppression can occur. So, when the Same is asymmetric towards the Others, the Same can also be relatively asymmetric towards itself in self-oppression, just as the Others can be likewise towards themselves. The subsequent oppressiveness is dualistic, as much by the Same against the Others and itself, as by the Others against the Same and themselves. Both oppression and self-oppression can be achieved by way of downgrading differences between the Same and the Others and of accentuating them.

This is true, even though not all forms of asymmetry have to be about oppression and self-oppression.

In addition, from Chapter Three of *FPHG*, symmetry is not perfect, to be understood in an approximate sense under many life circumstances. With this clarification in mind, hereafter are some empirical examples.

For instance, on the one hand, there is symmetry (e.g., the co-existence of different practices in science, be they "atomistic," "holistic," etc., in *Sec. 2.2.1*; both slates and granites can be used to build graves, in *Sec. 2.2.2*; the existence of different geological theories over time, like geosynclinal theory and the theory of plate tectonics, in *Sec. 2.3.1*; the existence of different ways to do geology, be they about the use of mineralogy, petrography, geochemistry, etc., in *Sec. 2.3.2*; the co-existence of different views about human relationship with nature, be they about harmonious co-existence or exploitative stewardship, in *Sec. 2.4.1*; the co-existence of different methods in hydrogeology, like numerical and analytical methods, in *Sec. 2.4.2*; the co-existence of different methods in the history of geology, in *Sec. 2.5.1*; the co-existence of different ways of classifying organisms in the history of paleontology, in *Sec. 2.5.2*; the co-existence of different views about the two genders in the history of geology, in *Sec. 3.2.1*; the co-existence of different views about the mind and nature in the history of ideas, in *Sec. 3.2.2*; the co-existence of different forms of ash, be they from burnng wood or from volcanic ejection, in *Sec. 3.3.1*; the occurrence of about 500,000 earthquakes per year, in *Sec. 3.3.2*; the existence of different versions of dating in geology, like relative dating and absolute dating, in *Sec. 3.4.1*; the chronic occurrences of different landslides in the U.S. over time, in *Sec. 3.4.2*; the existence of different geological beliefs in different historical eras of geology, in *Sec. 3.5.1*; and the existence of different views about geological time in geology, in *Sec. 3.5.2*).

On the other hand, there is also asymmetry (e.g., the acceptance of the "atomistic" practice in science in the older days—but the popularity of the "holistic" practice in science nowadays, like the use of gestalt psychology, in *Sec. 2.2.1*; granites are less susceptible than slates to decay due to weathering, in *Sec. 2.2.2*; the popularity of geosynclinal theory "in the late 19th and early 20th centuries"—but the popularity of the theory of plate tectonics nowadays, in *Sec. 2.3.1*; the popularity of using "mineralogy" and "petrography," for instance, in the older days to study "the composition and texture of rocks"—but the acceptance of using "geochemistry" and "geophysics," for instance, nowadays to study the "geochemical trends and cycles and the use of thermodynamic data and experiments to better understand the origins of rocks," in *Sec. 2.3.2*; the popularity of harmonious co-existence in the older days of antiquity, but the acceptance of exploitative stewardship towards nature in the modern days of industrialization, in *Sec. 2.4.1*; the popularity of numerical and analytical methods in our times—but the acceptance of hybrid methods combining both methods within some "academic and research groups" in our times, which means that these hybrid methods "have not seen as widespread use outside academic and research groups," in *Sec. 2.4.2*; the popularity of fieldwork in the older days of doing geology—but the acceptance of different competing methods in doing geology nowadays, in *Sec. 2.5.1*; the popularity of the creationist view of special creation in the older days—but the rise of the evolutionary versions of classifying organisms in our time, in *Sec. 2.5.2*; the popularity of the view that "women aren't as good at science as men" in the older days—but the popularity of the view that "women bring unique talents to geology" among the feminists in our time, in *Sec. 3.2.1*; the popularity of Newtonian physics without the consideration of the role of the mind in the older days—but the popularity of quantum mechanism with the consideration of the role of the mind nowadays, in *Sec. 3.2.2*; ash from burning wood is less abrasive than ash from volcanic ejection, in *Sec. 3.3.1*; some earthquakes are greater in magnitude than others, in *Sec. 3.3.2*; the usefulness of relative dating for a sequence of age but not for precise and absolute dates—but the usefulness of absolute dating for precise and absolute dates but not in cases where its limitations occur, like the limits when measured in number of half-lives or when the "fluctuating levels [of radiation] can skew results," in *Sec. 3.4.1*; "[t]he most costly landslide in U.S. history" happened in Thistle, Utah in the spring of 1983—whereas other landslides, by contrast, were less costly, in *Sec. 3.4.2*; the popularity of mystical beliefs about geological objects in antiquity—but the acceptance of an integrated approach towards uniformitarianism and catastrophism nowadays, in *Sec. 3.5.1*; and the focus on "biblical standards" in the view by Burnet—but the focus on "uniformity" in the view by Lyell, in *Sec. 3.5.2*).

And the reverse direction also holds true.

Again, for the last time, it should be stressed, however, that there are different shades of gray (or different degrees of truth) in the two opposites, and the classification is not necessarily mutually exclusive either. In fact, even when some combinations of the two occur, they only end up sharing the same dialectic relationship, but in a different degree.

Also, there are some other relationships (as a kind of family resemblance) which have something in common with the principle, but they are not exactly the same but only more or less comparable, which varies from case to case.

Good examples of family resemblance in relation to the principle are strength-weakness, potence-impotence, balance-extremity, reflexiveness-unreflexiveness, equality-inequality, harmoniousness-unharrmoniousness, and something like that.

14th Thesis: The Regression-Progression Principle

The fourteenth thesis is called the regression-progression princple (on the direction of history) in the category of "outcome" in existential dialectics (which was also already worked out in my previous books), in that neither the cyclical nor the linear views are adequate for explaining many phenomena at all levels. There is no regression without progression—and vice versa.

History progresses to more advanced forms, but with a regressive touch. Examples include no freedom without unfreedom, no equality without inequality, and no civilization without barbarity. This is not an inevitable law, but merely a highly likely empirical trend.

For instance, on the one hand, there is regression (e.g., the regression made in the application of gestalt psychology for geological fieldwork, as shown in the problems pointed out by the critics, in *Sec. 2.2.1*; the regression made in the use of slates to build cemeteries, like the problems of decayed slates and the geology of fear among visitors, in *Sec. 2.2.2*; the regression made in the theory of plate tectonics, as shown in the criticisms by the critics, in *Sec. 2.3.1*; the regression made in rock classification, as shown by the problems associated with it, in *Sec. 2.3.2*; the regression made in the use of Earth materials in economic geology, as shown in environmental and colonial problems pointed out by the critics, in *Sec. 2.4.1*; the regression made in hydrogeology, as shown in the problems pointed out by the critics, in *Sec. 2.4.2*; the regression made in field methods, as shown by the problems associated with them, in *Sec. 2.5.1*; the regression made in molecular systematics, as shown in the example about the inability to detect mitochondria in mammals along the "paternal" line, in *Sec. 2.5.2*; the regression made in the claim by Andrews, as shown in the problems pointed out by the critics, in *Sec. 3.2.1*; the regression made in the anti-reductionist view by Morowitz, as shown in the problems associated with it, in *Sec. 3.2.2*; the regression made in volcanic ejection, as shown by the undesirable consequences like deaths and property damages, in *Sec. 3.3.1*; the regression made in the study of earthquakes, as shown in the problems pointed out by the critics, in *Sec. 3.3.2*; the regression made in absolute dating, as shown in the problems pointed out by the critics, in *Sec. 3.4.1*; the regression made in engineering geology, as shown in the history of failures to protect "life and property against damage caused by geologic conditions," in *Sec. 3.4.2*; the regression made in uniformitarianism, as shown in the problems pointed out by Stephen Jay Gould, as already summarized in Chapter One, in *Sec. 3.5.1*; and the regression made in Gould's analysis, as shown by the criticisms pointed out by the criticis, in *Sec. 3.5.2*).

On the other hand, there is progress (e.g., the progress made in the application of gestalt psychology for geological fieldwork, as shown in the discovery of the HCS in Brazil, in *Sec. 2.2.1*; the progress made in knowing how to build more durable cemeteries, in *Sec. 2.2.2*; the progress made in the theory of plate tectonics for a better understanding of plate motions, in *Sec. 2.3.1*; the progress made in rock classification, for the better understanding of different types of rocks, in *Sec. 2.3.2*; the progress made in the use of Earth materials for the material enhancement of human life, in *Sec. 2.4.1*; the progress made in hydrogeology, for the better understanding of aquifer systems, in *Sec. 2.4.2*; the progress made in field methods for the collection of raw data in doing geology, in *Sec. 2.5.1*; the progress made in molecular systemtics, to help us understand the classification of organisms in relation to genetic details, in *Sec. 2.5.2*; the progress made in the claim by Andrews, in helping us understand the role of intuition in doing geology, in *Sec. 3.2.1*; the progress made in the anti-reductionist view by Morowitz, to help us better understand the role of the mind in understanding nature, in *Sec. 3.2.2*; the progression made in volcanic ejection, as shown by the formation of "highly fertile soil, which has made many volcanic regions densely cultivated and inhabited despite the inherent dangers," in *Sec. 3.3.1*; the progress made in the study of earthquakes, to help us better understand the occurrences of earthquakes and their prevention, in *Sec. 3.3.2*; the progress made in absolute dating for more precise dating in geology, in *Sec. 3.4.1*; the progress made in engineering geology, to help us understand geohazards and, under the right conditions, for the protection of life and property, in *Sec. 3.4.2*; the progress made in uniformitarianism, to help us understand better the role of gradual changes in Earth history, in *Sec. 3.5.1*; and the progression made in Gould's analysis, to help us better understand the nature of metaphorical construction in geology, in *Sec. 3.5.2*).

And the reverse direction also holds true.

Again, it should be stressed, however, that there are different shades of gray (or different degrees of truth) in the two opposites, and the classification is not necessarily mutually exclusive either. In fact, even when some combinations of the two occur, they only end up sharing the same dialectic relationship, but in a different degree.

Also, there are some other relationships (as a kind of family resemblance) which have something in common with the principle, but they are not exactly the same but only more or less comparable, which varies from case to case.

Good examples of family resemblance in relation to the principle are cost-benefit, undesirability-desirability, badness-goodness, risk-opportunity, and so forth.

15th Thesis: The Same-Difference Principle

The fifteen thesis refers to the same-difference principle (on the metamorphosis of change) in the category of "outcome" in existential dialectics (which was worked out in *ALD* and other books of mine), in that an entity, as it evolves over time, can be both different from and similar to its opposing alternatives and does not have to be solely more different from them over time. There is no similarity without difference—and vice versa.

Opposites are not absolute in a black-or-white fashion; so, an entity can become relatively more similar to (or more different from) its opposite over time.

This is further constrained by another principle, that is, the symmetry-asymmetry principle about the relationships among existents under the category about agency in existential dialectics, in that if there is symmetry (equality) between two entities, there is likewise asymmetry (inequality) emerging in a different way.

For instance, on the one hand, there is similarity in outcome (e.g., the contribution to the molding of beliefs and behaviors, regardless of whether this be done by way of the "atomistic" practice in science or by way of the "holistic" counterpart in gestalt psychology, in *Sec. 2.2.1*; the contribution of cemeteries to uneasy feelings among visitors, regardlesss of whether they are built with the use of slates, granites, or else, in *Sec. 2.2.2*; the contribution to the understanding of geolotical features on Earth, like existence of "mountain ranges," regardless of whether this is understood by way of the geosynclinal theory or the theory of plate tectonics, in *Sec. 2.3.1*; the contribution to the molding of beliefs about the world in a certain way, regardless of whether this be achieved by way of a neat rock classification or a more critical one, in *Sec. 2.3.2*; the contribution to the molding and control of beliefs and values, regardless of whether this be achieved by way of an instrumental rationality towards nature for exploitative domination or by way of a substantive rationality towards nature for harmonious co-existence, in *Sec. 2.4.1*; the contribution to the molding of beliefs and values, regardless of whether this be achieved by way of using analytical methods or numerical methods, in *Sec. 2.4.2*; the contribution to the molding and regulation of beliefs and values, regardless of whether this be done by way of field methods or by way of formal methods, in *Sec. 2.5.1*; the contribution to the molding and regulation of beliefs and values, regardless of whether this is achieved by way of creationist classification or by way of evolutionary classification, in *Sec. 2.5.2*; the contribution to the molding and control of beliefs, values, and behavior, regardless of whether this be achieved by way of feminism or by way of the opposite view, in *Sec. 3.2.1*; the contribution to the molding and regulation of beliefs and values, regardless of whether this is done by way of the tao of physics or by way of Newtonian physics, in *Sec. 3.2.2*; the occurrence of magma over time, regardless of whether it is due to intrusion, exrusion, or ejection, in *Sec. 3.3.1*; the contribution to the molding and control of beliefs and values, regardless of whether this is achieved by way of seismology or by way of ancient mythology about earthquakes, in *Sec. 3.3.2*; the contribution to the molding and control of beliefs and values, regardless of whether this is achieved by way of relative dating or by way of absolute dating, in *Sec. 3.4.1*; the contribution to the molding and control of beliefs and values, regardless of whether this is achieved by way of reductionistic engineering geological studies or by way of non-reductionistic engineering geological studies, in *Sec. 3.4.2*; the contribution to the molding and control of beliefs and values, regardless of whether this is achieved by way of mystical ideas about geological objects or by way of scientific ideas about them, in *Sec. 3.5.1*; and the contribution to the molding and control of beliefs and values, regardless of whether this is achieved by way of Burnet's work on geological time or by way of Gould's work on geological time, in *Sec. 3.5.2*).

On the other hand, there is difference in outcome (e.g., the contribution to the molding of beliefs and behaviors by way of the "atomistic" practice in science for the focus on individual parts of an object or a concept—but the contribution to the molding of beliefs and behaviors by way of the "holistic" counterpart in gestalt psychology for the focus on the whole instead, in *Sec. 2.2.1*; the contribution of cemeteries to uneasy feelings among visitors by way of the use of slates, with its "older is creepier" consequence—but the contribution of cemeteries to uneasy feelings among visitors by way of the use of granites, with its "dreary-in-dullness" consequence, in *Sec. 2.2.2*; the contribution to the understanding of geolotical features on Earth, like existence of "mountain ranges," by way of the geosynclinal theory, in that they are so because of "vertical crustal movement"—but the contribution to the understanding of geolotical features on Earth, like existence of "mountain ranges," by way of the theory of plate tectonics, in that they are the result of plate motions, in *Sec. 2.3.1*; the contribution to the molding of beliefs about the world by way of a neat rock classification into three types—but the contribution to the molding of beliefs about the world by way of a more critical analysis with more challenging questions without complete answers, in *Sec. 2.3.2*; the contribution to the molding and control of beliefs and values by way of an instrumental rationality towards nature for exploitative domination—but the contribution to the molding and control of beliefs and values by way of a substantive rationality towards nature for harmonious co-existence, in *Sec. 2.4.1*; the contribution to the molding of beliefs and values by way of using analytical methods for a more simplistic but exact way of understanding reality—but the contribution to the molding of beliefs and values by way of using numerical methods for a more general but approximate way of understanting reality, in *Sec. 2.4.2*; the contribution to the molding and regulation of beliefs and values by way of field methods for a more empiricist mindset to understand reality—but the contribution to the molding and regulation of beliefs and values by way of formal methods for a more analytical and numerical mindset to understand reality, in *Sec. 2.5.1*; the contribution to the molding and regulation of beliefs and values by way of creationist classification for a religious view of the world and its spiritual view of life—but the contribution to the molding and regulation of beliefs and values by way of evolutionary classification for a scientific view of the world and its instrumental way of life, in *Sec. 2.5.2*; the contribution to the molding and control of beliefs, values, and behavior by way of feminism for a more equalitarian way of life between the two genders—but the contribution to the molding and control of beliefs, values, and behavior by way of the opposite view of gender diffrences for a more inequalitarian way of life between the two genders, in *Sec. 3.2.1*; the contribution to the molding and regulation of beliefs and values by way of the tao of physics for a more harmonious relationship with nature—but the contribution to the molding and regulation of beliefs and values by way of Newtonian physics, for a more dominating relationship with nature, in *Sec. 3.2.2*; "[a]shfall [from magma] can become cemented together by heat to form a solid rock called tuff,…[and] breaks down over time, forming highly fertile soil"—whereas lava from magma becomes cooler and solidifies into igneous rocks over time, not soil, in *Sec. 3.3.1*; the contribution to the molding and control of beliefs and values by way of seismology for an instrumental/dominating way of dealing with nature—but the contribution to the molding and control of beliefs and values by way of ancient mythology about earthquakes for a substantive/reverent way of dealing with nature, in *Sec. 3.3.2*; the contribution to the molding and control of beliefs and values by way of relative dating for a relatively less certain view of the world—but the contribution to the molding and control of beliefs and values by way of absolute dating for a relatively more certain view of the world, in *Sec. 3.4.1*; the contribution to the molding and control of beliefs and values by way of reductionistic engineering geological studies for a relativley more simplistic and certain view of reality—but the contribution to the molding and control of beliefs and values by way of non-reductionistic engineering geological studies for a more complicated and uncertain view of reality, in *Sec. 3.4.2*; the contribution to the molding and control of beliefs and values by way of mystical ideas about geological objects for a relatively more reverent relation with nature—but the contribution to the molding and control of beliefs and values by way of of scientific ideas about geolotical objects for a relatively more dominating relation with nature, in *Sec. 3.5.1*; and the contribution to the molding and control of beliefs and values by way of Burnet's work on geological time for a stewardship of nature in relation to a supernatural being—but the contribution to the molding and control of beliefs and values by way of Gould's work on geological time for a relatively less certain view of nature, in *Sec. 3.5.2*).

And the reverse direction also holds true.

Again, it should be stressed, however, that there are different shades of gray (or different degrees of truth) in the two opposites, and the classification is not necessarily mutually exclusive either. In fact, even when some combinations of the two occur, they only end up sharing the same dialectic relationship, but in a different degree.

Also, there are some other relationships (as a kind of family resemblance) which have something in common with the principle, but they are not exactly the same but only more or less comparable, which varies from case to case.

Good examples of family resemblance in relation to the principle are homogeneity-heterogeneity, we-they, and so forth.

16th Thesis: The Post-Human Resettlement

And the sixteenth thesis is about the role of "post-humans," which I originally proposed in my previous books, starting with the first book titled *The Future of Human Civilization* in 2000 and all others afterwards.

As already pointed out in *Sec. 1.8*, I need to emphasize, as this is something that I used to repeat (*almost verbatim*) from my previous books, two clarifications here about the term "post-human" as a neologism in my works.

Firstly, the word "post-human" here should *not* be confused with another term which looks similar but has a totally different meaning in the literature of postmodernism, namely, "post-humanism"—which constitutes a critique of "humanism" as tradionally understood (especially, though not exclusively, in relation to the idea of progress in science and reason in the Enlightenment project). (WK 2008)

My works reject the project of "postmodernism" and propose the future world of what I orignally called "after-postmodernity" in *FHC* and *FCD*, for instance.

And secondly, the word "post-human" here should also *not* be confused with a similar term which is used to champion the ideology of technology for the future co-existence between humans and cyborgs in "trans-humanism." (WK 2008a)

Instead, my term "post-human" in relation to "posthuman-ism" also rejects "transhumanism" (especially, though not exclusively, in relation to the promises of technology) and refers to something else altogether, that is, the future extinction of humans and its post-human successors in deep space and beyond unto multiverses.

My critique of "transhumanism" was more extensively elabroated in *Sec. 2.4.1* of *BEPE*.

With these two clarifications in mind (as summarized in *Table 1.31*)—the post-human resettlement in geology can be addressed in relation to five main directions, based on the resettlement theory of geology as analysed in this book (together with my visions as already worked out in my previous books).

Firstly, the "post-human" resettlement in geology will learn from the resettlement theory of geology in relation to the dialectic context of dynamics and statics—especially in the larger dialectic context of no statics without dynamics (and vice versa).

Secondly, the "post-human" resettlement in geology will also learn from the resettlement theory of geology in relation to the four perspectives of the mind, nature, society, and culture.

Thirdly, the "post-human" resettlement in geology will also learn from the resettlement theory of geology in relation to the non-privilege of any specific theory over others in the literature and the non-integration of them all (as they are not necessarily compatible with each other).

Fourthly, the "post-human" resettlement in geology will have to confront the new challenges as posed by the long-term civilizational development of intelligent life in the distant future, both here on this planet Earth and elsewhere in deep space until multiverses.

As this is something that I regularly pointed out in my previous books for background information (and summarize here, *almost verbatim*), I already worked out, in my numerous books, what these new challenges will be and provide different original visions to meet them—especially, in regard to the future evolutions of the mind (e.g., "the contrarian personality," "the hyper-martial body," "the hyper-sexual body," "floating consciousness," "hyper-spatial consciousness," "thinking machines," "unfolding unconsciousness," "thinking robots," "genetically altered superior beings," "cyborgs," and others), of nature (e.g., "post-cosmology," "the alteration of space-time," "the creation of new matter-energy," "selective geometry," and the like), of society (e.g., "multifaceted war and peace," virtual organizations, different versions of "post-capitalism" and "post-democracy," the movement of "cyclical progression" at both structural and systemic levels, "ambivalent technology," and so forth), and of culture (e.g., the rise of "post-civilization," "transformative aesthetic experience," "contrastive mathematical-logic," and the whatnot) in history (e.g., the age of "after-postmodernity"), in the context of my approaches in relation to methodology (e.g., "sophisticated methodologial holism," "critical-dialectic formal science," etc,) and ontology (e.g., "existential dialectics," "contrastive rationality").

Of course, the examples (as listed above) are not exhaustive but illustrative, since my numerous books have worked out many other visions (as already roughly summarized in *Sec.1.8* and *Sec. 1.8*). Many, though not all, of my visions on the mind, nature, society, culture, history, methodology, and ontology (as cited above) are summarized in the tables as shown in Chapter One and Chapter Four (especially in *Table 4.43* about my original theories on numerous topics).

And finally, or fifthly, the "post-human" resettlement in geology will also learn the resettlement theory of geology in regard to the need to go beyond Earth geology (as further elaborated in the next section).

Beyond Earth Geology

This "post-human" resettlement in geology from the resettlement theory of geology to transcend both dynamics and statics beyond Earth geology can be understood in terms of three great future transformations of geology.

Three Great Future Transformations of Geology

The three great future transformations of geology can be explained hereafter, with the caveat, however, that the classification in terms of the number three is solely aesthetic and utilitarian, as there is no objective basis that it must be classified in term of the number three.

Solar Geology—and Geology of Other Solar Systems

With this caveat in mind—the first great future transformation of geology in the post-human era concerns what I originally call *solar geology* and *geology of other solar systems.*

In Chapter 9 of *The Future of Capitalism and Democracy* (*FCD*), I proposed that "civilizational history will continue into the following cyclical progression of expansion, before it is to be superseded (solely as a high probability, since humans might be destroyed sooner either by themselves or in a gigantic natural calamity) by posthumans at some distant point in after-postmodernity (which I already discussed in *FHC*)," all the way unto multiverses (different constellations of universes):

Local → Regional → Global →Solar →
Galactic → Clustery…→ Multiversal

In *BDPD*, this thesis of mine was referred to as "the theory of the cyclical progression of system integration and fragmentation," at the systemic level—in close relation to another thesis analyzed in Ch.9 of *FCD*, which was called, in the absence of better words, "the multifaceted theory of war and peace" in *FPHK* (and later in *FPHWP*).

If we are now in the age of "globalization" (another subject discussed in *FCD* in greater details), the next frontier will be "solar" (as shown in the diagram above).

This has implications for geology, in that the field of "Earth geology" will become obsolete in future eras, as there will be the need of new fields, and in the absence of better words, let me call them *solar geology* (for this solar system with our Sun as the center) and later, *geology of other solar systems* (for other solar systems beyond our Sun).

Thus, the word "resettlement" in my resettlement theory of geology is suggestive, because of the next frontier of resettlement for intelligent life on Earth to elsewhere in space and beyond.

These future humans (and then post-humans) will need to study the materials and resources on other planets in this solar system (and later in other solar systems) for survival, be they about the atmosphere, the cryosphere, the hydrosphere, the lithosphere, the biosphere, the pedosphere, and other spheres.

Galactic Geology—and Geology of Other Galaxies

The second great future transformation of geology in the post-human era concerns what I originally call *galactic geology* and *geology of other galaxies.*

If we are now in the age of "globalization" and later in the age of what I originally called *solarization*, the next frontier after them will be "galactic" (as shown in the diagram above).

This is where my theory of "post-humanity" at the systemic level becomes even more relevant, as new environments also lead to new species in evolution over time (or in what I called "exo-biology" in *FCD*), especially when combined with the advances in new technologies.

In *The Future of Human Civilizations* (2 volumes) and *The Future of Capitalism and Democracy*, for instance, I already went to great lengths to show the emergence of new post-humans species, as they migrate farther and farther away from Earth into deep space and beyond.

This has profound implications for geology, precisely because the new species will need to study the materials and resources in other galaxies for their own survival in new environments radically different from that on Earth.

This then means that the field of "Earth geology" will become obsolete in future eras, as there will be the need of new fields, and in the absence of better words, let me call them *galactic geology* (for this galaxy) and later, *geology of other galaxies* (for other galaxies beyond ours).

Cosmic Geology—and Geology of Other Cosmoses

And the third great future transformation of geology in the post-human era concerns what I originally call *cosmic geology* and *geology of other cosmoses.*

If we are now in the age of "globalization" and later in the age of what I originally called *solarization* (and then *galacticization*), the next frontier after them will be "multiversal" (as shown in the diagram above).

This is where my other theories in regard to the mind like "floating consciousness" and 'hyper-spatial consciousness," just to cite two examples, become even more useful (although they are already relevant to the earlier two frontiers too).

In *The Future of Capitalism and Democracy*, *The Future of Post-Human Consciousness*, *The Future of Post-Human Geometry*, and *Beyond Cosmology to Post-Cosmology*, just to cite a few examples, I already went to great lengths to show the emergence of new forms of conscousness, as they migrate farther and farther away from Earth into deep space unto multiverses.

This has profound implications for geology, precisely because the new forms of consciousness will need to study the materials and resources in other cosmoses for their own survival in new environments radically different from that on Earth.

This then means that the field of "Earth geology" will become obsolete in future eras, as there will be the need of new fields, and in the absence of better words, let me call them *cosmic geology* (for this cosmos) and later, *geology of other cosmoses* (for other cosmoses beyond ours).

Enormous Implications

These future great transformations of geology have enormous implications for some soul-searching questions which have bothered many fine minds in geological history—Does the earth have "no vestige of a beginning" and "no prospect of an end," as James Hutton once asked? Or, in the conversation between Burnet and Newton in 1681, is there a need for divine intervention for the understanding of the "days" of Genesis? (F. Sulloway 1987)

Questions like this will take whole new dimensions in post-human geology in the context of different planets, solar systems, galaxies, and cosmoses not imagined before in Earth history. Besides, in *Beyond Cosmology to Post-Cosmology* (2010), I already went to great lengths to answer them in the context of different worlds.

In any event, this will be the history of post-human geology which will be shaped by the post-humans in different places at different times.

Just as the study of geological thoughts in historical geology (in *Sec. 3.5.1*) reveals to us their cultural contingency, with each era having its own peculiar concern with particular ideas and holding particular values—the same can be said about post-human geology in other places at different times for "resettlement," in that, in each of those future eras, they too will have their own peculiar concern with particular ideas and holding particular values in the context of their own distinctive version of "solar geology," "galactic geology," and/or "cosmic geology."

Just as geology has contributed to the commerical and industrial uses of earth resources but in the end has brought neither utopia nor dystopia on Earth—it is not entirely nonsensical to think that a comparable fate will await post-humans, in that "solar geology," "galactic geology," and "cosmic geology" will also bring neither utopia nor dystopia in the distant future eras.

In the end, it is not up to us to judge whether or not, or to what extent, the post-human future of geology will be for better and for worse. Instead, it will be shaped by the values and beliefs of future post-humans in a way that the entire history of the human world hitherto existing has never known.

Table 4.1. Sophisticated Methodological Holism (Part I)

• "My methodological holism implies the partiality-totality principle in the ontology of existential dialectics (see the table on the partiality-totality principle for summary), which is against the varieties of (a) reductionism and (b) reverse-reductionism, in relation to (i) concept, (ii) theory, (iii) methodology, and (iv) ontology." (*FC*)

• "[M]y methodological holism here is not opposed to methodological individualism but includes it (and, for that matter, other methodologies too)…." (*FPHC*) For this reason (and others too, as summarized hereafter), my version of methodological holism is sophisticated—not vulgar as sometimes used by inapt scholars using the same term. (*FC*)

• "[M]y methodological holism does not democratically presume that all levels are equally valid, as all levels are not created equal. In other words, in relation to issue X, level A may be more relevant than level B, but in relation to Y, level B can be more relevant than level A instead." (*FPHC*) One excellent example of this vulgar democratic presumption is what I called in *BNN* "the compromise fallacy." (*FC*)

• My methodological holism does not presume that a lower level of analysis is more important than a higher level, solely because the former serves as the foundation for the latter—and vice versa, for that matter. One excellent example of this reductionistic presumption is what I called in *FPHST* "the foundation fallacy." (*FPHST*, *FC*)

(continued on next page)

Table 4.1. Sophisticated Methodological Holism (Part II)

• "[M]y methodological holism does not make any a-priori postulation that there must be a definite (and, for that matter, indefinite) number of levels" in any analysis. (*FPHC*) Nor does it dogmatically require that there must be a certain combination of levels of analysis in a given inquiry. (*FC*)

• "[M]ethodological holism, in my usage, does not assume that all levels…can necessarily be integrated, since methodological holism is not aimed to search for the holy grail of 'an integral theory…' (as is the case for Wilber). In other words, it allows that sometimes some levels may experience irreducible gaps between them, to be understood, at best, as empirical correlations, not as causal relations…." (*FPHC*)

• "[D]ifferent levels may overlap and even interact with each other in a given context (but sometimes may not), and the fact that I even proposed different ways of re-classifying the levels (whenever needed) in *FDC* reinforces this point….The dual danger here is either forcefully making different levels interact when they are just different (or, metaphorically speaking, apples and oranges) or inappropriately ignoring their interactions when some situations instead require them." (*FPHC*)

• "[T]o understand different levels from their own (unique) perspectives (as required by my methodological holism) is not the same as trying to reduce them to a preferred level in the process of learning from other levels. This second kind of multidisciplinary work is not genuine and does no justice to the unique complexities and merits inherent at each level." (*FPHC*)

(continued on next page)

Table 4.1. Sophisticated Methodological Holism (Part III)

• "[My] methodological holism walks a fine line between the artificial classification (separation) of levels and the simultaneous incorporation of them, if only for the sake of human scholarly endeavor. It should be reminded that nature does not impose upon itself the academic classification of the levels of analysis as humans have. The enterprise of classification is therefore anthropocentric." (*FPHC*)

• "[M]y methodological holism advocates neither epistemic subjectivism nor epistemic non-subjectivism (e.g., realism, idealism, and historicism), neither epistemic relativism (e.g., subjectivism, historicism) nor epistemic absolutism (e.g., realism, positivism), neither epistemic reductionism nor epistemic emergencism, and neither epistemic objectivism (e.g., realism, idealism) nor epistemic historicism....Neither does methodological holism, in my usage, accept the false meta-conceptual dichotomy between nominalism and realism....These false dichotomies...are to be transcended. In other words, methodological holism does not fully accept epistemic realism, positivism (a form of epistemic idealism), historicism, subjectivism, and reductionism in epistemology and philosophy of science but learns from the strengths and weaknesses of all of the opposing approaches without siding with any of them...." (*FPHC*)

• "Sophisticated methodological holism is subject to the constraints as imposed by the syntax of existential dialectics (e.g., the partiality-totality principle and the predictability-unpredictability principle). Even in predictability, outcomes are subject to uncertainty, the degree of which varies from case to case." (*FC*)

(continued on next page)

Table 4.1. Sophisticated Methodological Holism (Part IV)

• "Sophisticated methodological holism—when applied, especially though not exclusively, as illustrated in my numerous works—can enrich the understanding of reality in some distinctive ways. Here are three examples (as revealed in each of my books)." (*FPHCT*; *FPHL*)

—"Firstly, it provides a comprehensive analysis of a subject matter, from which much can be learned about reality," "both in relation to the perspectives of the mind, nature, society, and culture and also in relation to a new classification of the subject matter."

—"Secondly, it suggests some visions of the future in relation to the subject matter in question."

—"And thirdly, it proposes some insights on meta-theory (e.g., methodology and ontology) in general—with the clear understanding, however, of the dilemma of specific vs. general ontology (as shown in the table on the syntax of existential dialectics in the context of the dilemma of ontology). For this very reason, all of these ways are important, without reducing one into the analysis of another."

• "In the end, my meta-theory (both sophisticated methodological holism and existential dialectics) serves as a foundation to unify all domains of knowledge into *an unified theory of everything* (by way of some ontological principles and the comprehensive perspectives of the mind, nature, society, and culture). This is so, without commiting the sins of reductionism and reverse-reductionism (as often seen in many holistic approaches, with the fad of sysems approach—be it about systems theory, chaos theory, complexity theory, or else—as a most recent notorious example, which I debunked in *The Future of Complexity* and also in *The Future of Post-Human Formal Science*). My distinctive approach makes good use of different schools of thought without favoring any of them nor trying to integrate them (as they are not necessarily compatible with each other), so as to adjust for subjectivity, diversity, conflict, and complexity, for example. In this sense, the word "unified" does not have to mean integrative approach, in a narrow sense, nor systems approach, in a broad one, both of which I reject while learning from them." (*FPHFS*; *FPHS*)

Sources: A summary of *Sec.1.2* in *FPHC*—and also from *BNN*, *FPHST*, *ALD*, *FC*, *FPHCT*, and the rest of all other books of mine. See the books for more detail.

Table 4.2. On Reductionism and Reverse-Reductionism (Part I)

- **The Partiality-Totality Principle**
  —The partiality-totality principle in the ontology of existential dialectics targets against the varieties of reductionism and reverse-reductionism (as already worked out in my previous books).

- **Against the Varieties of Reductionism**
  —*Conceptual Reductionism*
  - Some illustrative instances involve myriad dualities like mind vs. body, self vs. world, democracy vs. non-democracy, and the like (as already addressed in *FHC*, *FPHC*, and *BDPD*, for instance).

  —*Theoretical Reductionism*
  - A fascinating case study concerns what I originally called "the foundation fallacy" in *FPHST*, in any attempt to naively understand space-time from the physical perspective as the foundation and, consequently, to dangerously dismiss other perspectives.
  - In *FAE*, I elaborated further these versions of reductionism in the literature on aesthetics (e.g., form vs. content, representation vs. expression, critics vs. artists, and externalism vs. internalism).
  - In *FIA*, I revealed other forms of reductionism in the literature on information architecture (e.g., the constructivist argument).
  - In *FPHU*, I showed the persistent legacy of reductionism, this time, in the literature on anomalous experience (e.g., the obsession with physics, chemistry, and biology for explaining anomalous experience).
  - In *FPHE*, I examined another case of reductionism in action, in the context of engineering (e.g., technical constraints vs. normative constraints).
  - In *FPHMM*, I elaborated one more version of reductionism, in relation to the three domains of communication (e.g., the competing views on sending, connecting, and receiving).
  - In *FPHCT*, I explored another version of reductionism, in relation to invention and innovation (e.g., the bio-psychological argument vs. the socio-cultural arguments).

(continued on next page)

Table 4.2. On Reductionism and Reverse-Reductionism (Part II)

- **Against the Varieties of Reductionism *(cont'd)***
  - *—Theoretical Reductionism (cont'd)*
    - In *FPHG*, I identify another version of reductionism, in relation to infinity, symmetry, and dimensionality (e.g., the Euclidean argument vs. the non-Euclidean arguments).
    - In *FPHUP*, I analyzed another version of reductionism, in relation to density and void (e.g., the engineering argument and the ecology argument).
    - In *FPHL*, I examined another version of reductionism, in relation to structure and context (e.g., the structuralist argument and the contextualist argument).
    - In *PFHO*, I explored another version of reductionism, in relation to communcation, decision-making, and leadership (e.g., the rational-system argument vs. the natural-system argument).
    - In *PFHMA*, I revealed another version of reductionism, in relation to the martial body and spirit (e.g., the spiritual argument vs. the materialist argument vs. the defensive argument).
    - In *PFHS*, I analyzed another version of reductionism, in relation to the sexual body and spirit (e.g., the naturalist argument vs. the constructivist argument).
    - In *PFHLAW*, I examined another version of reductionism, in relation to law (e.g., the necessity argument vs. the contengency argument).
    - In *FPHWP*, I showed another version of reductionism, in relation to war and peace (e.g., the aggressivist argument vs. the pacifist argument).
    - In *BEPE*, I analyzed another version of reductionism, in relation to morality and immorality (e.g., the objectivist argument vs. the non-objectivist argument vs. the skeptical argument).
    - In *BCOS*, I examined another version of reductionism, in relation to the contested beginnings and speculative ends of the universe (e.g., the scientific argument vs. the religious argument vs. the esoteric argument vs. the metaphysical argument).
    - In *FPHP*, I show another version of reductionism, in relation to normality and abnormality (e.g., the natural argument vs. the social argument vs. the cultural argument vs. the mental argument).

(continued on next page)

Table 4.2. On Reductionism and Reverse-Reductionism (Part III)

- **Against the Varieties of Reductionism** ***(cont'd)***
  - —*Theoretical Reductionism (cont'd)*
    - •And in *FPHGEOL*, I scrutinize another version of reductionism, in relation to statics and dynamics (e.g., the catastrophe argument vs. the uniformity argument vs. the revision argument).
  - —*Methodological Reductionism*
    - A good instance concerns the debate between different versions of qualitative and quantitative methods (as already analyzed in *FC* and also *FHC*). In *FPHML*, I examined similar reductionism, this time, in the literature on mathematical logic (e.g., the obsession with consistency, soundness, and completeness). And in *FPHFS*, I also explored the problems of reductionism in the context of formal science (e.g., the analytical argument).
  - —*Ontological Reductionism*
    - An excellent example is the debate between emergentism and reductionism in complexity theory and also in psychology (as elaborated in *FPHC*, in the context of Being and Becoming).

- **Against the Varieties of Reverse-Reductionism**
  - —*Theoretical Reverse-Reductionism*
    - There are numerous art and literary theories co-exist. Take the case of literary studies, as there are now Literary Structuralism, Marxist Literary Criticism, New Criticism, Phenomenology, Hermeneutics, Language-Game Literary Criticism, Feminist Literary Criticism, Reception Theory, Reader Response Criticism, Poststructuralism, Semiotics, Pyschoanalytic Literary Criticism, just to cite some well-known ones, with no one being said to be better than any others (as detailedly analyzed in Ch.4 of *FHC*). (S. Raman 1997) In *BNN*, I even introduced "the compromise fallacy" as another good illustration of theoretical reverse-reductionism, in misleadingly treating both genetic and environmental approaches as equally valid.

(continued on next page)

Table 4.2. On Reductionism and Reverse-Reductionism (Part IV)

- **Against the Varieties of Reverse-Reductionism *(cont'd)***
  —*Conceptual Reverse-Reductionism*
    - Any concept of "art" (e.g., fine art, cave art, outsider art, junk art) is deemed acceptable in postmodernism (as already addressed in Ch.4 of *FHC*).

  —*Methodological Reverse-Reductionism*
    - There is the "anything-goes" mentality in postmodernism (e.g., doing art without praxis, doing art with praxis, and doing art by sublation), as analyzed in Ch.4 of *FHC*. And in *FPHFS*, I also exposed the problems of reverse-reductionism in the context of formal science in relation to systems theory.

  —*Ontological Reverse-Reductionism*
    - There are likewise no privileged ontology, and the door is open for anything in postmodernism (e.g., the equal status of the ontology of Being vs. that of Becoming, as already addressed in Ch.4 of *FHC*—and also in *FPHC)*. In *FAE*, I also introduced another version of reverse-reductionism, that is, "the pluralist fallacy," in the context of understanding aesthetic experience, for instance—although this fallacy has been committed not exclusively in relation to the ontological level (but also at the conceptual, theoretical, and methodological ones).

Sources: From my previous books.

Table 4.3. The Concepton of Existential Dialectics (Part I)

- **Sets and Elements**
  - —Sets
    - Ex: the Same
    - Ex: the Others
  - —Elements
    - Ex: whites in 20th century America (in the set of "the Same")
    - Ex: Iraq during the U.S. invasion in 2003 (in the set of "the Others")

- **Relations, Operations, Functions**
  - —Relations (e.g., "belongs," "equals to," "is greater than")
    - Ex: symmetric interactions within the Same (or the Others)
    - Ex: asymmetric interactions between the Same and the Others
  - —Operations (e.g., "and," "or," "not," "if…then")
    - Ex: if the Same oppresses the Others, it will also oppress itself.
    - Ex: the Same is not the Others.
  - —Functions (e.g., goals)
    - Ex: the Same is hegemonic in relation to the Others.

- **Truth Values**
  - —"1" if True (in Symbolic Logic)
    - Ex: the proposition that imperial Japan was hegemonic to China during WWII
  - —"0" if False (in Symbolic Logic)
    - Ex: the proposition that Grenada invaded France in 2003
  - —"1" & "0" if Both True and False (in Dialectic Logic)
    - Ex: the proposition that the rabbit-duck picture refers to a duck
  - —"~1" & "~0" if Neither True Nor False (or N/A, in Dialectic Logic)
    - Ex: the proposition that God really exists

(continued on next page)

Table 4.3. The Conception of Existential Dialectics (Part II)

- **Axioms, Postulates, Theorems, Principles**
  - —Axioms
    - Ex: the reflexive axiom—"any quantity is equal to itself"
  - —Postulates
    - Ex: the SSS postulate—"if the three sides of a triangle are congruent to their corresponding parts, then the triangles are congruent"
  - —Theorems (and Principles) in Existential Dialectics
    - <u>In Relation to Method</u>
      - –#1: The formalness-informalness principle
      - –#2: The absoluteness-relativeness principle
      - –#3: The partiality-totality principle
      - –#4: The predictability-unpredictability principle
      - –#5: The explicability-inexplicability principle
    - <u>In Relation to Structure</u>
      - –#6: The finiteness-transfiniteness principle
      - –#7: The preciseness-vagueness principle
      - –#8: The simpleness-complicatedness principle
      - –#9: The openness-hiddenness principle
      - –#10: The denseness-emptiness principle
    - <u>In Relation to Process</u>
      - –#11: The change-constancy principle
      - –#12: The order-chaos principle
      - –#13: The slowness-quickness principle
      - –#14: The expansion-contraction principle
    - <u>In Relation to Agency</u>
      - –#15: The theory-praxis principle
      - –#16: The convention-novelty principle
      - –#17: The evolution-transformation principle
      - –#18: The symmetry-asymmetry principle
      - –#19: The softness-hardness principle
    - <u>In Relation to Outcome</u>
      - –#20: The regression-progression principle
      - –#21: The same-difference principle

(continued on next page)

Table 4.3. The Conception of Existential Dialectics (Part III)

Notes: The categories and examples in each are solely illustrative (not exhaustive). The comparison is also relative (not absolute), nor are they mutually exclusive. As generalities, they allow exceptions.
Sources: Starting from Ch.6 of *BCPC* and also from other books of mine

Table 4.4. The Syntax of Existential Dialectics I: The Principles (Part I)

- **In Relation to Method**

  —*The Formalness-Informalness Principle*
  (On the Formal Requirements of Logical Systems)
  - The formal requirements of a logical system (e.g., consistency, soundness, and completeness) have both usefulness and non-usefulness, to the extent that, if there are formal systems requiring them, there are alternative ones which do not. Thus, it does *not* exclude classical logics but simply goes beyond both classical and non-classical logics, while learning something from each. There is no formalness without informalness—and vice versa.
  - For instance, existential dialectics can make use of both classical logics under certain conditions (e.g., especially, though not exclusively, when they are clear-cut, etc.) and non-classical logics under alternative conditions (especially, though not exclusively, when they are "unknown," "irrelevant," "ambiguous," "possible," with "different degrees of truth," empirically inconsistent in a desirable way, etc.). (WK 2008u)
  - Family resemblance: e.g., logicalness-nonlogicalness, rationality-nonrationality, etc.
  - Sources: From *FPHML*. See also my later books.

  —*The Explicability-Inexplicability Principle*
  (On the Underlying Mechanisms of Things)
  - Both explicability and inexplicability are part of the understanding of things. There is no explicability without inexplicability—and vice versa.
  - This principle tells us the duality of the research dilemma, in that, if reality can be explained in some ways, it also has its other ways which are not quite explainable, at a given point in time.
  - Family resemblance: e.g., underlyingness-regularness, causation-regularness, causation-correlation, etc.
  - Sources: Especially from *FPHU*—and also *FPHC*. See also my later books.

(continued on next page)

Table 4.4. The Syntax of Existential Dialectics I: The Principles (Part II)

- **In Relation to Method** ***(cont'd)***

—*The Predictability-Unpredictability Principle (On the Occurrence of Events)*

- Both predictability and unpredictability have a major role to play in the occurrence of things, so that neither determinism nor indeterminism wins the centuries-old fight. There is no predictability without unpredictability—and vice versa.
- There are events which are predictable, just as there are those which are not. Or what is regarded as unpredictable at one point in time may turn out to be predictable later, and, conversely, what is deemed as predictable may turn out to not be so predictable. Even in predictability, outcomes are subject to uncertainty, the degree of which varies from case to case.
- Family resemblance: e.g., sureness-arbitrariness, etc.
- Sources: Especially from *FC*. See also my later books.

—*The Partiality-Totality Principle (On the Relationships between Whole and Parts)*

- The whole is not the sum of the parts. There is no partiality without totality—and vice versa.
- Any inquiry about a phenomenon in the work is to guard against the varieties of (a) reductionism and (b) reverse-reductionism.
- Reductionism and reverse-reductionism can be (i) conceptual, (ii) theoretical, (iii) methodological, and (iv) ontological.
- Family resemblance: e.g., individualisticness-holisticness, partness-wholeness, analysis-synthesis, etc.
- Sources: Especially from *FC*. See also my later books.

(continued on next page)

Table 4.4. The Syntax of Existential Dialectics I: The Principles (Part III)

- **In Relation to Method *(cont'd)***

—*The Absoluteness-Relativeness Principle*
(On the Multiplicity of Things)

  - There is the multiplicity of things in reality, be they about entities, qualities (or properties), and relationships, such that what is acceptable from one standpoint may not be so from another. For instance, if there is something absolute, there is likewise something relative. There is no absoluteness without relativeness—and vice versa.
  - Both absoluteness and relativeness here are also relevant to different modalities often cited in the literature on ontology, such as possibility (e.g., something "can" happen) and its opposite (e.g., impossibility), probability (e.g., something "will" happen) and its opposite (e.g., improbability), and necessity (e.g., something "should" happen) and its opposite (e.g., contingency).
  - Family resemblance: e.g., uniformity-diversity, internalness-externalness, immanence-transcendence, etc.
  - Sources: From *FPHK*. See also my later books.

(continued on next page)

Table 4.4. The Syntax of Existential Dialectics I: The Principles (Part IV)

- **In Relation to Structure**
  —*The Finiteness-Transfiniteness Principle*
  (On the Nature of Numbers)
  - If there are finite things, there are likewise transfinite ones. There is no finiteness without transfiniteness—and vice versa.
  - To avoid confusion, my usage of the word "transfinite" here differs radically from the one used by Cantor (and other mathematicians) for "relative" infinity—and is more limited, in light of the problems confronting any attempt to understand the idea of infinity, be it by intuition, imagination, and conception (as detailedly analyzed in *Sec. 2.2.3* of *FPHG*).
  - Instead, by "transfinity," I allow numbers which can be many times larger—or smaller, for that matter—than the finite things that we encounter in daily life, but they do not have to be related to the idea of infinity at all (which may not exist).
  - Of course, there may be some *borderline* cases, in which it is not clear whether the number in question is transfinite (in my usage) or simply a mathematical convenience. A good example of a borderline case is the Planck unit of length for "the smallest space possibly measured in nature," which is "less than billionths of trillionths of trillionths of an inch" (or something like 1.6×10^{-35} meters).
  - That said—my usage of "transfinity" can also resolve (or better, dissolve) an age-old problem in philosophy known as "Zeno's paradoxes" (as already explained in *Sec. 2.2.3* of *FPHG*).
  - Family resemblance: e.g., boundedness-quasiunboundedness, smallness-largeness, microness-macroness, etc.
  - Source: From *FPHG*. See also my later books.

(continued on next page)

Table 4.4. The Syntax of Existential Dialectics I: The Principles (Part V)

- **In Relation to Structure *(cont'd)***
  —*The Simpleness-Complicatedness Principle*
  (On the Interconnection among Things)
  - Both simpleness and complicatedness are vital, without favoring one over the other, in that each is utilized, depending on the basis of the perspectives of nature, the mind, culture, and society. And even when a combination of them is preferred, the dilemma is only shifted to a combinational degree of concern. There is no simpleness without complicatedness—and vice versa.
  - In relation to taxonomy, simpleness has its heuristic usefulness, just as complicatedness has its realistic representation, for instance. And in relation to network, simpleness has its economical attractiveness, just as complicatedness has its practical reliability, for instance.
  - Family resemblance: e.g., inflexibility-flexibility, standardization-specialization, imperfectness-perfectness, superficiality-depth, shallowness-deepness, economicalness-elaboratedness, plainness-circumspection, onesidedness-multisidedness, etc.
  - Sources: Especially from *FIA*. See also my later books.

  —*The Denseness-Emptiness Principle*
  (On the Distribution of Entities in Space)
  - Both density and void are needed, in relation to the mind, nature, culture, and society, albeit in different ways. There is no denseness without emptiness—and vice versa.
  - For clarity, the term "void" is used here only as an approximation of emptiness (depending on the degree of the lack of density), since, in physics, it is well known that "empty" space is not really empty all the way, because it can be full of energy (e.g., random quantum fluctuations at the sub-atomic level, and, for that matter, dark energy in the universe) and matter (e.g., different versions of sub-atomic particles, and, for that matter, dark matter in the universe).
  - Family resemblance: e.g., fullness-voidness, nearness-farness, concentration-dispersion, etc.
  - Sources: First worked out in *FPHUP*. See also my later books.

(continued on next page)

Table 4.4. The Syntax of Existential Dialectics I: The Principles (Part VI)

- **In Relation to Structure *(cont'd)***
  —*The Preciseness-Vagueness Principle*
  (On the Refinement of Things)
  - Both preciseness and vagueness are important, not that one is better than the other, but that both are used, in different degrees of preference, in accordance to the contextual application from the perspectives of nature, the mind, culture, and society. Even when both are used in a combination, the dilemma is shifted instead to one of combinational concern. There is no preciseness without vagueness—and vice versa.
  - In relation to taxonomy, preciseness has its taxonomic clarity, just as vagueness has its classificatory flexibility, for instance. And in relation to network, vagueness has its explorative liberty, just as preciseness has its conceptual definitiveness, for instance.
  - Family resemblance: e.g., clarity-ambiguity, directness-indirectness, quantitativeness-qualitativeness, describability-nondescribability, specificity-obscurity, specificity-generality, thickness-thinness, concretness-abstractness, etc.
  - Sources: Especially from *FIA*. See also my later books.

(continued on next page)

Table 4.4. The Syntax of Existential Dialectics I: The Principles (Part VII)

- **In Relation to Structure *(cont'd)***
  —*The Openness-Hiddenness Principle*
  (On the Detection of Things)
  - Reality has its hidden face, just as it is open to outside view in some other ways. There is no openness without hiddenness—and vice versa.
  - For instance, in the context of anomalous experience, certain aspects of reality can be open for examination (e.g., the experiment with SPECT images, the transmission of telepathemic bit, and the use of metonymies and metaphors). Other aspects, however, remain hidden, and examples include the elusive deeper nature of spiritual reality in the intangible realm, the undetectability of different branched universes, the underlying mechanisms of signals traveling between different folds of physical space-time, and the dependence on language and the bias for science and logic).
  - Family resemblance: e.g., overtness-covertness, publicness-privateness, openness-closedness, transparency-secrecy, openness-biasedness, etc.
  - Sources: Especially from *FPHU*. See also my later books.

- **In Relation to Process**
  —*The Change-Constancy Principle*
  (On the Alteration of Things)
  - Change occurs over time, although constancy is also allowed. There is no change without constancy—and vice versa.
  - Asymmetry undergoes changes over time, so does symmetry.
  - Old players fade away, and new ones emerges, with ever new causes and ever new forms.
  - Family resemblance: e.g., dynamicness-staticness, instability-stability, etc.
  - Sources: First named in *BCPC*. Especially from *FHC*, *FCD*, and *FPHC*. See also other books of mine (from that point on).

(continued on next page)

Table 4.4. The Syntax of Existential Dialectics I: The Principles (Part VIII)

- **In Relation to Process *(cont'd)***
  —*The Expansion-Contraction Principle*
  (On the Growth of Things)
  - Entities in the world can both expand in some ways and contract in other ones, as part of their nature. There is no expansion without contraction—and vice versa.
  - For instance, in the context of anomalous experience, one of its most fascinating expressions is none other than the analysis of different causes of the slow but steady expansion (and, furthermore, if put in a different way, unfolding) of unconsciousness.
  - Family resemblance: e.g., conquest-autarky, rise-fall (or up-down), spread-shrink, extendingness-shorteningness, etc.
  - Sources: Especially from *FPHU*. See also my later books.

  —*The Slowness-Quickness Principle*
  (On the Speed of Change)
  - Both slowness and quickness co-exist, with their own internal tension, to the extent that each fights for its own relevance with the other, in accordance to the perspectives of nature, the mind, culture, and society, without one being the victor and the other being the vanquished in the long haul. Even when both are chosen in other cases, this dilemma is only transferred into something else with a combinational character. There is no slowness without quickness—and vice versa.
  - In relation to taxonomy, quickness has its efficient usability, just as slowness has its aesthetic appeal, for instance. And in relation to network, quickness has its adventurous readiness, just as slowness has its risk-adverse convenience, for instance.
  - Family resemblance: e.g., inconvenience-convenience, passiveness-activenessness, gradualness-abruptness, deceleration-acceleration, etc.
  - Sources: Especially from *FIA*. See also my later books.

(continued on next page)

Table 4.4. The Syntax of Existential Dialectics I: The Principles (Part IX)

- **In Relation to Process *(cont'd)***
  —*The Order-Chaos Principle*
  (On the Pattern of Things)
  - Both order and chaos are vital in the process of change in the world. The preference for order is biased, since it does not give sufficient attention to the vital role of chaos in the transformation of the world (without somehow reducing it for the understanding of order). There is no order without chaos—and vice versa.
  - The scientific search for order in the world is often a hidden bias in its ontological obsession with order, since chaos is often treated as the "bad" guy, with order as the "good" guy (for the end goal of science).
  - Neither order nor chaos is the final end of the world, and one is not to be treated as the means for the other in the transformation of things. Both are fundamental in their recurrent dialectical interactions with each other over time, without reducing one for the other.
  - Family resemblance: e.g., lawfulness-disorder, order-disorder, etc.
  - Sources: Especially from *FC*. See also my later books.

- **In Relation to Agency**
  —*The Softness-Hardness Principle*
  (On the Force of Change)
  - This has to do with *the force of change* under the category about agency in existential dialectics, in that any change by an agent, be it organic (like humans) or non-organic (like natural objects), can occur in a forceful (aggressive) or gentle (pacific) way, which can come in all shapes and sizes, of course. There is no softness without hardness—and vice versa.
  - Family resemblance: e.g., peacefulness-violence, cooperation-competition, dovishness-hawkishness, reward-punishment, peace-war, pacificity-aggression, etc.
  - Sources: Especially from *ALD*. See also my later books.

(continued on next page)

Table 4.4. The Syntax of Existential Dialectics I: The Principles (Part X)

- **In Relation to Agency** ***(cont'd)***
  —*The Symmetry-Asymmetry Principle*
  (On the Relationships among Existents)
  - There is no symmetry without asymmetry—and vice versa.
  - For instance, the Same can be symmetric and asymmetric towards the Others. But in case of asymmetry, oppression and self-oppression can occur. So, when the Same is asymmetric towards the Others, the Same can also be relatively asymmetric towards itself in self-oppression, just as the Others can be likewise towards themselves.
  - The subsequent oppressiveness is dualistic, as much by the Same against the Others and itself, as by the Others against the Same and themselves.
  - Both oppression and self-oppression can be achieved by way of downgrading differences between the Same and the Others and of accentuating them.
  - This is true, even though not all forms of asymmetry have to be about oppression and self-oppression.
  - In addition, from Chapter Three of *FPHG*, symmetry is not perfect, to be understood in an approximate sense under many life circumstances.
  - Family resemblance: e.g., strength-weakness, balance-extremity, harmoniousness-unharrmoniousness, reflexiveness-unreflexiveness, equality-inequality, potence-impotence, etc.
  - Sources: From all my books, starting with *FHC*. First named in *BCPC*. See, for instance, *FCD* for more details on the Same and the Others.

(*continued on next page*)

Table 4.4. The Syntax of Existential Dialectics I: The Principles (Part XI)

- **In Relation to Agency** ***(cont'd)***

  —*The Theory-Praxis Principle*
  (On the Duality of Knowledge))
  - This has to do with the duality of knowledge in existential dialectics, in that, if there is theoretical construction, there is likewise its practical application, both technical and normative. There is no theory without praxis—and vice versa.
  - Family resemblance: e.g., discovery-application, knowledge-action, invention-innovation, etc.
  - Source: Especially from *FPHE*. See also my later books.

  —*The Evolution-Transformation Principle*
  (On the Multiple Kinds of Agency)
  - This principle (and the symmetry-asymmetry principle) are both about *the agency of change*. The word "agency," in a formal definition, refers to "a person or thing through which power is exerted or an end is achieved." (MWD 2007b) It therefore does not have to necessarily involve an intelligent lifeform.
  - Because of this dual meaning in agency, the evolution-transformation principle is more concerned with *the multiple kinds of agency*, that is, both about the *evolution* in the state of nature (e.g., an object of natural beauty) and the *transformation* in the world of intelligent lifeforms (e.g., a work of art, an air-conditioner, etc.). There is no evolution without transformation—and vice versa.
  - And the transformative part of the principle precisely refers to the other dimension in the dual meaning of agency, in giving technology a major role to play in the change of the world, which is something that I extensively analyzed in *FHC* in the context of the technophilic lifeworld, espeically though not exclusively since modern times.
  - Family resemblance: e.g., nonwillingness-willingness, naturalness-technologicalness, naturalness-nonnaturalness, inorganicness-volition, etc.
  - Sources: Especially from *FAE*. See also other books of mine.

(*continued on next page*)

Table 4.4. The Syntax of Existential Dialectics I: The Principles (Part XII)

- **In Relation to Agency *(cont'd)***
  —*The Convention-Novelty Principle*
  (On the Nature of Creative Thinking)
  - If there is conventional wisdom, there is likewise novel challenge, to the extent that both convergent and divergent thinking are part of life. There is no convention without novelty—and vice versa.
  - In addition, in this context of the convention-novelty principle, there are, in the absence of better words, what I want to call (a) *creative techniques* and (b) *creative traits*, which, when satisfied—in relation to the larger context of the mind, nature, society, and culture—can be used to enhance creative works. The list of creative techniques and traits (as summarized in *Table 4.1* of *FPHCT*) are not exhaustive, of course, but the examples here constitute a great beginning of understanding the structure of creative thinking.
  - Yet, lest the reader gets carried away by the euphoria about creative thinking in our time, it should be reminded that creative thinking has its own possibilities and limits (as shown in *Table 2.1* of *FPHCT* on invention), just as it has its own promises and pitfalls (as shown in *Table 3.1* of *FPHCT* on innovation). In the end, it should also be stressed that creative thinking has its own desirability and dark sides (as shown in *Table 4.2* of *FPHCT*).
  - This has important daily implications for an average individual, and the word "average" here allows different degrees of variation from one individual to another in a population, at any given point in history.
  - Family resemblance: e.g., normalness-nonnormalness, conformity-nonconformity, convergence-divergence, etc.
  - Sources: First from *FPHCT*. See also my later books.

(*continued on next page*)

Table 4.4. The Syntax of Existential Dialectics I: The Principles (Part XIII)

• **In Relation to Outcome**

—*The Regression-Progression Principle*
(On the Direction of History)

- Neither the cyclical nor the linear views are adequate for explaining many phenomena at all levels. There is regression without progression—and vice versa.
- History progresses to more advanced forms, but with a regressive touch. Examples include no freedom without unfreedom, no equality without inequality, and no civilization without barbarity.
- This is not an inevitable law, but merely a highly likely empirical trend.
- Family resemblance: e.g., cost-benefit, undesirability-desirability, badness-goodness, risk-opportunity, etc.
- Sources: From all my books, starting with *FHC*. First named in *BCPC*.

—*The Same-Difference Principle*
(On the Metamorphosis of Change)

- An entity, as it evolves over time, can be both different from and similar to its opposing alternatives and does not have to be solely more different from them over time. There is no similarity without difference—and vice versa.
- Opposites are not absolute in a black-or-white fashion; so, an entity can become relatively more similar to (or more different from) its opposite over time.
- Family resemblance: e.g., homogeneity-heterogeneity, we-they, etc.
- Sources: Especially from ALD. See also my later books.

Notes: The features in each principle are solely illustrative (not exhaustive). The comparison is also relative (not absolute), nor are they mutually exclusive. As generalities, they allow exceptions.

Table 4.5. The Syntax of Existential Dialectics II: The Principles as Short Cuts (Part I)

• The principles should be treated with caution, lest misunderstanding occurs, since they do not constitute rigid dualities (or dichotomies).

• The reason is that each pair in an ontological principle consists of two opposites, which are, however, merely short cuts both for multiple variations and degrees, as well as for different interactions with multiple other entities unlike them. In this light, each pair can end up having hundred (or even thousand, if not more) different versions, which interact with hundred (or even thousand, if not more) other entities. There are two clarifications here.

• Firstly, this conception of shortcuts is not mutually exclusive nor absolute, in that the opposites can come in all shapes and sizes, with different degrees. For instance, by analogy, just as there are different degrees of the two colors "white" and "black"—there are likewise different degrees of the opposites in each ontological pair, to the extent that there can be multiple entities (not only two) interacting with each other in each pair.

• And secondly, this conception of shortcuts do not ignore other possible entities in interacting with the two opposites (with their different versions) in each pair. For instance, by analogy, there are not only the two colors "white" and "black" as opposites, since there are other colors too besides them like "yellow," "green," "purple," or else. The same logic can be applied to each ontological pair, in that they also interact with other entities, not with only two of them (with their different versions).

• Therefore, with these two clarifications in mind—each pair in an ontological principle serves only as an abbreviation for something more complicated and, therefore, although it contains two opposites, it should not be confused as a duality (dualism).

(continued on next page)

Table 4.5. The Syntax of Existential Dialectics II:
The Principles as Short Cuts
(Part II)

• Existential dialectics rejects any dualism (or dichotomy) as too rigid and instead allows the multiplicity of entities, to the extent that between thc two opposites in each pair exist many other alternatives to choose from. They are named in that short form for aesthetic elegance, instead of listing all possible entities between the two opposites in the title.

• To be dialectic is to go beyond any rigid dichotomy and transcend into something different altogether in the long haul. One may be tempted to call the dialectic logic here with a different name like existential "multilectics" (instead of "dialectics"), but this naming is incorrect (or even misleading), for the two reasons aforestated.

• After all, each ontological pair come in all shapes and sizes (with different degrees) and do not exist by themselves but also interact with other entities unlike them. The virtue of revealing an ontological pair is to show how they relate within themselves (in multiple versions) and also interact with others unlike them (also in multiple versions).

• This will be clear in the table on "the dialectic constraints imposed by the principles."

Source: From Ch.1 of *FPHK*

Table 4.6. The Syntax of Existential Dialectics III: The Principles as Family Resemblances

• Each ontological principle is generic, with some other comparable ontological pairs to be put in the same family (like a *family resemblance*). Ludwig Wittgenstein (1953) once suggested the idea of "family resemblance" in explaining different games classified under the same family called *games*.

• Why should, for instance, playing football and chess as playing "games," when it is well understood that football is not the same as chess? The answer is that, although each game is different and has different rules, many of them (though not all) share, more or less, some commonalites (e.g., scoring as necessary for winning).

• And this is so, even though some games share more than some others in any given selection of critieria, and no two games are exactly identical. So, his point here is that there is no essential core which is common to all games, and the best that one can look for is some characteristics which are common to many (but not all) games. (A. Biletzki 2006)

• By the same logic—in the previous section on selection criteria, the flexibility-inflexibility pair can be put in the family resemblance of the simpleness-complicatedness principle, although the two pairs are not exactly identical. Likewise, the directness-indirectness pair can be put in the family resemblance of the preciseness-vagueness principle, although, again, the two pairs are not exactly identical.

• In this sense, which specific pair in a family should be used to designate the name of the family can be at times a bit arbitrary, but with good reason.

• For illustration, in the context of method, the partiality-totality principle can take the different form like individualisticness-holisticness, just as the explicability-inexplicability principle can take the different form like underlyingness-regularness—although each two pairs are not exactly identical and have slightly different meanings and usages. A more comprehensive listing of this family resemblance for all other principles is shown in the table on the syntax of existential dialectics.

Source: From Ch.1 of *FPHK*

Table 4.7. The Syntax of Existential Dialectics IV: The Dialectic Constraints Imposed by the Principles (Part I)

• **Co-Existent and Asymmetric**

—The principles, as they constitute the syntax of existential dialectics, are dialectic in character, such that, when they are applied, they impose dialectic constraints on how reality is to be understood. Consider, say, the symmetry-asymmetry principle as an illustration here, in order to summarize two main characters of the dialectic constraints in question.

—Firstly, to be dialectic here is to go beyond the narrow dichotomies (and, for that matter, any rigid multi-dimensional classificatory scheme), be they about "self" vs. "world," "freedom" vs. "unfreedom," "barbarity" vs. "civilization," "individuality" vs. "communality," and so on.

—One way to do so (to go beyond) is to consider them all in terms of co-existence (without favoring one over the rest). For instance, my theory of "post-civilization" (to be summarized later in the section on the pragmatics of existential dialectics) is to go beyond barbarity and civilization in terms of understanding barbarity and civilization as being co-existent. And the same logic can be said in relation to my theories of "post-democracy," "post-capitalsm," and others (also to be introduced later in the section on the pragmatics of existential dialectics), in regard to freedom vs. unfreedom, equality vs. inequality, communality vs. individuality, spirituality vs. materiality, and so on.

—But to consider them all (in the dichotomies—and, for that matter, in any rigid multi-dimensional classifactory schme) as co-existent is not the same as to imply that the opposites in any classificatory scheme are all equal, since, in accordance to the symmetry-asymmetry principle (as an illustration here), if they are equal in terms of being considered as co-existent, they are asymmetric in terms of being unequal in dialectic interaction (e.g., X can be more dominant than Y in case A, or Y is more dominant than X in case B).

(continued on next page)

Table 4.7. The Syntax of Existential Dialectics IV: The Dialectic Constraints Imposed by the Principles (Part II)

—For this reason, there are different versions of "post-democracy" and "post-capitalism" in my theories. As an illustration, in version I of the theory of post-democracy, freedom is more dominant than equality, whereas in version II of the theory of post-democracy, equality is more so than freedom.

—But this "X more than Y" has to be understood in the context of dialectic logic (not in conventional logic), in that both "X" and "Y" are important in post-democracy (in the context of dialectic logic), but in an asymmetry way. By contrast, in conventional logic, it often favors one over the other—be it in regard to privileging freedom over equality in Fascism, favoring freedom relatively more than equality in Liberal Democracy, or favoring equality relatively more than freedom in Socialist Democracy. In the latter two cases (about Liberal Democracy and Socialist Democracy), the difference between dialectic logic and conventional logic can be one in degree, not in kind—in this sense, albeit not in other senses.

—The same logic can be said about the relatonships between individuality and communality, between spirituality and materiality, and between formal legalism and informal legalism in the different versions of my theory of post-capitalism.

• **Transcendent**

—Secondly, to be dialectic is to go beyond the narrow dichotomies (and, for that matter, any rigid multi-dimensional classificatory scheme) in another way, this time, in a transcendent way, that is, in exploring other possibilities or even other issues not considered within the narrow confines of narrow dichotomies (and, for that matter, any rigid multi-dimensional classificatory scheme).

(continued on next page)

Table 4.7. The Syntax of Existential Dialectics IV: The Dialectic Constraints Imposed by the Principles (Part III)

—As an analogy, to go beyond the narrow color dichotomy of "black" and "white" is not just to choose both "black" and "white" (as in the first meaning) but also to explore other color options (e.g., "green," "purple," "blue," etc.—and, alternatively, "shade," "line," "curve," etc.). By the same logic, to go beyond "democracy" is to transcend democracy (as in version III of the theory of "post-democracy") and to explore other possibilties of lifeforms (e.g., floating consciousness, hyper-spatial consciousness, etc., to live beyond the narrow obsession with freedom and equality).

—This dialectic character of the principles in existential dialectics has important implications for the pragmatics of existential dialectics (as will be clear shortly, in the section on the pragmatics of existential dialectics).

Notes: The examples here are solely illustrative (not exhaustive). The comparison is also relative (not absolute), nor are they mutually exclusive. As generalities, they allow exceptions.

Sources: First explicitly stated in *ALD*. Also from all other books of mine.

Table 4.8. The Syntax of Existential Dialectics V: Further Clarifications (Part I)

• **Pioneering**
—Firstly, the total number of ontological principles is unknown, to be discovered later, as our knowledge of the world becomes more advanced.
—Consequently, the principles as introduced in my books are not exhaustive, with new ones being added, whenever more of them are discovered in later research.
—At least, future generations can pick up where I leave off and continue the discovery.
—In this sense, my work should be treated as a pioneering effort for the development of a systematic, comprehensive analysis of a new general ontology for the future of knowledge.

• **Flexible**
—Secondly, the principles are not rigidly classified, as they can be reclassified in a different way.
—For instance, the preciseness-vagueness principle is classified under the category of "structure" but can be reclassified under the category of "method," although in so doing, it has a different meaning in the context of method.
—The same logic applies to the same-difference principle under the category of "outcome," which can be reclassified under the category of "structure," although in so doing, once more, it has a different meaning in the context of structure.

• **Mutually Constraining**
—Thirdly, the principles are to be understood together, not that each prinicple is to be analyzed independently of others. After all, the principles are mutually constraining, in that they work together as a whole.

(continued on next page)

Table 4.8. The Syntax of Existential Dialectics V: Further Clarifications (Part II)

- **Selectively Useful**
  —Fourthly, the principles are relevant to all subject matters, but some principles are more useful to some subject matters than others—as implied in the symmetry-asymmetry principle. This is true, even if different studies of the same kind can yield different views about the degree of relevance for each ontological principle, depending on the specific nature of a research in question, needless to say.

- **Anti-Reductionistic**
  —And fifthly, the principles constitute only two levels of analysis, this time, at the ontological and methodological levels—while other levels of analysis (from the perspectives of the mind, nature, society, and culture) are also needed, in order to understand reality in its totality (as explained in the section on "sophisticated methodological holism," so as to avoid the dual dangers of reductionism and reverse-reductionism).

Notes: The examples here are solely illustrative (not exhaustive). The comparison is also relative (not absolute), nor are they mutually exclusive. As generalities, they allow exceptions.
Sources: From my previous books like *FHC*, *FCD*, *FPHC*, *BDPD*, *BCPC*, *BNN*, *FC*, *FAE*, and so on.

Table 4.9. The Syntax of Existential Dialectics VI: The Dilemma of Specific vs. General Ontology (Part I)

• Ontology is often more complicated and imposes some seemingly insurmountable difficulties concerning what constitutes a kind of ontology which can be valid enough to be accepted by the wider intellectual community in question. An excellent example concerns what I want to call *the dilemma of ontology* in relation to the relationship between "specific ontology" and "general ontology." (M. Bunge 1999; R. Corazzon 2007)

• On the one hand, "general ontology," as Mario Bunge (1999) put it, "studies all existents." But, on the other hand, "special ontology studies one genus of thing or process—physical, chemical, biological, social, etc." (M. Bunge 1999; R. Corazzon 2007)

• Consequently, the relationship between the former and the latter is that, as an illustration, "whereas general ontology studies the concepts of space, time, and event, the ontology of the social investigates such general sociological concepts as those of social system, social structure, and social change." (M. Bunge 1999; R. Corazzon 2007)

• With this formal definition in mind, the relationship between the two forms of ontology only poses *the dilemma of ontology*, in that general ontology is possible to the extent that its general contructs must be broad enough to serve as the "lowest common denominator" for all disciplines, but this character of the lowest common denominator is not only hard to find but also its usefulness is limited. (WK 2008b)

• But specific ontology fares no better either, in that its specific constructs primarily serve a specific discipline in question, although other disciplines can appropriate them for their own purposes, but in a different context (especially, though not exclusively, when used as a metaphor from one field to another). For instance, the Kantian category of understanding under the heading of "quantity" constitutes a type of specific ontology, which is specific to the discipline of mathematics, although other disciplines can make use of the categories in their own context.

(continued on next page)

Table 4.9. The Syntax of Existential Dialectics VI: The Dilemma of Specific vs. General Ontology (Part II)

• A solution lies in my proposal of "sophisticated methodological holism," which requires all levels of analysis in any subject matter. See the tables on sophisticated methodological holism, and on reductionism and reverse-reductionism, for more details.

Source: From *Sec. 1.4* of *FPHK*

Table 4.10. The Syntax of Existential Dialectics VII: Types of Inappropriate Family Resemblances

- **Inadequate Family Resemblances**
  —Ex: consistency-inconsistency: the formalness-informalenss principle
  —Ex: cognition-noncognition: the formalness-informalness principle

- **Compound Family Resemblances**
  —Ex: reality-nonreality: the formalness-informalness principles, the absoluteness-relativeness principle
  —Ex: normality-abnormality: the symmetry-asymmetry principle, the density-emptiness principle, the convention-novelty principle, the same-difference principle
  —Ex: structure-context: the simpleness-complicatedness principle, the theory-praxis principle
  —Ex: freedom-unfreedom: the symmetry-asymmetry principle, the expansion-contraction principle
  —Ex: morality-immorality: the denseness-emptiness principle, the symmetry-asymmetry, the same-difference principle
  —Ex: particularness-universalness: the preciseness-vagueness principle, the simpleness-complicatedness principle
  —Ex: pureness-mixedness: the absoluteness-relativeness principle, the simpleness-complicatedness principle, the same-difference principle
  —Ex: shortsightedness-foresightedness: the partiality-totality principle, the simpleness-complicatedness principle
  —Ex: shorttermness-longtermness: the partiality-totality principle, the simpleness-complicatedness principle, the slowness-quickness principle

Notes: The examples here are solely illustrative (not exhaustive). The comparison is also relative (not absolute), nor are they mutually exclusive. As generalities, they allow exceptions.
Source: First summarized in *BEPE* (and updated in later books)

Table 4.11. The Semantics of Existential Dialectics

- **Abstract Structure vs. Specific Meanings**
  —The syntax of existential dialectics so understood in terms of ontological principles only gives us the structure of ontology in the world, in an abstract (general) sense. These principles by themselves do not tell us the specific meanings in a given context.
  —In order to grasp the specific meanings of the principles in a given context, it is necessary to study the semantics of existential dialectics. And the analysis of the ontological principles as family resemblances in the section on syntax is only a starting point (and thus overlaps a bit with the semantics of existential dialectics here).
  —With this caveat in mind, the reason that I have often gone in great lengths in my previous books on different subjects is to explain the specific meanings of the principles when applied in different contexts.

- **Specific Meanings in Specific Fields**
  —For instance, in *FPHST*, I used the first three principles (i.e., the change-constancy principle, the regression-progression principle, and the symmetry-asymmetry principle) to propose "the perspectival theory of space-time," for a better way to understand space and time—especially, though not exclusively, in relation to future post-human history (as summarized in *Table 3.6*, *Table 3.7*, *Table 3.8*, and *Table 3.9*). In so doing, I had to introduce concepts and theories specific to the field of physics and other related fields (e.g., "absolute space" and "absolute time" in "classical mechanics" and "relative space-time" in "the theory of relativity").
  —In *BNN*, I also exploited the three principles to propose the "transcendent" approach to the study of genes and memes as a new way to understand the interaction between nature and nurture. In so doing, I had to explore concepts and theories in the world of evolutionary theory (e.g., "mutation," "variation," "adaptation," "selection," and "inheritance" in Darwinian evolutionary theory) and neural biology (e.g., "chromosome," "gene," "DNA," "RNA," "protein," "neuron," "neural network," and "behavior").

Notes: The categories and examples in each are solely illustrative (not exhaustive). The comparison is also relative (not absolute), nor are they mutually exclusive. As generalities, they allow exceptions.
Sources: First explicitly stated in *ALD*. Also from all other books of mine.

Table 4.12. The Pragmatics of Existential Dialectics (Part I)

- **The Two-Way Street Connecting Theory and Meta-Theory**
  —The pragmatics of existential dialectics is not a one-way street (that is, using the ontological principles for theoretical insights in praxis) but a two-way one, that is, (a) from meta-theory to theory, and (b) from theory to meta-theory.
  —(a) On one side of the street, the ontological principles can inspire some theoretical insights in praxis, that is, in relation to some specific fields.
  —(b) On the other (opposing) side of the street, however, the study of a subject matter in the specific fields in turn reveals some more hitherto unknown ontological principles to be discovered and identified. For this reason, three new principles were added in *FC*, one in *FAE*, and two in *ALD*, on top of the original three in *BCPC*—after some research on the specific subject matters.

- **Direct and Indirect Applications**
  —*Direct*
    - The logic of existential dialectics can shed some theoretical insights on diverse phenomena in the world, and good instances are the pertinent use of the principles of existential dialectics for the theoretical insights on the freedom/unfreedom dialectics, the equality/inequality dialectics, and the wealth/poverty dialectics in my previous works.
    - My latest books like *FPHST* and *BNN* also use the principles to reveal some theoretical insights on the perspectives of space and time (as in *FPHST*) and of nature and nurture (as in *BNN*).

  —*Indirect*
    - The theoretical insights can further be used to reveal other phenomena directly from them (viz., the theoretical insights) and therefore indirectly from the principles themselves. A good instance is the use of the theoretical insights on the freedom/unfreedom and equality/inequality dialectics for the understanding of the civilization/barbarity dialectics.

(continued on next page)

Table 4.12. The Pragmatics of Existential Dialectics (Part II)

- **Direct and Indirect Applications (*cont'd*)**
    - Even in indirect applications, however, a phenomenon under study can still be directly related back to the principles themselves. In the example as cited above, the civilization/barbarity dialectics can be directly related to the principles of existential dialectics without the intermediate role of the freedom/unfreedom and equality/inequality dialectics.

- **Multiple Levels of Application**

  —The theoretical insights can be applied to different levels of analysis, even though in a given example, it may refer to one level only. For instance, in the example concerning the freedom/unfreedom dialectics, it can be used at the structural level (e.g., in relation to the theory of cyclical progression of hegemony), but it can be exploited as well for other levels (e.g., the theory of post-capitalism at the institutional level).

Notes: The categories and examples in each are solely illustrative (not exhaustive). The comparison is also relative (not absolute), nor are they mutually exclusive. As generalities, they allow exceptions.
Sources: From Ch.6 of *BCPC*. See also other books of mine.

Table 4.13. The Freedom-Unfreedom Dialectics (Part I)

- **On Having**
  - —*In Relation to the Technological*
    - (1) if freer from submission to Nature, then less free from ecological degradation (Deep and Social Ecology), even if in a hi-tech form
    - (2) if freer from technological inconvenience / backwardness, then less free from technological control and the loss of privacy
    - (3) if freer from technological (material) backwardness, then less free from the abusive (barbaric) maltreatment of the primitive Others
  - —*In Relation to the Everyday*
    - (1) if freer from abject poverty, then less free from artificial needs/discontents (Frankfurt School)
    - (2) if freer from sensual suppression, then less free from violent sublimation (Freud)
    - (3) if freer from the snobbishness of high culture, then less free from the shabbiness (leveling-off effect) of mass culture (Tocqueville)
    - (4) if freer from the inefficiency of traditional "compassionate economy," then less free from the bondage of a "ruthless [competitive] economy" (Keynes)
    - (5) if freer from anarchy in the state of nature (system fragmentation), then less free from government regulations and controls in system integration

(continued on next page)

Table 4.13. The Freedom-Unfreedom Dialectics (Part II)

• **On Belonging**

—In Relation to the Good and the Just

(1) if freer from disciplinary society, then less free from society of control (Foucault)

(2) if freer from the tyranny of one or a few, then less free from the tyranny of the majority (or sometimes, minority veto)

(3) if freer from elitist decision making, then less free from political gridlock/cleavage

(4) if freer from arbitrary (discretionary) administration, then less free from bureaucratic irrationality (Weber) and legal trickery (loopholes)

• **On Being**

—In Relation to the True

(1) if freer from unscientific dogmas, then less free from instrumental abyss (nihilism). Or conversely, if freer from meaninglessness, then less free from dogmas.

(2) if freer from the bondage of partiality/partisanship (e.g., prejudice, discrimination), then less free from the danger of impartiality and neutrality (e.g., opportunism, unrealisticness, lack of compassion, inaction)

(3) if freer from making generalizations, then less free from being unable to understand much of anything

—In Relation to the Holy

(1) if freer from collective conscience, then less free from social loneliness

(2) if freer from religious absoluteness, then less free from spiritual emptiness

—In Relation to the Beautiful/Sublime

(1) if freer from artistic non-autonomy, then less free from aesthetic disillusion (deconstruction)

(continued on next page)

Table 4.13. The Freedom-Unfreedom Dialectics (Part III)

Notes: The examples in each category are solely illustrative (not exhaustive), and the comparison is relative (not absolute), nor are they necessarily mutually exclusive. And some can be easily re-classified elsewhere. As generalities, they allow exceptions.
Sources: A reconstruction from Ch.10 of *FCD*, based on *FHC*

Table 4.14. The Equality-Inequality Dialectics (Part I)

- **On Having**
  - —*In Relation to the Technological*
    - (1) if more equal in treating Nature with spiritual unity, then less equal in suppressing the dominant drive to transcend it altogether
  - —*In Relation to the Everyday*
    - (1) if more equal in building social plurality, then less equal in leveling-off effects (e.g., the subsequent relative intolerance of high/intellectual ethos in mass culture industry)
    - (2) if more equal in socioeconomic distribution beyond a certain point, then less equal in efficiency (e.g. resentment, the erosion of work ethics)
    - (3) if more equal in urging an affirmative action program, then less equal in creating victim mentality (in oneself), stigma (from others), reverse discrimination (against the once privileged), and mediocracy (against the more able)

- **On Belonging**
  - —*In Relation to the Good and the Just*
    - (1) if more equal in banning monarchic/oligarchic exclusion, then less equal in producing "the tyranny of the majority" or of "minority veto"
    - (2) if more equal in encouraging participatory decision making, then less equal in inducing political divisiveness (gridlock/cleavage in power blocs) and organizational oligarchy
    - (3) if more equal in institutionalizing a decentralized bureaucracy, then less equal in falling into more territorial/turf politics (intrigues)

(continued on next page)

Table 4.14. The Equality-Inequality Dialectics (Part II)

- **On Being**
  - —In Relation to the Beautiful / Sublime
    - (1) if more equal in accepting diverse styles ("anything goes" mentality), then less equal in artistic good quality (in leveling-off effects against the best)
  - —In Relation to the True
    - (1) if more equal in tolerating multiple viewpoints (no matter how extreme), then less equal in epistemic standards
  - —In Relation to the Holy
    - (1) if more equal in celebrating any cults and sects (no matter how questionable), then less equal in spiritual depth and authenticity

Notes: The examples in each category are solely illustrative (not exhaustive), and the comparison is relative (not absolute), nor are they mutually exclusive. And some can be easily reclassified elsewhere. As generalities, they allow exceptions.
Sources: A reconstruction from Ch.10 of *FCD*, based on *FHC*

Table 4.15. The Duality of Oppression in Existential Dialectics: Oppression and Self-Oppression (Part I)

- **From the Same to the Others and Itself**
  —The Oppression by the Same against the Others
  - *By way of downgrading differences*
    –Ex: on judiciary caprice for corporate crimes (*Sec.2.2.1.2.1*)
    –Ex: on the deceptive politics of liberation (*Sec.3.5*)
    –Ex: on the humanitarian mystique (*Sec.4.4*)
    –Ex: on the fad of emotional intelligence (*Sec.5.3*)
  - *By way of accentuating differences*
    –Ex: on the legal sophistry of self-defense (*Sec.2.3*)
    –Ex: on the legal semantics of proportionality (*Sec.2.4*)
    –Ex: on the tricky politics of external threat (*Sec.3.4*)
    –Ex: on the appeal of the Far Right for democracy (*Sec.5.4*)
    –Ex: on the democratic axis of evil (*Sec.5.5*)
    –Ex: on the democratic way of brutality and revenge (*Sec.5.6*)
    –Ex: on democratic autocracy (*Sec.6.4*)

  —The Oppression by the Same against Itself
  - *By way of downgrading differences*
    –Ex: on the politics of fear (*Sec.2.2*)
    –Ex: on the trickery of compassionate conservatism (*Sec.3.2*)
    –Ex: on the deceptive politics of patriotism (*Sec.3.3*)
  - *By way of accentuating differences*
    –Ex: on the caprice of due process on domestic suspects (*Sec.2.2*)
    –Ex: on the false security/freedom dilemma (*Sec.6.5.2*)

(continued on next page)

Table 4.15. The Duality of Oppression in Existential Dialectics: Oppression and Self-Oppression (Part II)

• **From the Others to the Same and Themselves**
—The Oppression by the Others against the Same
• *By way of downgrading differences*
–Ex: on judiciary caprice in the reverse direction (*Sec.2.2.1.2.2*)
–Ex: on equal pay (*Sec.6.2.1.1*)
–Ex: on equal representation (*Sec.6.2.1.2*)
–Ex: on affirmative action program (*Sec.6.3.1.1*)
–Ex: on same-sex marriage (*Sec.6.3.1.2*)
• *By way of accentuating differences*
–Ex: on sexual harassment (*Sec.6.2.2.1*)
–Ex: on physical violence (*Sec.6.2.2.2*)
–Ex: on sexual exploitation (*Sec.6.2.2.3*)
—The Oppression by the Others against Themselves
• *By way of downgrading differences*
–Ex: on the reverse-class mystique (*Sec.4.2*)
–Ex: on the reverse-black mystique (*Sec.4.3*)
–Ex: on self-discrimination by downgrading (*Sec.6.3.2.2*)
• *By way of accentuating differences*
–Ex: on self-discrimination by accentuating (*Sec.6.3.2.1*)

Notes: The examples are solely illustrative (not exhaustive), nor are they mutually exclusive. As generalities, they allow exceptions. Also, both forms of oppression co-exist in all of the examples, so the listing of them are only meant in a relative, not absolute, sense.
Source: A summary of the sections (as cited) in Chs.2-6 of *BDPD*. See text for more info and references.

Table 4.16. The Structure of Existential Dialectics I: The Freedom/Unfreedom and Equality/Inequality Dialectics

- Each freedom and equality produces its own unfreedom and inequality, regardless of whether the pair occurs in political society (with the nation-state), in civil society (with some autonomy from the state), or elsewhere (e.g., in the private sphere of individual homes)—and regardless of whether freedom and equality are understood as "negative" or "positive."

- Oppression is dualistic, as much by the Same against the Others and itself, as by the Others against the Same and themselves.

- Both forms of oppression and self-oppression can be achieved by way of downgrading differences (between the Same and the Others) and of accentuating them.

- The relationships are relatively asymmetric between the Same and the Others and relatively symmetric within them. This is true, even when the Same can be relatively asymmetric towards itself in self-oppression, just as the Others can be likewise towards themselves.

- Symmetry and asymmetry change over time, with ever new players, new causes, and new forms.

Notes: The examples in each category are solely illustrative (not exhaustive) nor necessarily mutually exclusive, and the comparison is relative (not absolute). As generalities, they allow exceptions. "Negative" freedom is freedom "from" (e.g., freedom from poverty), whereas "positive" freedom is freedom "to" (e.g., freedom to the state of enlightenment). "Negative" equality is "procedural" equality (e.g., equality of opportunity), while "positive" equality is "substantive" equality (e.g., equality of outcome). Existential dialectics impose constraints on freedom and equality in democracy, non-democracy, and post-democracy. There is no utopia, in the end; even should there be one, dystopia would exist within it.
Sources: From *Table 1.5* of *BDPD*—and also from *FHC*, *FCD*, and *FPHC*

Table 4.17. The Structure of Existential Dialectics II: The Wealth/Poverty Dialectics

• There is no wealth without poverty, just as there is no poverty without wealth.

• The wealth/poverty dialectics occurs in the realms of having, belonging, and being, in relation to the material, relational, and spiritual.

• The wealth/poverty dialectics also expresses itself at the multiple levels of analysis in accordance to methodological holism, be they about the micro-physical, the chemical, the biological, the psychological, the organizational, the institutional, the structural, the systemic, the cultural, and the cosmological.

• The wealth/poverty dialectics is a different manifestation of existential dialectics in general, subject to the principles in its logic of ontology—just as the freedom/unfreedom and equality/inequality dialectics are likewise.

• There is no economic utopia, in the end; even should there be one, dystopia would exist within it.

Notes: The main points here are solely illustrative (not exhaustive) nor necessarily mutually exclusive, and the comparison is relative (not absolute). As generalities, they allow exceptions.
Sources: From *BCPC*. See also *FCD* and *FHC*.

Table 4.18. The Structure of Existential Dialectics III: The Civilization/Barbarity Dialectics

---

• There is no civilization without barbarity.

• The civilization/barbarity dialectics applies in the four civilizing processes (e.g., the rationalizing process, the pacifying process, the stewardizing process, and the subliming process).

• The civilization/barbarity dialectics is another (different) manifestation of existential dialectics in general, subject to the principles in its logic of ontology—just as the freedom/unfreedom and equality/inequality dialectics and the wealth/poverty dialectics are likewise.

• There is no utopia, in the end; even should there be one, dystopia would exist within it.

Notes: The main points here are solely illustrative (not exhaustive) nor necessarily mutually exclusive, and the comparison is relative (not absolute). As generalities, they allow exceptions.
Sources: From *BCIV*. See also *FCD*, *FHC*, and *BDPD*.

Table 4.19. The Double Sides of Virtual Organizations (Part I)

- **Psychological**
  —Ex: virtual psychosis
  —Ex: impersonality and loneliness in quaternary social relations

- **Organizational**
  —Ex: the race for power and interests
  —Ex: the world of unequal successes
  —Ex: the bureaucratic life of its own
  —Ex: lesser accountability and transparency

- **Economic**
  —Ex: the economic divides
  —Ex: the erosive impact of commercialization

- **Political**
  —Ex: the anti-authoritarian myth
  —Ex: different power struggles among groups (e.g., Conservative, Reformist, Radical)

- **Structural**
  —Ex: different localities (e.g., climates, scenery)
  —Ex: divided domains (e.g., core, peripheral, sub-peripheral)
  —Ex: substitution effect of social capital

- **Systemic**
  —Ex: uneven advances in transportation and communications, and geographical migration

- **Cultural**
  —Ex: conflicting civilizational fabrics (e.g., Confucian, Islamic, Western)
  —Ex: the bias of the Liberal Democratic agenda
  —Ex: the addiction to moral fanaticism

(continued on next page)

Table 4.19. The Double Sides of Virtual Organizations (Part II)

Notes: These categories and examples are solely illustrative (not exhaustive), and some of the items can be reclassified somewhere else. Nor are they mutually exclusive. Since they are generalities, exceptions are expected.
Source: From Ch.7 of *FCD*

Table 4.20. Beyond the World of Titans, and the Remaking of the World Order

- **Hyper-Empires**
  —Ex: The Chinese Union
  —Ex: The Indian Union

- **Meso-Empires**
  —Ex: The European Union
  —Ex: The North American Union

- **Micro-Empires**
  —Ex: The Latin American Union
  —Ex: The Middle Eastern Union

- **The Rest of the World**
  —<u>Odd Powers</u>
    - Ex: Japan
    - Ex: Russia

  —<u>The Poor Club</u>
    - Ex: The African Union

  —<u>Ambivalent Regions</u>
    - Ex: Southeast Asia
    - Ex: Oceania
    - Ex: South Asia
    - Ex: Central Asia
    - Ex: Southern/Eastern Europe
    - Ex: North Africa
    - Ex: Central America
    - Ex: Others (e.g., the Korean peninsula)

Source: A summary of Chs.2-5 (of *BWT*)

Table 4.21. The Origins of Authoritarian Liberal Democracy

• **The Geopower of Nature (Ch.4 of *ALD*)**
—Ex: Power Character and Geographical Strategy
—Ex: Living Space and Territorial Expansion
—Ex: Strategic Heartland and Containment

• **The Biopsychology of the Mind (Ch.5 of *ALD*)**
—Ex: The Bell Curve and Mass Intelligence
—Ex: Group Analysis and Mass Knowledge
—Ex: Groupthink and Elite (Mis)calculation

• **The Disciplinary Control of Society (Ch.6 of *ALD*)**
—Ex: Social Organizations of Ruthlessness
—Ex: Social Institutions of Greed
—Ex: Social Structure of Exclusion
—Ex: Social Systems of Violence

• **The Molding Force of Culture (Ch.7 of *ALD*)**
—Ex: The Tradition of Conquest
—Ex: The Rationalization of Unreason

Notes: The examples in each category are solely illustrative (not exhaustive), and they are also relative (not absolute), nor are they mutually exclusive. As generalities, they allow exceptions.
Source: A summary of Chs.4-7 of *ALD*

Table 4.22. The Theory of Post-Democracy I: The Priority of Freedom over Equality (Part I)

- **Differences**
  - —*For the aggressive Lions (the strong Elitists)*
    - •Setting up rank distinctions among unequals (e.g., between inferior humans and superior post-humans, or later among inferior post-humans and superior ones, relatively speaking)
    - •Yearning for being not only distinguished from unequals, but also the first among equals (the best of the very best)
    - •Soul-searching for a high spiritual culture (not the trashy one for the masses). Mass culture is a dirty joke for them.
  - —*For the manipulative Foxes (the weak Counter-Elitists)*
    - •Seeking a gentle hegemony by way of more communitarian concerns (for inferior humans and, later, inferior post-humans)
    - •Being more sympathetic to less formal-legalistic institutions and values

- **Similarities**
  - —*For both Lions and Foxes*
    - •Exploring different spheres of non-human consciousness in the cosmos (something vastly superior than the human one)
    - •Recognizing the democratic illusions (e.g., no freedom without unfreedom, no equality without inequality, or simply no justice without injustice, and vice versa)

(continued on next page)

Table 4.22. The Theory of Post-Democracy I: The Priority of Freedom over Equality (Part II)

Notes: The two callings and examples in each category are solely illustrative (not exhaustive), since there will be many different post-human value ideals in the distant future of post-human civilization. The comparison is also relative (not absolute) towards post-democracy, so this is not just a version of free-market democracy (nor Fascism/Nazism, as shown in the table later on democracy, non-democracy, and post-democracy). Nor are they mutually exclusive. As generalities, they allow exceptions. And the specific forms of post-human post-democratic ideals need to be further developed in future after-postmodern history, as they will be different from the ones we now know. The point here is to solely give an extremely rough picture of a small part of the world to come that we have never known.
Source: From Ch.10 of *FCD*. Refer to text for more info and references.

Table 4.23. The Theory of Post-Democracy II: The Priority of Equality over Freedom

- **Hybrid Versions of**
  —Ex: the Trans-Feminine Calling
  —Ex: the Trans-Sinitic Calling
  —Ex: the Trans-Islamic Calling
  —Ex: the Trans-Outerspace Calling

- **Qualifications**
  —These four versions of post-capitalist value ideals need not automatically be post-democratic, just as capitalism does not necessarily mean democracy. They are two different entities—though closely related.
  —But up to a certain threshold of elevating equality at the farther expense of freedom, the democratic ideals will be overcome and cease to exist.
  —The overcome will not be socialist or communist, but post-democratic with no freedom without unfreedom and no equality without inequality, subject to the constraints of existential dialectics.

Notes: The callings are solely illustrative (not exhaustive), since there will be many different post-human value ideals in the distant future of post-human lifeforms. The comparison is also relative (not absolute), nor are they mutually exclusive. As generalities, they allow exceptions. And the specific forms of post-human post-democratic ideals need to be further developed in future after-postmodern history, as they will be different from the ones we now know. The point here is to solely give an extremely rough picture of a small part of the world to come that we have never known.
Source: From Ch.10 of *FCD*. Refer to text for more info and references.

Table 4.24. The Theory of Post-Democracy III: The Transcendence of Freedom and Equality (Part I)

- **Transcending Freedom in Floating Existence**
  —*Freedom*: seeking an ultimate elimination of the body. Being without the body. The aim is to transcend freedom in the end into a metaphysical state (i.e., beyond the physique).
  —*Unfreedom*: yet facing difficult trade-offs. The sacrifice of bodily existence and its joyfulness. An eternal boredom in floating existence in dark deep space, though with alternative pleasures. There is no free lunch even in the state of transcending freedom.

- **Transcending Equality in the Rivalry of Cosmic Hegemony**
  —*Inequality*: competing to outlast other lifeforms in floating existence, or just marginalizing them for one's hegemonic expansiveness in the rest of the cosmos (and even beyond). Universalism is only for the mediocre.
  —*Equality*: accepting only those of one's rank as equal partners in the vast spacetime for cosmic supremacy. Even here, the aim is to transcend equality into a metaphysical state.

(continued on next page)

Table 4.24. The Theory of Post-Democracy III: The Transcendence of Freedom and Equality (Part II)

Notes: Do not confuse this transcendence of freedom and equality (as one version of post-democracy) with the naïve temptation to transcend the freedom/unfreedom and equality/inequality dialectics. Existential dialectics hold true for freedom and equality in all cultures and societies—past, present, or future (i.e., democracy, non-democracy, and post-democracy), regardless of whether freedom and equality are conventionally understood as "negative" or "positive."
Also, the two features and examples in each are solely illustrative (not exhaustive), since there will be many different post-human value ideals in the distant future of post-human lifeforms. The comparison is also relative (not absolute), nor are they mutually exclusive. As generalities, they allow exceptions. And the specific forms of post-human ideals even for these radically alien floating lifeforms (and others unknown to us) need to be further developed in future after-postmodern history, as they will likely be different from the ones herein illustrated. The point here is to solely give a very rough picture of a small part of the extremely alien world to come that we have never known.
Source: From Ch.10 of *FCD*. Refer to text for more info and references.

Table 4.25. Democracy, Non-Democracy, and Post-Democracy (Part I)

• **Democracy**
—*Theoretical Constructs*
•The pursuit of freedom and equality (in various degrees), regardless of whether freedom and equality can be understood as "negative" or "positive"
(1) more equality than freedom: The relative priority of the good over the right
(2) more freedom than equality: The relative priority of the right over the good
—*Types*
•Only (1): Different versions of communitarian moral universalism
•Only (2): Different versions of liberal moral universalism
•(1) or (2): Different versions of anarchic (non-nation-state) moral universalism
•(1) or (2): Different versions of postmodern moral localism

• **Non-Democracy**
—*Theoretical Constructs*
•The focus on (1') equality or (2') freedom, but not both, regardless of whether freedom and equality can be understood as "negative" or "positive"
—*Types*
•Only (1'): Different versions on the Far Left (e.g., Stalinism, Robespierrianism)
•Only (2'): Different versions on the Far Right (e.g., Nazism, absolute monarchism)

(continued on next page)

Table 4.25. Democracy, Non-Democracy, and Post-Democracy (Part II)

- **Post-Democracy**
  - —*Theoretical Constructs*
    - •The priority of (1'') equality over freedom, or (2'') freedom over equality, or (3'') the transcendence of freedom and equality, regardless of whether freedom and equality are "negative" or "positive." In degree, (1'') or (2'') is less than (1') or (2') but more than (1) or (2)—respectively.
    - •Like democracy and non-democracy, post-democracy is also subject to the freedom/unfreedom and equality/inequality dialectics (or existential dialectics in general). Unlike them, post-democracy acknowledges the constraints of existential dialectics and no longer value freedom and equality as sacred virtues. There is no utopia, in the end; even were there one, dystopia would exist within it.
  - —*Types*
    - •(1''): Different versions of trans-Sinitic value ideals
    - •(1''): Different versions of trans-feminine value ideals
    - •(1''): Different versions of trans-Islamic value ideals
    - •(1''): Different versions of trans-outerspace value ideals
    - •(2''): Different versions of post-human elitist value ideals
    - •(3''): Different versions of the value ideals of floating consciousness (etc.)

(continued on next page)

Table 4.25. Democracy, Non-Democracy, and Post-Democracy (Part III)

Notes: The examples are solely illustrative (not exhaustive), nor are they mutually exclusive. As generalities, they allow exceptions. "Negative" freedom is freedom "from" (e.g., freedom from poverty), whereas "positive" freedom is freedom "to" (e.g., freedom to the state of enlightenment). "Negative" equality is "procedural" equality (e.g., equality of opportunity), while "positive" equality is "substantive" equality (e.g., equality of outcome). Existential dialectics impose constraints on freedom and equality in democracy, non-democracy, and post-democracy, regardless of whether freedom and equality can be understood as "negative" or "positive" in conventional discourse. Therefore, do not confuse the transcendence of freedom and equality in (3'') with the naïve temptation to transcend existential dialectics. There is no utopia, in the end; even should there be one, it would not exist without dystopia embedded within it.
Sources: A summary, based on my previous works, especially Ch.5 of *FHC*, Chs.5-10 of *FCD*, Chs.2-4 of *FPHC*, and Chs.1 & 7 of *BDPD*. The reader should consult the books for more analysis, as this is only a summary here.

Table 4.26. Multiple Causes of the Emergence of Post-Democracy (Part I)

- **At the Micro-Physical Level**
  —Ex: intelligent life without the human physical-chemical system
  —Sources: Ch.7 of *FHC*; Chs.9-10 of *FCD*; Ch.1 of *FPHC*

- **At the Chemical Level**
  —Ex: space radiation and toxins
  —Sources: Ch.7 of *FHC*; Chs.9-10 of *FCD*

- **At the Bio-Psychological Level**
  —Ex: exo-biological evolution in deep space
  —Ex: genetic engineering of new beings
  —Ex: limits of cognitive partiality
  —Ex: illusions of emotional neutrality
  —Ex: human biological inequality
  —Ex: the rise of unfolding unconsciousness
  —Sources: Ch.2 & Chs.9-10 of *FCD*; Ch.7 of *FHC*; Ch.4 of *BCPC*; *FPHU*

- **At the Institutional Level**
  —Ex: the flawed logic of equality
  —Ex: the conflicting nature of governance
  —Sources: Ch. 5 of *FHC*; Chs. 6 & 10 of *FCD*; Ch. 3 of *FPHC*; Chs.2-5 of *BDPD*

- **At the Organizational Level**
  —Ex: e-civic alienation
  —Ex: the dark sides of formal-legalistic routines
  —Sources: Ch.3 of *FHC*; Ch.7 of *FCD*; Ch.3 of *FPHC*

- **At the Structural Level**
  —Ex: ever new forms of inequities, at home and abroad
  —Ex: the emergence of China, women, and Islam as major actors
  —Sources: Chs.5-6 of *FHC*; Chs.7, 9 & 10 of *FCD*; Chs.4-5 of *BDPD*

(continued on next page)

Table 4.26. Multiple Causes of the Emergence of Post-Democracy (Part II)

- **At the Cultural Level**
  —Ex: freedom/unfreedom dialectics
  —Ex: equality/inequality dialectics
  —Ex: system fragmentation and integration
  —Sources: Ch.5 of *FHC*; Chs. 3, 9 & 10 of *FCD*; Ch.4 of *FPHC*; Ch.1 of *BDPD*; Ch.4 of *BCPC*

- **At the Systemic Level**
  —Ex: space habitats (in zero-gravity) and colonization
  —Ex: ultra advanced future info systems
  —Ex: qualitative demography
  —Sources: Ch.7 of *FHC*; Chs.9 &10 of *FCD*

- **At the Cosmological Level**
  —Ex: the colonization of multiverses
  —Ex: the alteration of space-time and the creation of new matter-energy
  —Ex: the expansion of floating consciousness
  —Ex: the spread of hyper-spatial consciousness
  —Sources: Ch.7 of *FHC*; Chs.9 &10 of *FCD*; Ch.4 of *FPHC*; *FPHST*

Notes: The examples in each category are solely illustrative (not exhaustive), and some of the items can be reclassified somewhere else. Nor are they always mutually exclusive. Since they are generalities, exceptions are expected.
Sources: Especially from *FHC*, *FCD*, *FPHC*, *BCPC*, *BDPD*, *FPHST*, and *FPHU*. See also other books and my perspectives on civilizational holism.

Table 4.27. Some Clarifications about Post-Capitalism and Post-Democracy (Part I)

• The prefix "trans-" in the first category of post-capitalism (with its four versions) refers to something "going beyond" (not "uniting" or "combining"). Ex: *Sec.10.3.3* of *FCD*; *Sec.2.4* & *Sec.4.4* of *FPHC*; *Sec.7.2* of *BCPC*

• Such terms like "post-democracy," "post-capitalism," "post-human elitist," "trans-feminine calling," and the like as used in my works are more for our current intellectual convenience than to the liking of future humans and post-humans, who will surely invent more tasteful neologisms to call their own eras, entities, and everything else, for that matter. But the didactic point here is to use the terms to foretell what the future might be like, not that its eras and entities must be called so exactly and permanently.
Ex: *Sec.11.1* of *FCD*; *Sec.7.2* of *BCPC*

• The four versions in the first category of post-capitalist value ideals need not automatically be post-democratic, just as capitalism does not necessarily mean democracy. They are two different entities—though closely related. But up to a certain threshold of elevating equality at the farther expense of freedom, the democratic ideals will be overcome and cease to exist. The same is true for the post-human elitist calling in the second category of post-capitalism in relation to post-democracy, depending on the extent to which freedom is elevated at the expense of equality.
Ex: *Sec.10.4.3.3* of *FCD*; *Table 3.9* of *FPHC*; *Table 7.6* of *BDPD*

• The comparison in each of the three realms of existence in all forms of post-capitalism is not absolute, but relative. Examples include "communal" vs. "individualistic," and the like.
Ex: Notes in *Table 10.8*, *Table 10.9*, *Table 10.10*, & *Table 10.11* of *FCD*; Chs.2-4 of *FPHC*; *Sec.7.2* of *BCPC*

(continued on next page)

Table 4.27. Some Clarifications about Post-Capitalism and Post-Democracy (Part II)

• The emergence of post-capitalism (and post-democracy, for that matter) has multiple causes (to not be reduced to one or only a few).
Ex: Ch.10 of *FCD*, Chs.2-4 of *FPHC*; *Sec.1.3* & *Sec.7.2* of *BCPC* (or *Table 1.8* & *Table 7.11*)

• The specific forms of post-capitalism (and post-democracy, for that matter) need to be further developed in future after-postmodern history, as they will be different from the ones we now know. The point here is to solely give an extremely rough sketch of a world to come that we have never known.
Ex: *Sec.10.3.3* & *Sec.10.4.3.3* of *FCD*; *Table 10.14* & *Table 10.15* of *FCD*; *Sec.7.2* of *BCPC*

• All forms of post-capitalism are not part of a "teleological law," but of "historical trends" only. The same is also true for all forms of post-democracy.
Ex: *Sec.7.1* of *FHC*; *Sec.9.5.3.2* & *Sec.10.3.4.2* of *FCD*; *Sec.7.2* of *BCPC*

• Post-capitalism is not better than capitalism in an "absolute" sense but only fits in better, on the basis of the historical contingency of culture, society, nature, and the mind in some future eras. The same is true for post-democracy in relation to democracy. The term "better" is historically relative.
Ex: *Sec.10.3.3* of *FCD*; *Sec.1.7* of *BDPD*; *Sec.1.5* of *BCPC*

(continued on next page)

Table 4.27. Some Clarifications about Post-Capitalism and Post-Democracy (Part III)

• All forms of post-capitalism and post-democracy are subject to the constraints of existential dialectics. In the process, the dialectic direction is to go beyond the conventional "either-or" dichotomies (e.g., freedom vs. unfreedom, equality vs. inequality, freedom vs. equality, individuality vs. communality, spirituality vs. materiality, formal legalism vs. informal legalism, etc.). As is true in post-civilization, to go beyond the dichotomies is to acknowledge the co-existence of both in each dichotomy, although the degree of scaling one over the over varies from case to case (e.g., the theory of post-capitalism I, the theory of post-capitalism II, the theory of post-democracy I, the theory of post-democracy II, etc.)—but is not to be extreme in largely favoring one over the other, *on average* (all things considered). There is no utopia to be had in the end; even should there be one, dystopia would exist within it.
Ex: Ch.5 of *FHC*; *Sec.10.4.4.2* of *FCD*; *Sec.1.5* of *BDPD*; *Sec.1.3* of *BCPC*; *BCIV*

• All forms of post-capitalism, however different from each other though they are, share one common feature, in that they all inspire for a higher spiritual culture. The same is also true for post-democracy.
Ex: *Sec.10.3*, *Sec.10.4* & *Sec.10.5* of *FCD*; *Chs.2-4* of *FPHC*; *Sec.7.2* of *BCPC*

• All forms of post-capitalism try to avoid the excess in capitalist consumerism by favoring more basic than artificial needs in having, but the quality and quantity of these "basic" needs will be measured by future standards, not by our current ones. Standards are historically relative.
Ex: *Sec.10.3*, *Sec.10.4* & *Sec.10.5* of *FCD*; Ch.2 of *FPHC*; *Sec.7.2* of *BCPC*

(continued on next page)

Table 4.27. Some Clarifications about Post-Capitalism and Post-Democracy (Part IV)

• All forms of post-capitalism make use of a different degree of political authority with advanced info systems in future history and strives for higher spiritual cultures (especially in the post-human age), while acknowledging the constraints of existential dialectics and no longer valuing free market (as in capitalism) and economic control (as in non-capitalism) as sacred virtues.
Ex: *Sec.10.3.4.2*, *Sec.10.3*, *Sec.10.4* & *Sec.10.5* of *FCD*; Chs.2-4 of *FPHC*; *Sec.1.5* of *BDPD*; *Sec.7.2* of *BCPC*

Notes:: The main points here are solely illustrative (not exhaustive) nor necessarily mutually exclusive, and the comparison is relative (not absolute). As generalities, they allow exceptions. The sections as cited are only illustrative (not exhaustive).
Sources: From *FHC*, *FCD*, *FPHC*, and *BDPD*

**Table 4.28. The Theory of Post-Capitalism I.1:
By Group—
Ex: Spiritual/Communal in the Trans-Feminine Calling**

• **More Communal Than Individual**
—*Sharing*: learning from others, as different ideas mutually enrich
—*Cooperative*: encouraging a sense of shared leadership and teamwork

• **More Informal-Legalistic Than Formal-Legalistic**
—*Specific*: listening more from the heart than from the head, to know a person as a concrete, not as an abstract, unit
—*Affective*: thinking and acting with others on a more affective tone. Business can mix with an emotional touch.
—*Ascriptive*: hiring (or firing) can be done on the basis of merit (or lack of it), but deep solidarity (sisterhood) is important too.
—*Particularistic*: making decisions on the basis of cost-benefit analysis, but a given group relationship is vital

• **More Spiritual Than Secular**
—*Long-Term Looking*: sharing for a long-term relationship (e.g., love, friendship), not just for a short-term gain
—*Loving/Caring*: showing compassion for the sufferings of others, without quickly blaming and pre-judging
—*Respectful*: showing acceptance about others' feelings (and thoughts)

Notes: The categories and examples are solely illustrative, since there can be different versions, and the comparison is relative (not absolute), nor are they mutually exclusive. As generalities, they allow exceptions. The specific forms of the trans-feminine version need to be further developed in future after-postmodern history, as they will be different from the ones we now know, since the prefix "trans-" here means going beyond or deconstructing the feminine values, while using them as the inspirational point at the beginning.
Source: From Ch.10 of *FCD*. Refer to text for more info and references.

Table 4.29. The Theory of Post-Capitalism I.2: By Nation-State— Ex: Spiritual/Communal in the Trans-Sinitic Calling

- **More Communal Than Individualistic**
  —*Centralized*: being more top-down in management
  —*Collective*: encouraging more group cooperation
  —*Social*: investing in trust and connection

- **More Informal-Legalistic Than Formal-Legalistic**
  —*Specific*: knowing more of those related or connected
  —*Affective*: behaving in a paternalistic, hierarchical way
  —*Ascriptive*: favoring family members and those related
  —*Particularistic*: building connection (guanxi) as imperative

- **More Spiritual Than Secular**
  —*Expansionist*: diffusing civilizational values (e.g., the superiority complex of civilizationalism)
  —*Holistic*: synthesizing things into a panoramic horizon
  —*Historical*: learning from the lessons of the ancient past
  —*Respectful*: deferential to elders and superiors

Notes: The categories and examples are solely illustrative, since there can be different versions, and the comparison is relative (not absolute), nor are they mutually exclusive. As generalities, they allow exceptions. The specific forms of the trans-Sinitic version need to be further developed in future after-postmodern history, as they will be different from the ones we now know, since the prefix "trans-" here means going beyond or deconstructing the Sinitic values, while using them as the inspirational point at the beginning.
Source: From Ch.10 of *FCD*. Refer to text for more info and references.

Table 4.30. The Theory of Post-Capitalism I.3: By Region— Ex: Spiritual/Communal in the Trans-Islamic Calling

- **More Communal Than Individualistic**
  —*Collective*: building the webs of relationships to bind individuals
  —*Sharing*: cultivating the established "wisdom" through common experience
  —*Cooperative*: stressing harmony, solidarity, and commonality

- **More Informal-Legalistic Than Formal-Legalistic**
  —*Specific*: making efforts to know well the participants (family and larger community) in matters of common concern
  —*Affective*: mixing work with language and ritual on explicit religious (Islamic) ideals, texts, stories, and examples
  —*Ascriptive*: privileging local history and custom on relationships among kinship groups
  —*Particularistic*: preferring an unbiased insider with ongoing connections to all parties

- **More Spiritual Than Secular**
  —*Historical*: learning from the lessons of the past as a source of stability and guidance
  —*Deferential*: showing respect for age, experience, status, and leadership in communal affairs
  —*Honorable*: emphasizing face, dignity, prestige, and fairness
  —*Compassionate*: giving mercy and charity ("Zahah") to others

Notes: The categories and examples are solely illustrative (not exhaustive), and the comparison is relative (not absolute), nor are they mutually exclusive. As generalities, they allow exceptions. The specific forms of the trans-Islamic version need to be further developed in future after-postmodern history, as they will be different from the ones we now know, since the prefix "trans-" here means going beyond or deconstructing the Islamic values, while using them as the inspirational point at the beginning.
Sources: From Ch.10 of *FCD*. Refer to text for more info and references, especially from the works by George Irani (2000) and C. Murphy (September 19, 2001).

Table 4.31. The Theory of Post-Capitalism I.4: By Universe— Ex: Spiritual/Communal in the Trans-Outerspace Calling

- **More Communal Than Individual**
  —*Cooperative*: requiring teamwork in small space habitats
  —*Sharing*: learning from, and enjoying being with, each other in a small group in outer space

- **More Informal-Legalistic Than Formal-Legalistic**
  —*Specific*: knowing more about each other to facilitate living and working together in space, both as fellow astronauts and space-mates
  —*Affective*: being friendly and social to each other as vital to working and living in small space quarters
  —*Ascriptive*: nurturing comaraderie among fellow astronauts as if they are family members over time
  —*Particularistic*: building work relationship with enduring memory in a space mission

- **More Spiritual Than Secular**
  —*Long-Term*: looking beyond selfish materialistic concerns in a precarious space environment with potential life or death
  —*Loving/Caring*: cultivating deep bondage for the success of a long term space mission
  —*Transcendent*: searching for life meaning in outer space

Notes: The calling and examples in each category are solely illustrative (not exhaustive), since there will be many different outer-space value ideals in the distant future of space colonization. The comparison is also relative (not absolute), nor are they mutually exclusive. As generalities, they allow exceptions. And the specific forms of trans-outer-space calling need to be further developed in future after-postmodern history, as they will be different from the ones we now know, since the prefix "trans-" here means going beyond or deconstructing the current outer-space values, while using them as the inspirational point at the beginning. The point here is to solely give an extremely rough picture of a small part of the world to come that we still do not know much about.
Source: From Ch.10 of *FCD*. Refer to text for more info and references.

Table 4.32. The Theory of Post-Capitalism II: Spiritual/Individualistic in the Post-Human Elitist Calling (Part I)

- **More Individualistic Than Communal**
  —Setting up rank distinctions among unequals (e.g., between inferior humans and superior post-humans, or later among inferior post-humans and superior ones, relatively speaking)
  —Yearning for being not only distinguished from unequals, but also the first among equals (the best of the very best)
  —Recognizing the constraints of equality/inequality dialectics (or existential dialectics in general)

- **More Spiritual Than Secular**
  —Soul-searching for a high spiritual culture (not the trashy one for the masses). Mass culture is a dirty joke for them.
  —Exploring different spheres of non-human consciousness in the cosmos (something vastly superior than the human one)
  —Recognizing the constraints of freedom/unfreedom dialectics (or existential dialectics in general)

- **Qualifications**
  —Although post-human elitist post-democracy is comparable to post-human elitist post-capitalism in some respects, the former does not necessarily imply the latter (post-human elitist post-capitalism), just as democracy does not have to entail capitalism. They are two different (though related) entities.
  —But up to a certain threshold of incorporating government intervention with advanced info systems in future civilizations for higher spiritual concerns at the expense of the free market and materialist pursuit, the capitalist ideal will be overcome.
  —The overcome will not be Fascist or feudalistic, but post-capitalist, subject to the constraints of existential dialectics.

(continued on next page)

Table 4.32. The Theory of Post-Capitalism II: Spiritual/Individualistic in the Post-Human Elitist Calling (Part II)

Notes: The calling and examples in each category are solely illustrative (not exhaustive). The comparison is also relative (not absolute), nor are they mutually exclusive. As generalities, they allow exceptions. And the specific forms of post-human elitist post-capitalism need to be further developed in future after-postmodern history, as they will be different from the ones we now know, while using them as the inspirational point at the beginning. The point here is to solely give an extremely rough picture of a small part of the world to come that we still do not know much about.
Sources: From Ch.10 of *FCD* (and also *FPHC*, *BDPD*, and *BCPC*). Refer to the text for more info and references.

Table 4.33. Capitalism, Non-Capitalism, and Post-Capitalism (Part I)

- **Capitalism**
  - —*Theoretical Constructs*
    - •Allocation of scarce resources among alternative wants largely by free market for competition (whose characteristics in its ideal form include, for instance, no barrier to entry or exit, homogeneity, perfect information, a large number of buyers/sellers, and perfect factor mobility)
    - •More formal-legalistic than informal-legalistic, more individualistic than communal, and more material (secular) than spiritual
    - •Either (1) minimal government or (2) relatively active government
  - —*Types*
    - •Only (1): Different versions of market capitalism (e.g., the U.S.)
    - •Only (2): Different versions of welfare capitalism (e.g., Sweden)

- **Non-Capitalism**
  - —*Theoretical Constructs*
    - •Allocation of scarce resources among alternative wants mainly by political authority for policies (which can be regulative, redistributive, symbolic, and participatory)
    - •More informal-legalistic than formal-legalistic
    - •Either (1') more individualistic (for the elites), often (though not always) for material (secular) concerns, or (2') more communal (for the masses), often (though not always) for spiritual concerns
  - —*Types*
    - •Only (1'): Different versions on the Right (e.g., Fascist corporate-state economy for the glory of the new Rome, medieval lord-vassal-serf economy for the power of the feudalistic order)
    - •Only (2'): Different versions on the Left (e.g., Soviet command economy for the creation of the New Socialist Man)

(continued on next page)

Table 4.33. Capitalism, Non-Capitalism, and Post-Capitalism (Part II)

• **Post-Capitalism**
  —*Theoretical Constructs*
    •Allocation of scarce resources among alternative wants largely by political authority with advanced info systems in future civilizations, subject to existential dialectics. In degree of allocating by authority, post-capitalism is more than capitalism but less than non-capitalism.
    •More spiritual than secular (material)
    •Either (1'') more individualistic or (2'') more communal
    •Like capitalism and non-capitalism, post-capitalism is also subject to the freedom/unfreedom and equality/inequality dialectics (or existential dialectics in general). There is no utopia, in the end; even were there one, dystopia would exist within it.
    •Unlike capitalism and non-capitalism, post-capitalism makes use of a different degree of political authority with advanced info systems in future civilizations and strives for higher-spiritual cultures (especially in the post-human age), while acknowledging the constraints of existential dialectics and no longer valuing free market (as in capitalism) and economic control (as in non-capitalism) as sacred virtues.
  —*Types*
    •Only (1''): Different versions of post-human elitist value ideals
    •Only (2''): Different versions of trans-Sinitic value ideals
    •Only (2''): Different versions of trans-feminine value ideals
    •Only (2''): Different versions of trans-Islamic value ideals
    •Only (2''): Different versions of trans-outerspace value ideals

---

(continued on next page)

Table 4.33. Capitalism, Non-Capitalism, and Post-Capitalism (Part III)

Notes: The calling and examples in each category are solely illustrative (not exhaustive). The comparison is also relative (not absolute), nor are they mutually exclusive. As generalities, they allow exceptions. And the specific forms of each calling need to be further developed in future after-postmodern history, as they will be different from the ones we now know, while using them as the inspirational point at the beginning. The point here is to solely give an extremely rough picture of a small part of the world to come that we still do not know much about.
Source: From Ch.10 of *FCD*. Refer to the text for more info and references.

Table 4.34. Multiple Causes of the Emergence of Post-Capitalism (Part I)

- **At the Micro-Physical Level**
  —Ex: intelligent life without the human physical-chemical system
  —Ex: mastering of quantum mechanics, electromagnetism, and other fields for the understanding of a broad range of anomalous experiences and the application for artificial intelligence for spiritual quest
  —Sources: Ch.7 of *FHC*; Chs.9-10 of *FCD*; Ch.1 of *FPHC*

- **At the Chemical Level**
  —Ex: space radiation and toxins
  —Sources: Ch.7 of *FHC*; Chs.9-10 of *FCD*

- **At the Bio-Psychological Level**
  —Ex: exo-biological evolution in deep space
  —Ex: genetic engineering of new beings
  —Ex: limits of human cognition
  —Ex: the rise of unfolding unconsciousness
  —Sources: Ch.2 & Chs.9-10 of *FCD*; Ch.7 of *FHC*; *FPHU*

- **At the Institutional Level**
  —Ex: the flawed logic of the free market
  —Ex: the need of a post-autistic economics
  —Sources: Ch.10 of *FCD*

- **At the Organizational Level**
  —Ex: the dark sides of formal-legalistic routines
  —Sources: Ch.3 of *FHC*; Ch.7 of *FCD*; Ch.3 of *FPHC*

- **At the Structural Level**
  —Ex: ever new forms of inequities, at home and abroad
  —Ex: the emergence of China, women, and Islam as major actors
  —Sources: Chs.5-6 of *FHC*; Chs.7, 9 & 10 of *FCD*; Chs.4-5 of *BDPD*

(continued on next page)

Table 4.34. Multiple Causes of the Emergence of Post-Capitalism (Part II)

- **At the Cultural Level**
  —Ex: freedom/unfreedom dialectics
  —Ex: equality/inequality dialectics
  —Sources: Ch.5 of *FHC*; Chs.3 & 10 of *FCD*; Ch.4 of *FPHC*; Ch.1 of *BDPD*

- **At the Systemic Level**
  —Ex: space habitats (in zero-gravity) and colonization
  —Ex: ultra advanced future info systems
  —Ex: qualitative demography
  —Sources: Ch.7 of *FHC*; Chs. 9 & 10 of *FCD*

- **At the Cosmological Level**
  —Ex: the colonization of multiverses
  —Ex: the alteration of space-time and the creation of new matter-energy
  —Ex: the expansion of floating consciousness
  —Ex: the spread of hyper-spatial consciousness
  —Sources: Ch.7 of *FHC*; Chs. 9 & 10 of *FCD*; Ch.4 of *FPHC*; *FPHST*

Notes: The examples in each category are solely illustrative (not exhaustive), and some of the items can be reclassified somewhere else. Nor are they always mutually exclusive. Since they are generalities, exceptions are expected.
Sources: Especially from *FHC*, *FCD*, *FPHC*, *BCPC*, *BDPD*, *FPHST*, and *FPHU*. See also other books and my perspectives on civilizational holism.

Table 4.35. The Theoretical Debate on Civilization

- **The Progressive Theory of Civilization**
  —*Thesis*: The "civilizing" process is "good," as opposed to the "barbarizing" process as something "bad," relatively speaking.
  —*Discourse*: Especially, though not exclusively, in the Enlightenment era and a bit before. Example: Thomas Hobbes—in that the tribes in primitive societies were "savages."

- **The Romantic Theory of Barbarity**
  —*Thesis*: The "civilizing" process is "bad," as opposed to the "barbarizing" process as something "good," relatively speaking.
  —*Discourse*: Especially, though not exclusively, in the Counter-Enlightenment circle. Example: Jean-Jacques Rousseau—in that civilization "corrupts" men, and the "savages" are in fact "noble."

- **The Moderate Theory of Civilization**
  —*Thesis*: The "civilizing" process is "good," but there is a price to pay, especially in systematic (compulsive) self-control.
  —*Discourse*: Especially, though not exclusively, in some late modern and postmodern circles. Example: Norbert Elias—in that social manners become more refined in the civilizing process, but self-control also becomes more systematic.

- **The Theory of Post-Civilization**
  —*Thesis*: The civilizing process is as evil and good as barbarity, and each cannot exist without the other, to be eventually superseded by post-civilization unto the post-human age.
  —*Discourse*: Proposed by Peter Baofu. See the rest of *BCIV* for more analysis.

Source: From *BCIV* on the theoretical debate

Table 4.36. No Freedom Without Unfreedom in the Civilizing Processes (Part I)

- **The Rationalizing Process (at the Level of Culture)**
  —if freer from the dominance of unreason (as in barbarism) in the civilizing process, then less free from the rationalizing process (be it in the form of the principle of either transcendence or immanence)
  —if freer from the principle of immanence in the rationalizing process, then less free from the inclination to commit terror in the name of reason and the relative underdevelopment of non-reason (e.g., in relation to yoga and meditation)
  —if freer from the principle of transcendence in the rationalizing process, then less free from the relative underdevelopment of reason (e.g., in relation to systematic methodology) and the occurrence of oppression in the name of non-reason

- **The Pacifying Process (at the Level of Society)**
  —if freer from the dominance of pillage (as in savagery) in the civilizing process, then less free from the pacifying process (be it in the form of external control or self-control)
  —if freer from self-control in the pacifying process, then less free from the temptation of expansionist oppression and rebellious mindset in external control
  —if freer from external control in the pacifying process, then less free from the gruesome psychological self-torture and conformism in self-control

- **The Stewardizing Process (at the Level of Nature)**
  —if freer from the dominance of nature (as in the state of nature) in the civilizing process, then less free from the stewardizing process (be it in the form of the stewardship of creation or the covenant with nature)
  —if freer from the stewardship of creation in the stewardizing process, then less free from material underdevelopment, relatively speaking, and spiritual exclusion in the covenant with nature
  —if freer from the covenant with nature in the stewardizing process, then less free from ecological degradation and spiritual disconnection from nature in the stewardship of creation

(continued on next page)

Table 4.36. No Freedom Without Unfreedom in the Civilizing Processes (Part II)

- **The Subliming Process (at the Level of the Mind)**
  —if freer from the dominance of spontaneity (as in the wild state of the mind) in the civilizing process, then less free from the subliming process, be it in the form of (cyclical-centric) self-refinement or (linear-centric) self-discipline
  —if freer from (cyclical-centric) self-refinement in the subliming process, then less free from the (linear-centric) self-regimen (as a form of neurosis)
  —if freer from (linear-centric) self-discipline in the subliming process, then less free from the (cyclical-centric) self-torture (equally as a form of neurosis)

Notes: The examples in each category are solely illustrative (not exhaustive), and the comparison is relative (not absolute), nor are they necessarily mutually exclusive. And some can be easily re-classified elsewhere. As generalities, they allow exceptions.
Sources: From *BCIV*. See also *FHC*, *FCD*, *FPHC*, *BDPD*, and *BCPC*.

Table 4.37. No Equality Without Inequality in the Civilizing Processes (Part I)

- **The Rationalizing Process (at the Level of Culture)**
  —if more equal for the role of rationalization in the rationalizing process (of civilizational making), then less equal for that of mythicization (as in barbarism)
  —if more equal for the principle of transcendence in (linear-centric) rationalizing process, then less equal for the principle of immanence
  —if more equal for the principle of immanence in (cyclical-centric) rationalizing process, then less equal for the principle of transcendence

- **The Pacifying Process (at the Level of Society)**
  —if more equal for pacification in civilizational making, then less equal for the institution of pillaging and others (as in savagery)
  —if more equal for external control, relatively speaking, in pacifying process, then less equal for self-control
  —if more equal for self-control, relatively speaking, in pacifying process, then less equal for external-control

- **The Stewardizing Process (at the Level of Nature)**
  —if more equal for stewardship in the stewardizing process (of civilizational making), then less equal for reverent (submissive) existence (as in barbarism)
  —if more equal for the stewardship of creation in (linear- centric) stewardizing process, then less equal for the (cyclical-centric) covenant with nature for harmonious co-existence
  —if more equal for the (cyclical-centric) covenant with nature in the stewardizing process, then less equal for the (linear-centric) stewardship of nature for domination

(continued on next page)

Table 4.37. No Equality Without Inequality in the Civilizing Processes (Part II)

- **The Subliming Process (at the Level of the Mind)**
  —if more equal for the role of reason in the subliming process, then less equal for that of unreason (as in the natural state of wildness)
  —if more equal for the primacy of reason in (linear-centric) subliming process, then less equal for other faculties (e.g., intuition, existential feelings, and analogous thinking) in cyclical-centric one
  —if more equal for the exercise of other faculties (e.g., intuition, existential feelings, and analogous thinking) in cyclical-centric subliming process, then less equal for the role of reason in linear-centric counterpart

---

Notes: The examples in each category are solely illustrative (not exhaustive), and the comparison is relative (not absolute), nor are they mutually exclusive. And some can be easily reclassified else-where. As generalities, they allow exceptions.
Sources: From *BCIV*. See also *FHC*, *FCD*, *FPHC*, *BDPD*, and *BCPC*.

Table 4.38. Five Theses on Post-Civilization

• Post-civilization no longer treats civilization as good and barbarity as evil (relatively speaking), nor does it nostalgically regard barbarity as good and civilization as evil (relatively speaking again). Civilization is as evil and good as barbarity.

• Post-civilization also no longer accepts the dichotomy between civilization and barbarity. Civilization cannot exist without barbarity. It is no longer necessary to preserve civilization, any more than it is imperative to destroy barbarity. To go beyond civilization and barbarity is to acknowledge the co-existence of both, although the degree of scaling one over the over varies from case to case—but is not to be extreme in largely favoring one over the other, *on average* (subject to the constraints of existential dialectics).

• Post-civilization is thus subject to the constraints of existential dialectics. There is no freedom without unfreedom, and no equality without inequality, for instance. There will be no utopia; even should there be one, there would be dystopia embedded within it.

• Post-civilization will eventually replace civilization (as a form of life settlement), to be dominated by post-capitalist and post-democratic lifeforms here on earth and in deep space (besides other alien lifeforms that we have never known), unto the post-human age in multiverses. Those few post-humans who keep civilization will live in a "post-human civilization," while the rest (the majority), who choose post-civilization, will evolve towards the state of "post-human post-civilization." One therefore should not confuse "post-human civilization" with "post-human post-civilization," as the two are not the same.

• Post-civilization will confront psychosis as a primary problem in the culture of virtuality unto the post-human age, just as civilization has neurosis as a primary one of its own (although both neurosis and psychosis are major problems in both).

Notes: The comparison in each category is relative (not absolute), nor are they necessarily mutually exclusive. And some can be easily re-classified elsewhere. As generalities, they allow exceptions.
Sources: From *BCIV*. See also *FHC*, *FCD*, *FPHC*, *BDPD*, and *BCPC*.

Table 4.39. Barbarity, Civilization, and Post-Civilization

- **The Rationalizing Process (at the Level of Culture)**
  —*Barbarity*
    - More mythicizing than rationalizing, relatively speaking
  —*Civilization*
    - More rationalizing than mythicizing, relatively speaking
  —*Post-Civilization*
    - Beyond the dichotomy, subject to existential dialectics

- **The Pacifying Process (at the Level of Society)**
  —*Barbarity*
    - More pillaging than pacifying, relatively speaking
  —*Civilization*
    - More pacifying than pillaging, relatively speaking
  —*Post-Civilization*
    - Beyond the dichotomy, subject to existential dialectics

- **The Stewardizing Process (at the Level of Nature)**
  —*Barbarity*
    - More revering than stewardizing, relatively speaking
  —*Civilization*
    - More stewardizing than revering, relatively speaking
  —*Post-Civilization*
    - Beyond the dichotomy, subject to existential dialectics

- **The Subliming Process (at the Level of the Mind)**
  —*Barbarity*
    - More impulsing than subliming, relatively speaking
  —*Civilization*
    - More subliming than impulsing, relatively speaking
  —*Post-Civilization*
    - Beyond the dichotomy, subject to existential dialectics

Notes: The comparison in each category is relative (not absolute), nor are they necessarily mutually exclusive. And some can be easily re-classified elsewhere. As generalities, they allow exceptions.
Sources: From *BCIV*. See also *FHC*, *FCD*, *FPHC*, *BDPD*, and *BCPC*.

Table 4.40. Types of Super Civilization in the Cosmos (Part I)

- **Type I**
  —a civilization which gains control of and uses the total energy output "falling on its planet from its sun for interstellar communication" (or, in general, space colonization). For N. Kardashev, who proposed the first three types, human civilization is currently Type Zero (Type O), which is below even Type I, since its present energy consumption for all purposes, let alone for interstellar communication, is still 10,000 times less.

- **Type II**
  —a civilization which gains control of and uses directly the total energy output of its sun for interstellar communication (or, in general, space colonization).

- **Type III**
  —a civilization which gains control of and uses the total energy output of its galaxy for interstellar communication (or, in general, space colonization).

- **Type IV**
  —a civilization which gains control of and uses the total energy output of its cluster of galaxies for interstellar communication (or, in general, space colonization).

- **Type V**
  —a civilization which gains control of and uses the total energy output of its supercluster of galaxies for interstellar communica-tion (or, in general, space colonization).

(continued on next page)

Table 4.40. Types of Super Civilization in the Cosmos (Part II)

- **Type…n**
  —So continues the series in what I call the cyclical progression of hegemony in the cosmos and beyond.

Notes: The Russian astrophysicist Nikolai Kardashev proposed the first three types of super civilization in terms of total energy out-put for interstellar communication. (CSM 1979) I extend his argument further to propose Type IV, Type V, Type VI, and Type…n, in the context of my claim about the cyclical progression of he-gemony in the cosmos and beyond.
Sources: From *Table 9.4* of *FCD.* See *FHC*, *FCD*, and *FPHC* for more info.

Table 4.41. The Civilizational Project from Pre-Modernity to After-Postmodernity (Part I)

	Pre-Modern	*Modern*	*Postmodern*	*After-Postmodern*
Main narratives	•Sacralness •Courtliness •Vitalism •Animism	•Freedom •Equality •Fraternity	•Multiplicity •Hybridization	•Naked contingency •Cyclical progression of hegemony
Main institutions	•Monarchy •Aristocracy •Feudalism •Holy order •Primitivism	•Capitalism •Liberalism •Socialism •Nazism •Fascism	•Capitalism •Liberalism •Postmodern politics of difference	•Post-Capitalism •Post-Democracy •Others
Main techno-logical and economic revolutions	•Agricultural	•Service •Industrial	•Informational	•Biological •Material •Energy •Space •Others

(continued on next page)

Table 4.41. The Civilizational Project from Pre-Modernity to After-Postmodernity (Part II)

	Pre-Modern	*Modern*	*Postmodern*	*After-Postmodern*
Main agents	•Males •Upper strata •Mini-states	•Males •Upper strata •Whites •Empires	•Males •Upper strata •Whites •Others •Supra-states •IO's	•Post-humans •Humans •Others
Main impacts	•Local	•Inter-national	•Global	•Outer-space •Multiverse
Main outcomes	•Towards moderntiy •Rise of linear- & cyclical-centric civiliza-tions	•Towards post-modernity •Dominance of linear-centric civilization	•Towards after-post-moderntiy •Linear-centric civiliza-tion in crisis	•Towards human (& maybe post-human) extinction •Rise of post-civiliza-tion, especially in post-human forms of space-time

Notes: The examples in each category are solely illustrative (not exhaustive) nor necessarily mutually exclusive, and the comparison is relative (not absolute). As generalities, they allow exceptions.
Sources: From *Table 10.16* of *FCD*—and also from *BCIV* on post-civilization (and *FPHST*)

Table 4.42. Civilizational Holism (Part I)

- **At the Micro-Physical Theoretical Level**
  —Ex: Mastering of quantum mechanics, electromagnetism, and other fields for the understanding of a broad range of anomalous experiences and theapplication for artificial intelligence (*Sec.1.4.1* of *FPHC*)

- **At the Chemical Theoretical Level**
  —Ex: Unprecedented expansion of (and violence to) the mind through ever new forms of drugs (and virtual technologies, for that matter) (Ch.9 of *FCD*)

- **At the Biological Theoretical Level**
  —Ex: Humans are not biologically equal, on the basis of race, gender, ethnicity, age, and whatnot. (*Sec.2.6* & Ch.10 of *FCD*; *BNN*) And post-humans will experience the same fate, in an even more amazing way.

- **At the Psychological Theoretical Level**
  —Ex: Human cognitive impartiality and emotional neutrality are quite limited. (*Secs.2.4-2.5* of *FCD*)
  —Ex: Rise of Floating Consciousness (Ch.10 of *FCD*; Chs.1 & 4 of *FPHC*) and Unfolding Unconsciousness (*FPHU*)

- **At the Organizational Theoretical Level**
  —Ex: Administrative colonization of deep space, with less legal-formalism in some corners. (Chs.9-10 of *FCD*)

- **At the Institutional Theoretical Level**
  —Ex: Both capitalism and democracy will not last, to be superseded by different versions of post-capitalism and post-democracy in after-postmodernity. (Ch.10 of *FCD*)

(continued on next page)

**Table 4.42. Civilizational Holism
(Part II)**

• **At the Structural Theoretical Level**
—Ex: Social stratification reappears in ever new forms, also with new causes and new players in the cyclical progression of hegemony. (Chs.8-10 of *FCD*)
—Ex: The world of hyper-empires, and the union of the unions (*BWT*)

• **At the Systemic Theoretical Level**
—Ex: Outerspace expansion: local → regional → global → solar → galactic → clustery → multiversal (Ch.9 of *FCD*)
—Ex: Demographic transition: human extinction, and the rise of post-humans (e.g., cyborgs, thinking machines, thinking robots, genetically altered superior beings, floating consciousness, hyper-spatial consciousness) (Ch.4 of *FPHC*; Ch.10 of *FCD*; & Ch.7 of *FHC*)
—Ex: New technological forces in material sciences, electronic and communication sciences, energy sciences, biosciences, manufacturing and engineering sciences, and space sciences (Ch.10 of *FCD* & Ch.7 of *FHC*)
—Ex: Systematic dominance towards nature for space colonization (Chs.9-10 of *FCD*; Chs.2 & 7 of *FHC*)

• **At the Cultural Theoretical Level**
—Ex: The post-human transcendence of freedom and equality (Ch.10 of *FCD*)
—Ex: Methodological Holism (Ch.1 of *FCD*; Ch.1 of *FPHC*; *Sec.2.1* & *Sec.2.5* of *BCPC*)
—Ex: The Evolution from Barbarity to Post-Civilization (*BCIV*)

• **At the Cosmological Theoretical Level**
—Ex: Mastering of dark matter and dark energy, and the exploration of multiverses (Ch.4 of *FPHC*; Ch.10 of *FCD*; & Ch.7 of *FHC*)
—Ex: Alternation of space-time (*FPHST*)
—Ex: The emergence of hyper-spatial consciousness (*FPHC*)

(continued on next page)

Table 4.42. Civilizational Holism (Part III)

- **At Other Levels**
  —Ex: Historical: pre-modernity → modernity → postmodernity → after-postmodernity (human distinction, and the rise of post-humans, including floating consciousness, hyper-spatial consciousness, and unfolding unconsciousness) (Ch.7 of *FHC*; Ch.10 of *FCD*; *FPHC*; *FPHU*)

Notes: These examples are solely illustrative (not exhaustive), and some of the items can be reclassified somewhere else. Nor are they always mutually exclusive. Since they are generalities, exceptions are expected. And the comparison is relative, not absolute.
Sources: From *Table 5.1* of *FPHC*—with details from *FHC*, *FCD*, and the rest of my books. Not every aspect in each category as discussed in all my books are presented here, since there are too many issues. For more info, also consult the table on theories on civilizational holism and, of course, the books themselves.

Table 4.43. Theories on Civilizational Holism (Part I)

I. Theories in Relation to Nature

—*At the Macro-Physical (Cosmological) Theoretical Level*

- 44. Resettlement Theory of Geology (Peter Baofu) (*FPHGEOL*)
- 43. Theory of Post-Cosmology (Peter Baofu) (*BCOS*)
- 42. Theory of Hyper-Spatial Consciousness (Peter Baofu) (Ch.4 of *FPHC*; *FPHG*)
- (• 40). Selective Theory of Geometry (Peter Baofu) (*FPHG*)
- (• 39). Perspectival Theory of Space-Time (Peter Baofu) (*FPHST*)
- (• 38). Dialectic Theory of Complexity (Peter Baofu) (*FC*)
- (• 9). Theory of Floating Consciousness (Peter Baofu) (Ch.10 of *FCD*; Chs.1 & 4 of *FPHC*)
- 41. Theory of the Geopower of Nature (Peter Baofu) (Ch.4 of *ALD*)

—*At the Micro-Physical Theoretical Leve*

- 40. Selective Theory of Geometry (Peter Baofu) (*FPHG*)
- 39. Perspectival Theory of Space-Time (Peter Baofu) (*FPHST*)
- 38. Dialectic Theory of Complexity (Peter Baofu) (*FC*)

(continued on next page)

Table 4.43. Theories on Civilizational Holism (Part II)

II. Theories in Relation to Culture

—At the Cultural Theoretical Level

- • 37. Theory of Post-Ethics (Peter Baofu) (*BEPE*)
- • 36. Dualistic Theory of Mass Culture (Peter Baofu) (Ch.2 of *FHC*)
- • 35. Comparative Theory of Religion (Peter Baofu) (Ch.3 of *FHC*; Ch.9 of *FCD*; Ch.1 of *FPHK*)
- • 34. Theory of Post-Civilization (Peter Baofu) (*BCIV*)
- • 33. Theory of the Trinity of Modernity to Its After-Postmodern Counterpart (Peter Baofu) (*FHC*; Ch.10 of *FCD*)
- • 32. Transformative Theory of Aesthetic Experience (Peter Baofu) (*FAE*)
- (• 17). Theory of Post-Capitalism (Peter Baofu) (Ch.10 of *FCD*; Chs.2 & 4 of *FPHC*; *BCPC*)
- (• 16). Theory of Post-Democracy (Peter Baofu) (Ch.10 of *FCD*; Chs.3 & 4 of *FPHC*; *BDPD*)
- (• 5). Theory of Existential Dialectics, or the Holistic Theory of Knowledge (Peter Baofu) (*FHC*; *FCD*; *FPHC*; *BDPD*; *FC*; *FAE*; *ALD; FIA; FPHK*; etc,)
- (• 4). Contrastive Theory of Rationality (Peter Baofu) (*FPHML*)
- (• 2). Theory of Methodological Holism (Peter Baofu) (Ch.1 of *FCD*; Ch.1of *FPHC*; *Sec.2.1* & *Sec.2.5* of *BCPC*; *FC*; *FPHK*; etc.)

(continued on next page)

Table 4.43. Theories on Civilizational Holism (Part III)

III. Theories in Relation to the Mind

—At the Biological Theoretical Level

- 31. Theory of Contrastive Advantages (Peter Baofu) (*Sec.2.6* & Ch.10 of *FCD; BNN*)
- (• 24). Theory of Floating Consciousness (Peter Baofu) (Ch.10 of *FCD*; Chs.1 & 4 of *FPHC*)

—At the Psychological Theoretical Level

- 30. Contrarian Theory of Personality (Peter Baofu) (*FPHP*)
- 29. Theory of Virtual Sexuality (Peter Baofu) (*FPHS*)
- 28. Expansive-Contractive Theory of Martial Arts (Peter Baofu) (*FPHMA*)
- 27. Multilogical Theory of Learning (Peter Baofu) (*FPHL*)
- 26. Comprehensive Theory of Creative Thinking (Peter Baofu) (*FPHCT*)
- 25. Theory of Unfolding Unconsciousness—also known as the Unfolding Theory of Anomalous Experience (Peter Baofu) (*FPHU*)
- 24. Theory of Floating Consciousness (Peter Baofu) (Ch.10 of *FCD*; Chs.1 & 4 of *FPHC*)
- 23. Theory of Cognitive Partiality (Peter Baofu) (*Sec.2.4* of *FCD*; *Sec.4.5.1.1* of *BCPC*)
- 22. Theory of Emotional Non-Neutrality (Peter Baofu) (*Sec.2.5* of *FCD*; *Sec.4.5.2* of of *BCPC*)
- 21. Theory of Behavioral Alteration (Peter Baofu) (*Sec.4.5.3* of *BCPC*)

Table 4.43. Theories on Civilizational Holism (Part IV)

IV. Theories in Relation to Society

—*At the Organizational Theoretical Level*

- 20. Theory of E-Civic Alienation (Peter Baofu) (Ch.7 of *FCD*)
- 19. Combinational Theory of Organization (Peter Baofu) (*FPHO*; Ch.6 of *ALD*)

—*At the Institutional Theoretical Level*

- 18. Reconstruction of Theory of Law (Peter Baofu) (*FPHLAW*)
- 17. Theory of Post-Capitalism (Peter Baofu) (Ch.10 of *FCD*; Chs.2 & 4 of *FPHC*; *BCPC*)
- 16. Theory of Post-Democracy (Peter Baofu) (Ch.10 of *FCD*; Chs.3 & 4 of *FPHC*; *BDPD*)
- 15. Dynamic Theory of Comparative Political Systems (Peter Baofu) (*ALD*)

—*At the Systemic Theoretical Level*

- 14. Contingent Theory of Urban Planning (Peter Baofu) (*FPHUP*)
- 13. Totalistic Theory of Communication (Peter Baofu) (*FPHMM*; *FCD*; *FHC*)
- 12. Ambivalent Theory of Technology (Peter Baofu) (*FPHE*; *FCD*; *FHC*)
- 11. Multifaceted Theory of War and Peace (Peter Baofu) (Ch.9 of *FCD*; Ch.1 of *FPHK*)
- 10. Theory of Post-Humanity (Peter Baofu) (Ch.7 of *FHC*; Chs.3, & 10 of *FCD*; Chs.1, 3 & 4 of *FPHC*; and other books of mine)
- 9. Theory of the Cyclical Progression of System Integration and Fragmentation (Peter Baofu) (Chs.9-10 of *FCD*)
- 8. Synthetic Theory of Information Architecture (Peter Baofu) (*FIA*)

(continued on next page)

Table 4.43. Theories on Civilizational Holism (Part V)

IV. Theories in Relation to Society (*cont'd*)

—*At the Structural Theoretical Level*

- 7. Theory of the Cyclical Progression of Hegemony (Peter Baofu)
  (Chs.9-10 of *FCD*; Chs.1, 3 & 4 of *FPHC*; *BDPD*)
- 6. Theory of the Cyclical Progression of Empire-Building (Peter Baofu)
  (*BWT*)

V. Meta-Theories (in Relation to Theories)

—*At the Ontological Meta-Theoretical Level*

- 5. Theory of Existential Dialectics, or the Holistic Theory of Knowledge (Peter Baofu)
  (*FHC*; *FCD*; *FPHC*; *BDPD*; *FC*; *FAE*; *ALD; FIA; FPHK*; etc,)
- 4. Contrastive Theory of Rationality (Peter Baofu)
  (*FPHML*)

—*At the Methodological Meta-Theoretical Level*

- 3. Critical-Dialectic Theory of Formal Science (Peter Baofu)
  (*FPHFS*)
- 2. Theory of Methodological Holism (Peter Baofu)
  (Ch.1 of *FCD*; Ch.1of *FPHC*; *Sec.2.1* & *Sec.2.5* of *BCPC*; *FC*; *FPHK*; etc.)

VI. Theories in Relation to the Rest

—*At Other Levels (Historical)*

- 1. Theory of the Evolution from Pre-Modernity to After-Postmodernity (Peter Baofu)
  (*FHC*; Ch.9-10 of *FCD*; *FPHC*)

Notes: All these theories are my constructions, as some of the main contributions of my grant project on civilization and its future. These examples are solely illustrative (not exhaustive), and some of the items can be reclassified somewhere else. Nor are they always mutually exclusive. Since they are generalities, exceptions are expected.
Sources: From my previous books.

BIBLIOGRAPHY

Ager, Derek V. 1993. *The Nature of the Stratigraphical Record.* Chichester, New York: Brisbane.

Andrews, Sarah. 2010. "What Women Bring to Geology." <http://www.sarahandrews.net/Women_Geology.pdf>.

Bally, A. W, ed. 1987. *Atlas of Seismic Stratigraphy.* Tulsa, OK: American Association of Petroleum Geologists.

Baofu, Peter. 2010. *Beyond Ethics to Post-Ethics: A Preface to a New Theory of Morality and Immorality.* Charlotte, NC: Information Age Publishing.

______.2010a. *Beyond Cosmology to Post-Cosmology: A Preface to a New Theory of Different Worlds.* Cambridge, England: Cambridge International Science Publishing, Ltd.

______.2010b. *The Future of Post-Human Personaltiy: A Preface to a New Theory of Normality and Abnormality.* Cambridge, England: Cambridge International Science Publishing, Ltd.

______.2010c. *The Future of Post-Human War and Peace: A Preface to a New Theory of Aggression and Pacificity.* Cambridge, England: Cambridge Scholars Publishing.

______.2010d. *The Future of Post-Human Formal Science: A Preface to a New Theory of Abstraction and Application.* Cambridge, England: Cambridge Scholars Publishing.

______.2010e. *The Future of Post-Human Law: A Preface to a New Theory of Necessity, Contingency, and Justice.* Cambridge, England: Cambridge Scholars Publishing.

______.2010f. *The Future of Post-Human Sexuality: A Preface to a New Theory of the Body and Spirit of Love Makers.* Cambridge, England: Cambridge Scholars Publishing.

______,2009. *The Future of Post-Human Martial Arts: A Preface to a New Theory of the Body and Spirit of Warriors.* Cambridge, England: Cambridge Scholars Publishing.

______.2009a. *The Future of Post-Human Organization: A Preface to a New Theory of Communication, Decision-Making, and Leadership.* Cambridge, England: Cambridge Scholars Publishing.

_____.2009b. *The Future of Post-Human Language: A Preface to a New Theory of Structure, Context, and Learning*. Cambridge, England: Cambridge Scholars Publishing.

_____.2009c. *The Future of Post-Human Urban Planning: A Preface to a New Theory of Density, Void, and Sustainability*. Cambridge, England: Cambridge Scholars Publishing.

_____.2009d. *The Future of Post-Human Geometry: A Preface to a New Theory of Infinity, Symmetry, and Dimensionality*. Cambridge, England: Cambridge Scholars Publishing.

_____.2009e. *The Future of Post-Human Creative Thinking: A Preface to a New Theory of Invention and Innovation*. Cambridge, England: Cambridge Scholars Publishing.

_____.2009f. *The Future of Post-Human Mass Media: A Preface to a New Theory of Communication*. Cambridge, England: Cambridge Scholars Publishing.

_____.2009g. *The Future of Post-Human Engineering: A Preface to a New Theory of Technology*. Cambridge, England: Cambridge Scholars Publishing.

_____.2008. *The Future of Post-Human Mathematical Logic: A Preface to a New Theory of Rationality*. Cambridge, England: Cambridge Scholars Publishing.

_____.2008a. *The Future of Post-Human Knowledge: A Preface to a New Theory of Methodology and Ontology*. Oxford, United Kingdom: Chandos Publishing (Oxford) Ltd.

_____.2008b. *The Future of Post-Human Unconsciousness: A Preface to a New Theory of Anomalous Experience*. Cambridge, England: Cambridge Scholars Publishing.

_____.2008c. *The Future of Information Architecture: Conceiving a New Way to Understand Taxonomy, Network, and Intelligence*. Oxford, United Kingdom: Chandos Publishing (Oxford) Ltd.

_____.2007. *The Rise of Authoritarian Liberal Democracy: A Preface to a New Theory of Comparative Political Systems*. Cambridge, England: Cambridge Scholars Publishing.

_____.2007a. *The Future of Aesthetic Experience: Conceiving a Better Way to Beauty, Ugliness, and the Rest*. Cambridge, England: Cambridge Scholars Publishing.

_____.2007b. *The Future of Complexity: Conceiving a Better Way to Understand Order and Chaos*. London, United Kingdom: World Scientific Publishing Co.

_____.2007c. *Beyond the World of Titans, and the Remaking of World Order: A Preface to a New Logic of Empire-Building*. Cambridge, England: Cambridge Scholars Publishing.

_____.2006. *Beyond Nature and Nurture: Conceiving a Better Way to Understand*

Genes and Memes. Cambridge, England: Cambridge Scholars Publishing.

_____.2006a. *The Future of Post-Human Space-Time: Conceiving a Better Way to Understand Space and Time*. New York: Peter Lang Publishing, Inc.

_____.2006b. *Beyond Civilization to Post-Civilization: Conceiving a Better Model of Life Settlement to Supersede Civilization*. New York: Peter Lang Publishing, Inc.

_____.2005. *Beyond Capitalism to Post-Capitalism: Conceiving a Better Model of Wealth Acquisition to Supersede Capitalism*. New York: The Edwin Mellen Press.

_____.2004. *The Future of Post-Human Consciousness*. New York: The Edwin Mellen Press.

_____.2004a. *Beyond Democracy to Post-Democracy: Conceiving a Better Model of Governance to Supersede Democracy*. Volume 1. New York: The Edwin Mellen Press.

_____.2004b. *Beyond Democracy to Post-Democracy: Conceiving a Better Model of Governance to Supersede Democracy*. Volume 2. New York: The Edwin Mellen Press.

_____.2002. *The Future of Capitalism and Democracy*. Maryland: The University Press of America.

_____.2000. *The Future of Human Civilization*. Volume 1. New York: The Edwin Mellen Press.

_____.2000a. *The Future of Human Civilization*. Volume 2. New York: The Edwin Mellen Press.

Barrow, John D. 1998. *Impossibility: The Limits of Science and the Science of Limits*. Oxford: Oxford University Press.

BBC. 2010. "The Eruption that Changed Iceland Forever" (April 16). <http://news.bbc.co.uk/2/hi/uk_news/magazine/8624791.stm>.

Biletzki, Anat; & Anat Matar. 2006. "Ludwig Wittgenstein" (Nov. 17). *Stanford Encyclopedia of Philosophy*. <http://plato.stanford.edu/entries/wittgenstein/>.

Blatt, Harvey; & Robert J. Tracy. 1996. *Petrology*. W. H. Freeman.

Bowler, Peter J. 2003. *Evolution: The History of an Idea*. Berkeley, CA: University of California Press.

British Geological Survey (BGS). 2006. "Studying Earth Sciences." Natural Environment Research Council. <http://www.bgs.ac.uk/vacancies/studying.htm>.

Bunge, Mario. 1999. *Dictionary of philosophy*. Amherst, MA: Prometheus Books.

Burger, H. Robert; Anne F. Sheehan; & Craig H. Jones. 2006. *Introduction to Applied Geophysics: Exploring the Shallow Subsurface*. New York: W.W. Norton.

Burnet, Thomas. 1965. *The Sacred Theory of the Earth*. Basil Willey, intro.

London, England: Centaur Press.

Chase, Scott. 1993. "Tachyons" (March). *Physics FAQ.* <http://math.ucr.edu/home/baez/physics/ParticleAndNuclear/tachyons.html>.

Columbia Encyclopedia, the (CE). 2007. "Uniformitarianism." New York: Columbia University Press.

Compton, Robert. 1985. *Geology in the Field.* New York: Wiley.

Condomines, M. 1995. "Magma Dynamics at Mt Etna: Constraints from U-Th-Ra-Pb Radioactive Disequilibria and Sr Isotopes in Historical Lavas." *Earth and Planetary Science Letters,* 132: 25.

Conrad, Clinton P; & Carolina Lithgow-Bertelloni. 2002. "How Mantle Slabs Drive Plate Tectonics." *Science,* 298 (5591): L45. <http://www.soest.hawaii.edu/GG/FACULTY/conrad/resproj/forces/forces.html>.

Corazzon, Raul. 2007. "Definitions of Ontology: From Nicolai Hartmann to the Present Time." <http://www.formalontology.it/section_4_two.htm>.

Corbett, Dan. 2008. "Introduction to String Theory" (November 23). *ThinkQuest.* <http://library.thinkquest.org/27930/stringtheory1.htm>.

Cowen, R. 2000. *History of Life.* Malden, MA: Blackwell Science.

Dahlen, F. A. 1990. "Critical Taper Model of Fold-And-Thrust Belts and Accretionary Wedges." *Annual Review of Earth and Planetary Sciences*, 18: 55.

de Castellejo, Irene Claremont. 1980 *Knowing Woman: A Feminine Psychology.* Boston: Shambala.

Encyclopedia of World Environmental History (EWEH). 2003. "Earthquakes."

Everett, Hugh. 1957. "'Relative State' Formulation of Quantum Mechanics" (July). *Reviews of Modern Physics*, vol. 29, no.3. <http://www.univer.omsk.su/omsk/Sci/Everett/paper1957.html>.

Faure, Gunter. 1998. *Principles and Applications of Geochemistry: A Comprehensive Textbook for Geology Students.* Upper Saddle River, NJ: Prentice-Hall.

Favera, Jorge Carlos Della; & Marco André Malmann Medeiros. 2007. "Gestalt Psychology and the Recognition of Complex Sedimentary Structures in Geology." *Revista Brasileira de Geociências*, 37(4): 841-847. <http://ojs.c3sl.ufpr.br/ojs2/index.php/rbg/article/viewFile/11449/7980>.

Fedonkin, M.A.; J. Gehling; K. Grey; G. Narbonne; & P. Vickers-Rich. 2007. *The Rise of Animals: Evolution and Diversification of the Kingdom Animalia.* Baltimore, MD: Johns Hopkins University Press.

Fernández, O. 2004. "Three-Dimensional Reconstruction of Geological Surfaces: An Example of Growth Strata and Turbidite Systems from the Ainsa Basin (Pyrenees, Spain)." *AAPG Bulletin*, 88: 1049.

Firehammer, Reginald. 2005. "Basic Principles of Ontology" (January 05).

<http://usabig.com/autonomist/philosophy/ontology2.html>.

Gish, Duane. 1995. *Evolution: The Fossils Say No*. Dallas, Texas: Master Books.

Gott, J. Richard. 2002. *Time Travel in Einstein's Universe*. Boston : Houghton Mifflin.

Gould, Stephen Jay. 1987. *Time's Arrow, Time's Cycle: Myth and Metaphor in the Discovery of Geological Time*. Cambridge, MA: Harvard University Press.

______.1965. "Is Uniformitarianism Necessary?" *American Journal of Science*, 263: 223–228.

Guth, Alan. 1997. *The Inflationary Universe: The Quest for a New Theory Of Cosmic Origins*. New York: Helix Books.

Gutscher, M. 1998. "Material Transfer in Accretionary Wedges from Analysis of a Systematic Series of Analog Experiments." *Journal of Structural Geology*, 20: 407.

Hancock, Paul L; Brian J. Skinner; & David L. Dineley. 2000. *The Oxford Companion to The Earth*. New York: Oxford University Press.

Heiken, G.; & K. Wohletz. 1985. *Volcanic Ash*. Berkeley, CA: The University of California Press.

Hess, H. H. 1962. "History of Ocean Basins" (November 1). *IN: Petrologic Studies: A volume in Honor of A. F. Buddington*. A. E. J. Engel, Harold L. James, and B. F. Leonard, eds. Geological Society of America: 599–620.

Hodell, David A. 1994. "Magnetostratigraphic, Biostratigraphic, and Stable Isotope Stratigraphy of an Upper Miocene Drill Core from the Salé Briqueterie (Northwestern Morocco): A High-Resolution Chronology for the Messinian Stage." *Paleoceanography*, 9: 835.

Hooykaas, Reijer. 1963. *Natural Law and Divine Miracle: The Principle of Uniformity in Geology, Biology, and Theology*. Leiden: EJ Brill.

Huang, Zhen Shao. 1997. "Speed of the Continental Plates." *The Physics Factbook*. <http://hypertextbook.com/facts/ZhenHuang.shtml>.

Hubbard, Bryn; & Neil Glasser. 2005. *Field Techniques in Glaciology and Glacial Geomorphology*. Chichester, England: J. Wiley.

Hutton, James. 1795. *Theory of the Earth*. Two volumes. Edinburgh: Creech.

Jedburgh Online (JO). 2010. "Hutton's Unconformity" (July 19). <http://www.jedburgh-online.org.uk/aroundjedburgh.asp>.

Jensen, Dennis. 1999. "Pain, Pleasure and Evolution: An Analysis of Paul Draper's Critique of Theism." *Perspectives on Science and Christian Faith*, 51.1 (March): 40-46.

Kaku, Michio. 1994. *Hyperspace: A Scientific Odyssey Through Parallel Universes, Time Warps, and The Tenth Dimension*. Illustrations by Robert O'Keefe. Oxford: Oxford University Press.

King, Clarence. 1877. "Catastrophism and Evolution." *The American Naturalist*,

11 (8): 449–470.

Kious, Jacquelyne; & Robert Tilling. 1996. "Developing the Theory" (February). *This Dynamic Earth: The Story of Plate Tectonics*. Martha Kiger & Jane Russel, eds. Reston, Virginia: United States Geological Survey. <http://pubs.usgs.gov/gip/dynamic/understanding.html>.

_____ & Robert Tilling. 1996a. "Understanding Plate Motions." *This Dynamic Earth: The Story of Plate Tectonics*. Martha Kiger & Jane Russel, eds. Reston, Virginia: United States Geological Survey. <http://pubs.usgs.gov/gip/dynamic/understanding.html>.

Koons, P. O. 1995. "Modeling the Topographic Evolution of Collisional Belts." *Annual Review of Earth and Planetary Sciences*, 23: 375.

Krumbein, Wolfgang. 1978. *Environmental Biogeochemistry and Geomicrobiology*. Ann Arbor, MI: Ann Arbor Science Publishers.

Kuhn, Thomas S. 1962. *The Structure of Scientific Revolutions*. Chicago, IL: University of Chicago Press.

Laudan, R. 1992. "What's so Special about the Past?" *History and Evolution*. M. Nitecki and D. Nitecki. New York: SUNY Press.

Levin, Harold. 2003. *The Earth through Time*. Hoboken, New Jersey: John Wiley & Sons.

Lovett, Richard A. 2006. "Moon Is Dragging Continents West, Scientist Says." *National Geographic News*. <http://news.nationalgeographic.com/news/2006/01/0124_060124_moon.html>.

Lyman, Marian. 2009. "The Geology of Fear" (October 07). <http://www.freeradicalsmag.com/2009/10/07/the-geology-of-fear/>.

Madrigal, Alexis. 2008. "Top 5 Ways to Cause a Man-Made Earthquake" (June 04). *Wired News*. <http://blog.wired.com/wiredscience/2008/06/top-5-ways-that.html>.

McAleer, Neil. 1987. *The Mind-Boggling Universe*. New York: Doubleday & Co.

McDougall, Ian; & T. Mark Harrison. 1999. *Geochronology and Thermochronology by the $^{40}Ar/^{39}Ar$ Method*. New York: Oxford University Press.

Macfarlane, Robert. 2003. "Glimpses into the Abyss of Time" (September 13). *The Spectator*.

McGowan, Christopher. 2001. *The Dragon Hunters*. Cambridge, MA: Persus Publishing.

MSNBC. 2010. "How Long Will Volcano Ash Hit Europe Flights?" (April 15). <http://www.msnbc.msn.com/id/36563626>.

Mendia-Landa, Pedro. 2007. "Myths and Legends on Natural Disasters: Making Sense of Our World." <http://www.yale.edu/ynhti/curriculum/units/2007/4/07.04.13.x.html>.

Merriam-Webster's Collegiate Dictionary Online (MWD). 2010. "Geology" (July 05). <http://www.merriam-webster.com/dictionary/geology>.

______.2010a. "Statics" (July 05). <http://www.merriam-webster.com/dictionary/ statics>.

______.2010b. "Dynamics" (July 05). <http://www.merriam-webster.com/ dictionary/dynamics>.

______.2010c. "Ashfall" (July 15). <http://www.merriam-webster.com/ dictionary/ Ashfall>.

______.2010d. "Tuff" (July 15). <http://www.merriam-webster.com/dictionary/ Tuff>.

Moore, George W. 1973. "Westward Tidal Lag as the Driving Force of Plate Tectonics." *Geology*, 1: 99–100.

Morowitz, Harold J. 1981. *The Mind's I*. Toronto: Bantam Books.

Morton, Andrew C. 1985. "A New Approach to Provenance Studies: Electron Microprobe Analysis of Detrital Garnets from Middle Jurassic Sandstones of the Northern North Sea." *Sedimentology*, 32: 553.

Musson, Roger. 2003. "Seismicity and Earthquake Hazard in the UK" (June 15). *British Geological Survey*. <http://www.quakes.bgs.ac.uk/hazard/Hazard_ UK.htm>.

NationalAtlas.gov (NAG). 2009. "Geologic Hazards" (November 17). <http://www.nationalatlas.gov/articles/geology/a_geohazards.html>.

National Research Council (NRC). 2003. "5. Earthquake Physics and Fault-System Science." Living *on an Active Earth: Perspectives on Earthquake Science*. Washington D.C.: National Academies Press. <http://www.nap.edu/ openbook.php?record_id=10493&page=282>.

Nesse, William. 1991. *Introduction to Optical Mineralogy*. New York: Oxford University Press.

Nova Online (NO). 2000. "Time Can Vary." <http://www.pbs.org/ wgbh/nova/time/thinktime.html>.

Olsen, Paul E. 2001. "Steno's Principles of Stratigraphy." *Dinosaurs and the History of Life*. New York, NY: Columbia University Press. <http://rainbow.ldeo.columbia.edu/courses/v1001/steno.html>.

PlateTectonics.com (PT). 2010. "Plate Boundaries." <http://www.platetectonics. com/book/page_5.asp>.

Poulsen, Chris J. 1998. "Three-Dimensional Stratigraphic Evolution of the Miocene Baltimore Canyon Region: Implications for Eustatic Interpretations and the Systems Tract Model." *Geological Society of America Bulletin*, 110: 1105.

Raman, Shankar. 1997. Lectures on Literary Theory. Cambridge, MA: MIT (spring).

Rocchio, Laura. 2010. "The Landsat Program" (July 16). *National Aeronautics and Space Administration*. <http://landsat.gsfc.nasa.gov>.

Rollinson, Hugh. 1996. *Using Geochemical Data Evaluation, Presentation, Interpretation*. Harlow, England: Longman.

Rudwick, Martin J. S. 1972. *The Meaning of Fossils*. Chicago, Il: The University of Chicago Press.

Scoppola, B.; D. Boccaletti; M. Bevis; E. Carminati; & C. Doglioni. 2006. "The Westward Drift of the Lithosphere: A Rotational Drag?" *Geological Society of America* Bulletin, 118: 199–209.

Scottish Geology (SG). 2010. "Glen Tilt" (July 19). <http://www.scottishgeology.com/outandabout/classic_sites/locations/glen_tilt.html>.

Shlain, Leonard. 1991. *Art and Physics: Parallel Visions in Space,Time, and Light*. New York: William Morrow and Co., Inc.

Skwirk. 2010. "Volcanic Mountains and Living in Volcanic Zones" (July 15). <http://www.skwirk.com.au/p-c_s-57_u-183_t-490_c-1796/NT/7/Volcanic-mountains-and-living-in-volcanic-zones/Mountains-and-rivers/Global-environments-part-A/SOSE-Geography/>.

Smith, Gary A.; & Aurora Pun. 2006. *How Does the Earth Work?* Upper Saddle River, NJ: Pearson Prentice Hall.

Stenson, James. 1984. "Evolution: A Catholic Perspective." *Catholic Position Papers*, Series A (116). Japan: Seido Foundation for the Advancement of Education.

Sulloway, Frank. 1987. "The Metaphor and the Rock: A Review of *Time's Arrow, Time's Cycle* by Stephen Jay Gould." *The New York Review of Books*, 34 (May 28): 37-40.

Thomson, Keith Stewart. 2001. "Vestiges of James Hutton." *American Scientist* online, 89 (3): 212. <http://www.americanscientist.org/issues/id.3275,y.0,no.,content.true,page.2,css.print/issue.aspx>.

Today in Science History (TSH). 2010. "Tablets" (July 06). <http://www.todayinsci.com/QuotationsCategories/T_Cat/Tablet-Quotations.htm>.

Torrans, C.; & N. Dabbagh. 1999. "Gestalt and Instructional Design." <http://chd.gse.gmuedu/immrsion/knowledgebase/strategies/cognitivism/gestalt/gestalt>.

Toscano, M. 1999. "Submerged Late Pleistocene Reefs on the Tectonically-Stable S.E. Florida Margin: High-Precision Geochronology, Stratigraphy, Resolution of Substage 5a Sea-Level Elevation, and Orbital Forcing." *Quaternary Science Reviews*, 18: 753.

Trembath, Brendan. 2007. "Researcher Claims Mining Triggered 1989 Newcastle Earthquake" (January 09). *Australian Broadcasting Corporation*. <http://www.abc.net.au/am/content/2007/s1823833.htm>.

Tucker, R. D.; D. Bradley; C. Ver Straeten; A. Harris; J. Ebert; & S. McCutcheon. 1998. "New U–Pb Zircon Ages and the Duration and Division of Devonian time." *Earth and Planetary Science Letters*, 158: 175.

Turcotte, Donald L.; & G. Schubert. 2002. *Geodynamics*. Cambridge, England: Cambridge University Press.

Turner, Christine. 2000. "Messages in Stone--Field Geology in the American West." *Earth Matters—The Earth Sciences, Philosophy, and the Claims of Community*. Robert Frodeman. New Jersey: Prentice Hall.

United States Geological Survey (USGS). 2009. "USGS Topographic Maps" (January 02). <http://topomaps.usgs.gov/>.

_____.2009a. "What's It like during an Ash Fall?" (February 03). <http://volcanoes.usgs.gov/ash/ashfall.html>.

_____.2009b. "Earthquake Facts" (October 27). <http://earthquake.usgs.gov/learn/facts.php>.

_____.2009c. "Historic Earthquakes and Earthquake Statistics: Where Do Earthquakes Occur?" (October 27). <http://earthquake.usgs.gov/learning/faq.php?categoryID=11&faqID=95>.

_____.2009d. "Visual Glossary—Ring of Fire" (November 03). <http://earthquake.usgs.gov/learning/glossary.php?termID=150>.

_____.2005. "M7.5 Northern Peru Earthquake of 26 September 2005" (September 26). <ftp://hazards.cr.usgs.gov/maps/sigeqs/20050926/20050926.pdf>.

_____.1999. "Understanding Plate Motions." <http://pubs.usgs.gov/gip/dynamic/understanding.html>.

University of California Museum of Paleontology (UCMP). 2010. "Alfred Wegener (1880-1930)." <http://www.ucmp.berkeley.edu/history/wegener.html>.

_____.2010a. "What is Paleontology?" <http://www.ucmp.berkeley.edu/faq.php#paleo>.

Walker, Dan. 2007. "When a Killer Cloud Hit Britain" (January 19). <http://news.bbc.co.uk/2/hi/science/nature/6276291.stm>.

Watson, John; & Kathie Watson. 1998. "Volcanoes and Earthquakes" (January 07). *United States Geological Survey*. <http://pubs.usgs.gov/gip/earthq1/volcano.html."

Weber, Max. 1930. *The Protestant Ethic and the Spirit of Capitalism*. Talcott Parsons, tr. New York: Scribner.

Whewell, William. 1832. "*Principles of Geology*, by Charles Leyell." *Quart. Rev.*, 47: 103-123.

Wikipedia (WK). 2010. "Geology" (July 04). <http://en.wikipedia.org/wiki/Geology>.

_____.2010a. "Sediment" (July 05). <http://en.wikipedia.org/wiki/Sediment>.

_____.2010b. "Rock Cycle" (July 05). <http://en.wikipedia.org/wiki/Rock_cycle>.

_____.2010c. "Stratigraphy" (July 05). <http://en.wikipedia.org/wiki/Stratigraphy>.

_____.2010d. "Petrology" (July 05). <http://en.wikipedia.org/wiki/Petrology>.

_____.2010e. "Catastrophism" (July 06). <http://en.wikipedia.org/wiki/Catastrophism>.

_____.2010f. "Uniformitarianism" (July 06). <http://en.wikipedia.org/wiki/Uniformitarianism>.

_____.2010g. "Plate Tectonics" (July 08). <http://en.wikipedia.org/wiki/Plate_tectonics>.

_____.2010h. "Rock (Geology)" (July 08). <http://en.wikipedia.org/wiki/Rock_(geology)>.

_____.2010i. "Formation of Rocks" (July 08). <http://en.wikipedia.org/wiki/Formation_of_rocks)>.

_____.2010j. "List of Rock Types" (July 09). <http://en.wikipedia.org/wiki/List_of_rock_types>.

_____.2010k. "Economic Geology" (July 10). <http://en.wikipedia.org/wiki/Economic_geology>.

_____.2010l. "Mineral Resource Classification" (July 10). <http://en.wikipedia.org/wiki/Mineral_resource_classification>.

_____.2010m. "Mineral" (July 10). <http://en.wikipedia.org/wiki/Mineral>.

_____.2010n. "Ore Genesis" (July 10). <http://en.wikipedia.org/wiki/Ore_genesis>.

_____.2010o. "Natural Resource" (July 10). <http://en.wikipedia.org/wiki/Natural_ resource>.

_____.2010p. "Ores" (July 10). <http://en.wikipedia.org/wiki/Ores>.

_____.2010q. "Earth Sciences" (July 10). <http://en.wikipedia.org/wiki/Earth_sciences>.

_____.2010r. "Hydrogeology" (July 10). <http://en.wikipedia.org/wiki/Hydrogeology>.

_____.2010s. "Hydraulic Head" (July 10). <http://en.wikipedia.org/wiki/Hydraulic_head>.

_____.2010t. "Field Work" (July 11). <http://en.wikipedia.org/wiki/Field_work>.

_____.2010u. "Field Experiment" (July 11). <http://en.wikipedia.org/wiki/Field_experiment>.

_____.2010v. "Laboratory" (July 11). <http://en.wikipedia.org/wiki/Laboratory>.

_____.2010w. "Experiment" (July 11). <http://en.wikipedia.org/wiki/Experiment>.

_____.2010x. "Paleontology" (July 12). <http://en.wikipedia.org/wiki/Paleontology>.

_____.2010z. “Biological Classification” (July 12). <http://en.wikipedia.org/wiki/Biological_classification>.

_____.2010aa. “Phylogenetic Nomenclature” (July 12). <http://en.wikipedia.org/wiki/Phylogenetic_nomenclature>.

_____.2010bb. “Common Descent” (July 12). <http://en.wikipedia.org/wiki/Common_descent>.

_____.2010cc. “Special Creation” (July 12). <http://en.wikipedia.org/wiki/Special_creation>.

_____.2010dd. “Molecular_phylogenetics” (July 12). <http://en.wikipedia.org/wiki/Molecular_phylogenetics>.

_____.2010ee. “Charles Lyell” (July 12). <http://en.wikipedia.org/wiki/Sir_Charles_Lyell>.

_____.2010ff. “Magma” (July 15). <http://en.wikipedia.org/wiki/Magma>.

_____.2010gg. “Volcanic Ash” (July 15). <http://en.wikipedia.org/wiki/Volcanic_ash>.

_____.2010hh. “Earthquake” (July 16). <http://en.wikipedia.org/wiki/Earthquake>.

_____.2010ii. “Principle of Original Horizontality” (July 18). <http://en.wikipedia.org/wiki/Principle_of_original_horizontality>.

_____.2010jj. “Absolute Dating” (July 18). <http://en.wikipedia.org/wiki/Absolute Dating>.

_____.2010kk. “Engineering Geology” (July 18). <http://en.wikipedia.org/wiki/Engineering_geology>.

_____.2010ll. “Historical Geology” (July 19). <http://en.wikipedia.org/wiki/Historical_geology>.

_____.2010mm. “Nicolaus Steno” (July 19). <http://en.wikipedia.org/wiki/Nicolaus_Steno>.

_____.2010nn. “The History of Geology” (July 19). <http://en.wikipedia.org/wiki/History_of_geology>.

_____.2008. “Posthumanism” (October 20). <http://en.wikipedia.org/wiki/Posthumanism>.

_____.2008a. “Transhumanism” (October 20). <http://en.wikipedia.org/wiki/Transhumanism>.

_____.2008b. “Lowest Common Denominator” (March 26). <http://en.wikipedia.org/ wiki/Least_common_denominator>.

_____.2008c. “Infinity” (November 11). <http://en.wikipedia.org/wiki/Infinity>.

_____.2008d. “Transfinite Number” (November 10). <http://en.wikipedia.org/wiki/Transfinite_ number>.

_____.2008e. “Quantum Mechanics” (November 26).

<http://en.wikipedia.org/wiki/Quantum_mechanics>.

_____.2008f. "Copenhagen Interpretation" (November 26). <http://en.wikipedia.org/wiki/Copenhagen_interpretation >.

_____.2008g. "Many Worlds Interpretation" (November 26). <http://en.wikipedia.org/wiki/Many_Worlds_Interpretation>.

_____.2008h. "Hyperspace (Science Fiction)" (November 23). <http://en.wikipedia.org/wiki/Hyperspace_(science_fiction)>.

_____.2008i. "Time Travel" (November 26). <http://en.wikipedia.org/wiki/Time_travel#Time_dilation>.

_____.2008j. "Wormhole" (November 25). <http://en.wikipedia.org/wiki/Wormhole>.

_____.2007. "Linguistics" (March 04). <http://en.wikipedia.org/wiki/Linguistics>.

Wikiversity (WKV). 2008. "Why 10 Dimensions" (November 23). <ttp://en.wikiversity.org/wiki/Why_10_dimensions>

Winchester, Simon. 2001. *The Map that Changed the World.* New York: HarperCollins.

Wilson, Leonard G. 1973. "Charles Lyell." *Dictionary of Scientific Biography.* Charles Coulston Gillispie, ed. PA: Charles Scribner's Sons.

Winchester, Simon. 2001. *The Map that Changed the World: William Smith and the Birth of Modern Geology*. New York: HarperCollins Publishers.

Wittgenstein, Ludwig.1953. *Philosophical Investigations*. G. E. M. Anscombe and R. Rhees, eds. G.E.M. Anscombe, tr. Oxford: Blackwell.

Woods, Thomas. 2005. *How the Catholic Church Built Western Civilization.* Washington, DC: Regenery.

Zananas. 2008. "Saint-Pierre Martinique: Pelée Mountain and Eruption of 1902." <http://www.zananas-martinique.com/en-saint-pierre-martinique/>.

Zheng, Y. 2003. "Stable Isotope Geochemistry of Ultrahigh Pressure Metamorphic Rocks from the Dabie–Sulu Orogen in China: Implications for Geodynamics and Fluid Regime." *Earth-Science Reviews*, 62: 105.

INDEX

•A•

•B•

•C•

•D•

•E•

•F•

•G•

•H•

•K•

•L•

•M•

•N•

•O•

•P•

•Q•

•R•

•T•

•U•

•V•

•W•

•X•

•Y•

•Z•